W9-CMN-959

The Politics of Space Security

The Politics of Space Security

*Strategic Restraint and
the Pursuit of National Interests*

JAMES CLAY MOLTZ

STANFORD SECURITY STUDIES

An Imprint of Stanford University Press

Stanford, California 2008

Stanford University Press
Stanford, California
©2008 by the Board of Trustees of the Leland Stanford Junior
University. All rights reserved.

No part of this book may be reproduced or transmitted in any form
or by any means, electronic or mechanical, including photocopy-
ing and recording, or in any information storage or retrieval system
without the prior written permission of Stanford University Press

Library of Congress Cataloging-in-Publication Data

Moltz, James Clay.
    The politics of space security : strategic restraint and the pursuit of
national interests / James Clay Moltz.
        p. cm.
    Includes bibliographical references and index.
    ISBN 978-0-8047-5914-4 (cloth : alk. paper) —
    ISBN 978-0-8047-6010-2 (pbk. : alk. paper)
    1. Astronautics—International cooperation—History.
2. Astronautics—Government policy—United States—History.
3. Astronautics—Government policy—Soviet Union—History.
4. Outer space—Civilian use—Political aspects—History.   5. Outer
space—Exploration—Security measures.   6. Space warfare—
Prevention.   7. Space race—History.   I. Title.
TL788.4.M65   2008
358'.8—dc22   2007049990

Printed in the United States of America on acid-free, archival-quality
paper

Typeset at Stanford University Press in 10/13 Minion

Special discounts for bulk quantities of Stanford Security Studies
are available to corporations, professional associations, and other
organizations. For details and discount information, contact the
special sales department of Stanford University Press. Tel: (650)
736–1783, Fax: (650) 736–1784

Contents

Acknowledgments

This book is the product of twenty years of studying and analyzing problems in the field of space security. My interest in these issues began in the mid-1980s during the sharp U.S.-Soviet debates over the Soviet missile threat and the U.S. Strategic Defense Initiative. I completed a doctoral dissertation on the subject in 1989, entitled *Managing International Rivalry on High Technology Frontiers: U.S.-Soviet Competition and Cooperation in Space*. For their help and guidance at the time, I again want to recognize my advisors at the University of California, Berkeley—Ernst B. Haas (formerly Robson research professor at Berkeley, now deceased), George W. Breslauer (now the provost at Berkeley), and John Holdren (now the Theresa and John Heinz professor of Environmental Policy at Harvard University). I also benefited from the help of Roald Sagdeev, then-director of the Space Research Institute in Moscow (now a distinguished professor of Physics at the University of Maryland), who facilitated a series of difficult-to-arrange interviews with officials across the Soviet space program in the spring of 1988. The Social Science Research Council and the Berkeley-Stanford Program on Soviet Studies provided grants in support of this earlier research, and dozens of U.S. and Soviet space analysts and officials kindly agreed to be interviewed for that project.

After this work, however, I largely set my interests in the space field aside in the midst of the changes in the Soviet Union and its eventual collapse. Space security questions seemed to be less pressing amidst the scaling back of the U.S. Strategic Defense Initiative and the breakup of the Soviet Union. New problems had emerged related to weapons proliferation, loose nuclear materials, and the spread of delivery systems. From July 1993 to June 2007, I worked in various positions at the Center for Nonproliferation Studies of the Monterey Institute of International Studies, where I conducted research, ran projects, supervised nonproliferation training programs, and taught about nuclear risks in the former Soviet Union, as well as questions related to missile proliferation. My space research percolated along on the back burner, as part of the question of economic conversion in Russia and Ukraine and, within the U.S. context, questions of theater missile defense. In the late 1990s, when the issue of national missile defense returned to the forefront of security debates, I began to return more directly to questions of space security, particularly in the context of the increasing number of players in space, their interactions in the commercial, civil, and military sectors, and emerging problems of space management and possible conflict. Fortunately, with successive grants from the W. Alton Jones Foundation, the Carnegie Corporation of New York, the Ploughshares Fund, and the John D. and Catherine T. MacArthur Foundation, I was able to work on a series of interesting projects on space and missile defense issues with partners in Russia (the Committee on Critical Technologies and Nonproliferation and the PIR Center), Ukraine (the Scientific and Technical Centre for Export and Import of Special Technologies, Hardware, and Materials and the Research Center for Nonproliferation Problems), the United Kingdom (the University of Southampton's Mountbatten Centre), and the United States (the Henry Stimson Center, Pugwash USA, the Center for Defense Information, the Nautilus Institute, and the Space Policy Institute at George Washington University). At various times, I also had the chance to contribute to some of the work at the United Nations and the Conference on Disarmament, as well as to conduct space-related technical seminars for members and staff in the U.S. Congress. As my work coalesced into a book manuscript, I benefited from a small travel grant from the Monterey Institute in the winter of 2005 that funded a weeklong research trip to Washington, D.C., for interviews, as well as from additional grants from the Ploughshares Fund to participate in several related conferences on space security issues. Finally, a cooperative arrangement with the Secure World Foundation allowed me to begin writing up this large study. The people who made this support possible were Paul Carroll and Naila Bolus at the Ploughshares Fund and, at the Secure World Foundation, program officer Joe Bock (now at Haverford College) and, especially, its president Cynda Collins Arsenault. I gratefully thank them all.

My research benefited from the help of many individuals who assisted me at various points along the way. In 1995, archivist Diane Barrie at the Ronald Reagan Library proved both cordial and helpful as I searched for understanding regarding a complex, contradictory, and yet very important period in U.S. space policy. I found that the Reagan years have been frequently misunderstood by recent authors—from both the political right and left—who did not have access to these materials. In 2002, Yuri Belitskiy assisted me with a smaller but still useful set of materials from the Russian Foreign Ministry Archives. Other materials came from a trip to the National Security Archives at George Washington University. In addition, I benefited greatly from a series of interviews I conducted in January 2006 (mentioned earlier) in Washington, D.C., with the following individuals (and some additional officials who asked not to be named): Ralph Braibanti, director of the Office of Space and Advanced Technology at the Department of State; Richard DalBello, vice president of INTELSAT; Steven A. Mirmina, a senior attorney at NASA's Commercial and International Law Division; Clayton Mowry, president of Arianespace USA; Roy Pettis, then-science advisor for Strategic and Theater Defenses at the state department; Joan Rolf, then-acting director of the Science Division in NASA's Office of External Relations; Marcia Smith, then just finishing a long and distinguished career as a senior space analyst at the Congressional Research Service of the Library of Congress and moving on to a new position; and Vann Van Diepen, then-director of the Office of Missile Threat Reduction at the state department. None of these individuals are quoted directly in the book, but useful background information from these interviews (and others conducted strictly on a non-attribution basis) is used. Also, Wolfgang (Pief) Panofsky, a Manhattan Project participant and technical advisor on a number of U.S. government studies in the late 1950s and early 1960s regarding nuclear weapons testing in space, kindly agreed to be interviewed in March 2006 in his office at the Stanford Linear Accelerator. His death a few weeks before the manuscript's completion in mid-October 2007 made me especially grateful for his contribution. Nicholas Johnson, NASA's chief scientist on orbital debris management and a longtime participant in the formation of national and international debris guidelines, provided extremely useful (and just plain interesting) information on particular points in the little-known history of debris discussions within the U.S. government and with other countries, especially from the 1980s. I greatly appreciated his generosity with his time. In addition, I want to thank John Logsdon, director of the Space Policy Institute at George Washington University, whose work I have known and respected for a long time, for sharing his tremendous knowledge and contacts as part of our collaboration on several projects in the past few years, including forwarding me copies of some important and newly declassified U.S. documents from the National Security Archives collection to supplement my research on the Kennedy administration.

Among those persons who provided comments on the manuscript itself, I am very grateful to Joan Johnson-Freese of the Naval War College and my colleague James Wirtz of the Naval Postgraduate School for providing positive feedback and useful suggestions to improve the manuscript. Special thanks go to Dr. Pete Hays of the National Security Space Office and the Science Applications International Corporation for his encouragement and for his detailed (and highly informed) comments on a variety of historical, thematic, and technical issues. His care and interest improved the accuracy of the book significantly. In addition, I thank an anonymous reviewer at Stanford University Press.

A series of highly qualified research assistants—Caitlin Baczuk, Rebecca Schauer, Charlotte Savidge, Josh Levinger, Erik Quam, and Adam Williams— have helped me gather materials on space security developments during the past four years, while I was working on the book. Adam Williams deserves particular recognition for having proofread the whole manuscript and for providing useful substantive comments. My former colleague at the Monterey Institute, Dan Pinkston (now with the International Crisis Group in Seoul), commented on two chapters as well.

In terms of the publication process, this book benefited from the careful shepherding provided by Amanda Moran and Geoffrey Burn at Stanford University Press. Their support of this project helped convince me that I had a worthwhile story to tell and allowed me to bring it to a much wider audience through their press's resources.

Lastly, I want to give credit—in different ways—to two people who helped motivate me to complete this book. The first is Maxim V. Tarasenko, a Russian physicist, space enthusiast, and valued professional colleague at the Moscow Physical-Technical Institute. Maxim died in an automobile accident outside of Moscow in May 1999 at the age of thirty-seven. He was an already accomplished figure in the Russian space field, with incredible knowledge and energy. His untimely death deprived the space community of a person deeply devoted to understanding space security and to promoting U.S.-Russian cooperation. The second person is my wife, Sarah J. Diehl, a lawyer, poet, pianist, and nonproliferation affairs expert. I owe her a great debt of gratitude for her patience regarding this second great passion of mine, including the many weekends and vacations missed, which she had to endure while I worked on it (especially in 2006–2007). In addition, Sarah read several of the critical draft chapters—when I could not bear to look at them again—and helped point out mistakes and areas for improvement that greatly strengthened the manuscript.

In the end, all of this collective wisdom and assistance will certainly not prevent me from making mistakes in these pages, for which I am alone responsible. I should also note that nothing in these pages should be interpreted as the official policy of the U.S. Navy, the U.S. Department of Defense, or the U.S. government, but are solely my opinions. Nevertheless, I hope that this book will promote the efforts of states, companies, national militaries, nongovernmental organizations, and individuals to live and work together cooperatively in space and to preserve it for many generations to come.

The Politics of Space Security

Introduction

Fifty years after the opening of space in 1957, the United States finds itself in a unique historical period of overwhelming global strength and influence. A major reason for this power is U.S. space technology. Satellites speed our personal, business, and military communications around the world, transferring tremendous amounts of data nearly instantaneously. Global positioning system (GPS) spacecraft track our products and save lives by locating ships, keeping planes from colliding, and delivering weapons with uncanny precision, reducing casualties and collateral damage. Weather and remote-sensing satellites boost agricultural production and warn of coming disasters. Military early-warning and reconnaissance satellites enforce treaties, help track foreign armies and navies, and provide advance information on missile launches. No other country enjoys the advantages that the United States currently reaps from space, and no other country has made such an investment in space technology.

Despite this power, some U.S. officials and policy analysts fear that space is an Achilles' heel, an environment on which the United States is uniquely dependent but also in which it is highly vulnerable to possible attack.[1] They argue that the increasing number of countries (such as Iran and North Korea) now acquiring ballistic missiles might be able to place objects into space—including crude weapons—to put our assets at risk. There are also fears that space powers with more developed capabilities, such as China, will develop weapons that could eventually hold a number of U.S. satellites "hostage" in a crisis. As a consequence, these officials and analysts argue that substantial U.S. space de-

---

[1] For example, see the "Report of the Commission to Assess United States National Security Space Management and Organization," Pursuant to Public Law 108–65, January 11, 2001, online via DefenseLINK at <http://www.defenselink.mil/pubs/space20010111.html> (accessed September 24, 2006).

fenses will be needed to protect access to critical military assets in orbit or risk a conflict where these now-essential communications, reconnaissance, and targeting spacecraft are denied to U.S. warfighters. For these reasons, the United States since 2001 has resumed active consideration of space-based, anti-ballistic missile (ABM) defenses and anti-satellite (ASAT) weapons, programs it had largely moved to the back burner of national priorities during the 1990s. It has also tested a sea-based ASAT. Some air force officials have called for space-based "Global Strike" capabilities[2] using constellations of military spacecraft capable of hurling high-speed tungsten rods down on rogue states or terrorist facilities harboring weapons of mass destruction (WMD).[3] From their perspective, newly threatening conditions in space provide the rationale for the need to act now.

On the other side of this equation is a perspective that views space as a valuable "sanctuary" from deployed weapons and active military conflict, one that we disturb only at our peril.[4] Supporters of this viewpoint note that the United States and the Soviet Union did not station weapons in space; they question whether threats today are actually worse than during the Cold War. As one study points out, "If the United States detected a missile that appeared to be attacking a satellite, even a relatively small maneuver could essentially eliminate the probability of an intercept."[5] This school argues that the active testing and deployment of space defenses and offensive weapons would alter dramatically a system of tacit and formal restraint in space behavior established during the Cold War, harming U.S. security in the long run.[6] They argue that the further testing of kinetic-kill weapons in orbit (particularly like China's high-altitude destruction of a satellite in January 2007, which produced considerable long-lasting orbital debris) could ruin space for other purposes and might stimulate

[2] For the origins of this concept, see Gen. (USAF) John P. Jumper, "Global Strike Task Force: A Transforming Concept, Forged by Experience," *Aerospace Power Journal*, Vol. 15, No. 1 (Spring 2001), at <http://www.airpower.au.af.mil/airchronicles/apj/apj01/spr01/jumper.htm> (accessed September 24, 2006).

[3] On U.S. plans, see Tim Weiner, "Air Force Seeks Bush's Approval for Space Weapons," *New York Times*, May 18, 2005, p. A1.

[4] See, for example, "Space Security or Space Weapons? A Guide to the Issue," Space Security Project, Henry L. Stimson Center, Washington, D.C., 2005, online at <http://stimson.org/space/pdf/issueguide.pdf> (accessed July 20, 2006).

[5] David Wright, Laura Grego, and Lisbeth Gronlund, *The Physics of Space Security: A Reference Manual* (Cambridge, Mass.: American Academy of Arts and Sciences, 2005), p. 164. Also, as the United States found with its own ground-based ASAT program in the 1960s, one of the problems of such a system is that targets have to pass close enough to the basing site in order for it to be used.

[6] See, for example, Nancy Gallagher, "Towards a Reconsideration of the Rules for Space Security," in John M. Logsdon and Audrey M. Schaffer, eds., *Perspectives on Space Security* (Washington, D.C.: Space Policy Institute, George Washington University, December 2005), p. 35.

a global arms race. Their answer to the security challenge lies instead in crafting new treaties or "rules of the road" to safeguard space against weapons,[7] drawing on the transparency of space for the necessary verification. They point out that the overwhelming majority of states at the United Nations are on record condemning the notion of an arms race in outer space[8] and that only a handful of countries maintain military space programs of any sort, mostly for reconnaissance and communications.

Given the importance of this debate to future U.S. national security and the lack of a comprehensive study of the first fifty years of space security, this book analyzes the period from 1957 to the present in hopes of drawing some practical lessons. The aim is not to describe every launch or mission (which has been done admirably by other authors[9]) or to provide new data on civilian cooperation,[10] but instead to focus on space security issues and turning points in the management of *military* space threats as experienced to date.[11] The book traces the main trends in military space developments—including weapons tests and deployments, arms control treaties, and less formal cooperative agreements— their underlying causes, and the factors that are likely to influence their future course. It is intended primarily for a scholarly audience (particularly students and analysts in the fields of space policy and security studies), but it also may be hoped to reach interested members of the policymaking community, the media, and the general public. Although it provides some background and arguments (particularly in Chapters One and Two) from the academic literature, the book keeps jargon to a minimum so that it may provide an accessible framework for addressing practical problems in the area of space security policymaking.

[7] See Michael Krepon (with Christopher Clary), *Space Assurance or Space Dominance? The Case Against Weaponizing Space* (Washington, D.C.: Henry L. Stimson Center, 2003).

[8] For example, the yearly U.N. resolution (60/54) calling for the "Prevention of an Arms Race in Space" passed on December 8, 2005, with a vote of 180–2. Only the United States and Israel opposed the measure.

[9] For example, on the U.S. side, see William E. Burrows, *This New Ocean: The Story of the First Space Age* (New York: Random House, 1998). On the Soviet side, see the two-volume history by Asif A. Siddiqi, *Sputnik and the Soviet Space Challenge* (Gainesville: University of Florida Press, 2003) and *The Soviet Race with Apollo* (Gainesville: University of Florida Press, 2003).

[10] For detailed studies on U.S.-Soviet and U.S.-Russian civilian space cooperation, see Matthew J. Von Bencke, *The Politics of Space: A History of U.S.-Soviet/Russian Competition and Cooperation* (Boulder, Colo.: Westview Press, 1997); and Susan Eisenhower, ed., *Partners in Space: US-Russian Cooperation After the Cold War* (Washington, D.C.: Eisenhower Institute, 2004).

[11] Given the book's focus on the first fifty years of space security, it does not deal extensively with deep-space issues or with questions related to near-Earth objects (i.e., planetary defense). Both of these topics are likely to become of greater interest as the space age progresses, but they did not factor in the 1957–2007 period as central space security issues, which focused almost exclusively on Earth-orbital space (from geostationary to low–Earth orbits).

Outline of the Chapters

This book is organized into three parts. Part I ("Explaining Space Security: Concepts and Historical Comparisons") covers the existing literature, its strengths and weaknesses, and possible alternative explanations for space outcomes. Chapter One focuses in particular on historical analogies and underlying assumptions among analysts of space security in the four existing schools of thought. Chapter Two provides an alternative explanation, stressing the role of the space environment and gradual learning regarding such problems as man-made electromagnetic pulse radiation and orbital debris in explaining the cooperative outcomes seen since 1957. The argument also shows that these developments were not inevitable. Situational factors, communication breakdowns between leaders, and attempts to assert unilateral advantages could have led already distrustful U.S. and Soviet officials to adopt policies of heightened confrontation rather than strategic restraint.

Following this groundwork, Part II ("Reassessing Twentieth-Century Space Security") provides a detailed history of U.S.-Soviet space security relations, focusing in particular on how more limited forms of competition emerged from initially hostile, open-ended, and military-led space programs. The Cold War evidence presented in Chapters Three through Five shows how and why the two sides gradually accepted mutual constraints on deployable weapons in return for safe access to the space environment for military reconnaissance, weather forecasting, tracking, early warning, and a range of civilian uses. This cooperation proved exceptionally durable despite the periodic rise of political hostilities, such as during the early to mid-1980s in the context of the Strategic Defense Initiative.

Next, Chapter Six examines the end of the Cold War in 1991 and how norms of self-restraint and negotiated space security came under question as Russia's space capabilities declined and as the United States emerged as the dominant player. Nevertheless, the U.S. government made a strategic decision not to exploit this asymmetry in space power and to continue—and even expand—prior forms of cooperative space relations. The data for Chapters Three through Six are drawn from primary source materials in English and in Russian (including declassified U.S. and Soviet government documents), secondary sources, and personal interviews with past and present participants in the two space programs, with industry representatives, and with military officials.

Finally, Part III ("Considering Twenty-First-Century Space Security") examines the new dynamics that have emerged in international space activity since 2001. Chapter Seven discusses the major shift in military space policy under President George W. Bush, including the U.S. decision to withdraw from the ABM Treaty in 2002 and to examine seriously the placement of missile defenses and other weapons in space, thus returning to the military-led direction of U.S. policies typical of the late 1950s and early 1960s. It analyzes the factors behind this shift, reactions from other states, and the stability of the broader space restraint regime in the face of this challenge. China's rise as a major space power—including in the military sector—is also traced. Finally, this chapter highlights the emergence of new commercial actors and activities, as well as the greater practical demands of multilateral (compared to bilateral) space management.

With this background, Chapter Eight looks to the future. It considers the contradictory trends in military and commercial space activities, one toward increasing nationalism and one toward greater internationalism. A major concern is whether weapons deployments could stunt the development of new commercial applications and also threaten passive military assets because of the worsening problem of orbital space debris. Yet, new means for addressing the vulnerability of space assets may emerge out of enhanced communications, interactions, and transparency among space actors, as well as from strategies allowing for the diversification of space platforms (making individual targets less attractive). The book concludes with the elaboration of alternative space futures, ranging from atomized and state-centric to highly integrated and transnational.

Themes and Overall Argument

Ultimately, given the destructive powers of modern states and the particular fragility of the space environment, this book argues that there is a compelling logic to the exercise of military restraint by all actors in space because of their shared national interest in maintaining safe access to critical regions of space—especially low–Earth orbit (from around 60 to 1,000 miles in altitude).[12] During the Cold War, the United States and the Soviet Union (and, indeed, a handful of other nuclear and missile powers of the time) had the potential to render space unusable. The launch of even a dozen nuclear weapons or the dispersal of large amounts of speeding debris into critical low–Earth orbits could have ruined near-Earth space for any significant commercial, scientific, or passive military uses for an indefinite period of time. Today, there are about two dozen states that could do so with nuclear or conventional weapons, and a few states

---

[12] Space itself can be defined as the area beginning at roughly 60 miles above the Earth's surface. Above this altitude, the Earth's atmosphere dissipates to a degree that orbital flight becomes possible, although higher orbits are more favorable because of the further lessening of the effects of atmospheric drag.

among those—with more sophisticated space tracking networks—that have the capability to hit *specific* space targets. This is the threat that some military analysts point to when they argue that advanced space defenses are needed.[13] However, as the history of the Cold War shows, vulnerability in space cannot be erased through military means alone. As University of California physicist Joel R. Primack argues, space "is the most fragile environment that exists because it has the least ability to repair itself."[14] Using orbital physics, he makes the case that "any kind of space warfare will put all satellites at risk."[15] Interestingly, the recent growth of the debris problem has enlisted some unlikely allies to this school's perspective, including at the Pentagon. Air Force Undersecretary for Space Programs Gary Payton argued in 2006 for a more sophisticated U.S. policy of rejecting debris-producing weapons, explaining, "We'd be fools to actually get into the kinetic energy anti-satellite business."[16] But other U.S., Chinese, and perhaps additional foreign officials still remain supportive of keeping destructive weapons available as an option. The U.S. decision to destroy an ailing National Reconnaissance Office satellite packed with hydrazine fuel in February 2008 indicated the Bush administration's willingness to use space weapons in at least certain prescribed conditions—high perceived threat and low orbital altitude (thus minimizing debris consequences). Whether new international norms will be developed to ensure the adoption of these restrictive criteria by other military space powers remains to be seen.

As noted earlier, another factor that may change space activity over time is the emergence of new commercial actors. While "revolutions" in space commerce have long been overpredicted, trends that are making satellite technology, manned spacecraft, and low-cost launchers more accessible are finally beginning to alter the military-dominated nature of the space age. Space's second fifty years may look very different because of this greater diversity of actors and their impact on practical dynamics. As then–Commander in Chief of U.S. Space Command General Howell M. Estes predicted in a speech in April 1997, "It is not the future of military space that is critical to the United States—it is

[13] Critics of "space control" theories raise the logical objection: even if such weapons are developed, other countries will still retain the ability to negate U.S. space assets through *asymmetric* military means—including resort to space debris and radiation, jamming, and attacks on ground stations.

[14] Joel R. Primack, "Debris and Future Space Activities," in James Clay Moltz, ed., *Future Security in Space: Commercial, Military, and Arms Control Trade-Offs*, Occasional Paper No. 10 (Monterey, Calif.: Center for Nonproliferation Studies, Monterey Institute of International Studies, July 2002), p. 18.

[15] Ibid., p. 21.

[16] Quoted in Jeremy Singer, "USAF Interest in Lasers Triggers Concerns About Anti-Satellite Weapons," *Space News*, May 1, 2006, p. A4.

the continued commercial development of space that will provide continued strength for our great country in the decades ahead."[17]

In this context, effective coordination among a range of actors and activities may be the most serious emerging space challenge. This is a fundamentally *political* task. The aim of this book, therefore, is first to analyze the past connections among national politics, the space environment, and the practice of space security; then, taking into account the influence of emerging changes, it seeks to project these lessons forward in order to develop meaningful guidelines for the future. The main conclusion is that the most useful framework for analyzing the past, present, and future of these issues in space is not a traditional military-strategic one, but instead the interdependent concept of environmental security.

[17] Quoted in Lt. Col. (USAF) Peter L. Hays, *United States Military Space: Into the Twenty-First Century*, Occasional Paper No. 42 (Colorado Springs: U.S. Air Force Academy, Institute for National Security Studies, September 2002), p. 14.

PART I

Explaining Space Security

Concepts and Historical Comparisons

CHAPTER ONE

The Dynamics of Space Security
Existing Explanations

The concept of national security describes the relationship between a country's capabilities and the challenges posed by the surroundings in which it must operate. When a country is *secure*, it enjoys the ability to conduct its activities free from harm. Although we normally view security as reliant solely on military power, it is also influenced by a variety of other factors: alliances, economic strength, treaty memberships, political stance (such as declared neutrality), social cohesion, and even perceived moral authority.

In space, the attainment of security involves the task of overcoming both man-made and natural threats, given the extreme hostility of the space environment. Since orbital dynamics require a certain level of interaction with other actors, the behavior of all space-faring entities (states, companies, universities, private citizens, and international consortia) inevitably affects the security of others, more so than in other realms. In general, we can define "space security" as *the ability to place and operate assets outside the Earth's atmosphere without external interference, damage, or destruction.* By this definition, all actors have enjoyed a high level of space security for most of the space age, with very few exceptions, as will be discussed later. Unfortunately, challenges to space security are increasing today, particularly as space becomes more crowded. Arguably, at least three alternatives exist: (1) space actors can assume the worst and prepare for eventual warfare; (2) they can hedge their bets with weapons research and begin efforts at better coordination and conflict avoidance; or (3) they can reject military options altogether and heighten their efforts to build new cooperative mechanisms for developing space jointly.

During the Cold War, the behavior of the Soviet Union and the United States dominated space security considerations. These two sides conducted well more than 95 percent of space activities during the Cold War. Although Russian activity has declined significantly since 1991, even in 2005 the combined total of U.S. and Russian activities still made up 50 percent of all commercial space launches, 63 percent of civil launches,[1] and fully 68 percent of military launches.[2] For each of these two countries, achieving space security was for many years primarily a matter of understanding the policies of the other side and trying to reach consensus on how to manage disputes and prevent hostile acts. As discussed in this book, space security evolved during the Cold War in two primary stages: the 1957–62 period (characterized by military-led approaches) and the 1963–91 period (characterized mainly by military "hedging" and negotiated approaches). With the end of the Cold War in 1991, space became a realm led mainly by the United States. For a decade, Washington continued a policy of negotiated space security, in close cooperation with the Russian Federation (drawing on the third option listed above). After 2001, however, a new U.S. leadership, focusing on emerging foreign missile threats and eventual U.S. space vulnerabilities, shifted course back toward a military-led strategy in the belief that hostile actors would arise among new space powers and create threats requiring military solutions. In part for this reason, it withdrew from one of the main, negotiated space security arrangements of the Cold War—the 1972 Anti-Ballistic Missile Treaty—and also developed more space-specific military plans for defensive operations. However, the United States did not abandon the 1967 Outer Space Treaty or a number of other cooperative agreements.

Some analysts believed that the George W. Bush administration's moves had finally paved the way for an historically inevitable process of space's weaponization and the occurrence of direct military conflict, which had been delayed by political and technological factors. As Steven Lambakis of the National Institute of Public Policy in a 2001 book complained, regarding the behavior of past U.S. space policies: "If freedom of space is our guidestar, what is being done to nurture and protect it? Are not U.S. policy makers setting a bad precedent by unilaterally restricting national activities in the force-application and space-control areas, limiting in effect the country's freedom to exploit space?"[3] For others, these developments marked a sharp and negative change from wise policies by past presidents that had helped create the foundations for U.S. space preeminence. As the Center for Defense Information's Theresa Hitchens argued in 2003: "Unfortunately, this [Bush] administration has done little thinking . . .

---

[1] "Civil" space refers to launches for the purpose of noncommercial, nonmilitary activities, including primarily human exploration and space science.

[2] Spacesecurity.org, *Space Security 2006* (accessed July 2006), pp. 78, 95, 113.

[3] Steven Lambakis, *On the Edge of Earth: The Future of American Space Power* (Lexington: University Press of Kentucky, 2001), p. 276.

about the potential for far-reaching military, political and economic ramifications of a U.S. move to break the taboo against weaponizing space."[4] Whether this outcome and the U.S. movement toward the deployment of space-based defenses is somehow historically predetermined or instead related mainly to the specific policy preferences of the Bush administration remains a subject of debate among space experts.

These developments and the prospect of space-based defenses and offensive weapons raise a series of important questions: Is the deployment of space-based weapons somehow unavoidable, or can space actors prevent it through rules of the road, treaties, or tacit avoidance? Is there such a thing as the "partial" weaponization of space? Are there definable cut-off lines among systems and could they be enforced? Or, could weapons in space be used to prevent an arms race through some form of "space hegemony"?[5] Skeptics believe that any form of weaponization would be a slippery slope, likely to result in a multilateral arms race and a reversion by states to military-led solutions. But there is also the possibility that recent military trends are an epiphenomenon and instead that the expansion of commercial actors in space will change priorities in Washington and other capitals and lead human space developments away from conflictual, weapons-based scenarios.

In seeking guidance on these questions, we might observe that space is but one of many new frontiers visited by states over the past several centuries; to better understand the dynamics of space, we can start with this history. Indeed, many analysts of space security have attempted to draw lessons from historical rivalries on new physical frontiers. This chapter begins by summarizing some of the key dynamics involved in policies of "expansionist security." It then focuses on the three most often mentioned historical analogies for space security—the settling of the New World, the development of sea and air power in the late 1800s and early 1900s (taken together), and negotiations over Antarctica in the late 1950s—examining "how parallel" their dynamics actually are in regard to space. The analysis then turns to the four main schools of existing thought regarding space security and where their strengths and weaknesses lie. The chapter concludes with an argument for a new approach: environmentally influenced learning.

---

[4] Theresa Hitchens, "Weapons in Space: Silver Bullet or Russian Roulette? The Policy Implications of U.S. Pursuit of Space-Based Weapons," in John M. Logsdon and Gordon Adams, eds., *Space Weapons: Are They Needed?* (Washington, D.C.: Space Policy Institute, George Washington University, October 2003), p. 88.

[5] On this point, see Everett Carl Dolman, "Space Power and US Hegemony: Maintaining a Liberal World Order in the 21st Century," in Logsdon and Adams, *Space Weapons*.

The Past as Precedent: Three Analogies

Debates on the future of international relations in space revisit a long history of great power competition on new frontiers, coincident with the rise of the modern nation-state. Advances in maritime technology (sails, rudders, and portable chronometers[6]) allowed countries to seize and control distant lands with the aim of achieving strategic, military, political, and economic advantages over their rivals for the purposes of maintaining or advancing their security. At the domestic level, powerful coalitions often pushed these enterprises in order to promote self-interested aims,[7] with the prizes being profitable new lands, their populations, and their natural resources. Frederick Jackson Turner argued in the late 1800s that expansionism offered states a natural and psychologically necessary release from domestic tensions, and might even be *required* for the continued stability and development of major nation-states.[8]

By the twentieth century, however, competing Western countries had seized all of the most readily accessible regions of the world that could not be defended by resident populations, leaving only unpopulated areas: the seabed, the polar icecaps, and, finally, space. These new frontiers required a combination of technological innovations and considerable funding to enable human beings to navigate, operate in, and make use of their more hostile environments.[9]

Part of the motivation for states to enter new frontiers has to do with national reputation. As political leaders have long recognized, international influence at any given point in history is a product not only of a country's economic and military power but also of its perceived *momentum* as a state.[10] As seen in

---

[6] Portable clocks were critical to the calculation of longitudinal coordinates, which could finally be linked with previously accessible latitudinal information from sextants. See Martin van Creveld, *Technology and War: From 2000 B.C. to the Present* (New York: Free Press, 1991), p. 128.

[7] For a study of expansionism drawing on the combined forces of coalition building and ideology, see Jack Snyder, *Myths of Empire: Domestic Politics and International Ambition* (Ithaca, N.Y.: Cornell University Press, 1991).

[8] Frederick Jackson Turner, "The Significance of the Frontier in American History," speech before the American Historical Association, Chicago, July 12, 1893, reprinted in Frederick Jackson Turner, *The Frontier in American History* (New York: Henry Holt, 1921).

[9] Looking ahead, the frontiers of other dimensions (such as nanotechnology and even time travel) may yet create future forms of international competition.

[10] The importance of national "reputation" in international relations has long been recognized both by game theorists and by those working in the area of deterrence theory. For a recent summary of these issues, see especially chapters by Jack Levy and by Stern, Axelrod, Jervis, and Radner in Paul C. Stern, Robert Axelrod, Robert Jervis, and Roy Radner, eds., *Perspectives on Deterrence* (New York: Oxford University Press, 1989). See also Robert Wilson, "Reputations in Games and Markets," in G. F. Feiwel, ed., *Game-Theoretic Models of Bargaining* (New York: Cambridge University Press, 1985). For a two-level view analyzing both the creation of state reputations and their *interpretation*

the troubles of the Ottoman Empire in the late 1800s, the malaise of the United States in the late 1970s, and the stagnation of the Soviet Union in the late 1980s, when a major world government fails to maintain a national image of power, efficacy, and forward technological progress, it can be perceived as weak by its adversaries and, in the eyes of its population, even as questionable in its claim of legitimacy.

For countries and corporations alike, however, deciding *when* and *where* to compete is not easy given the limits of available resources and the presence of risks.[11] While offering great opportunities for those who succeed, costly failures in frontier struggles can destabilize national governments and make them liable to domestic or external subversion. As Paul Kennedy, Richard Rosecrance, and Jack Snyder have observed, this struggle to achieve expansionist versions of security has had many losers resulting from the unexpected effects of frontier competitions on geopolitics, trade, political affairs, and military alliances.[12]

The New World Analogy

The opening of space by the Soviet Union and the United States in the late 1950s and 1960s shares certain characteristics with the competition between Spain and Portugal over the New World in the late fifteenth and early sixteenth centuries.[13] For this reason, the New World analogy has been frequently referred to by officials, analysts, and authors on space since the 1950s.[14] As in space, the effort to develop new sea routes to India and the eastern islands required the utmost secrecy and involved technologies crucial to national security.[15] The actu-

---

by other states, see Barry Nalebuff, "Rational Deterrence in an Imperfect World," *World Politics*, Vol. 43, No. 3 (April 1991), pp. 314–16.

[11] President Dwight D. Eisenhower, for example, believed that racing in space made no sense. He was eventually overwhelmed by public and congressional pressure.

[12] Paul Kennedy, *The Rise and Fall of the Great Powers* (New York: Vintage, 1989); Richard Rosecrance, *The Rise of the Trading State: Commerce and Conquest in the Modern Age* (New York: Basic, 1986); Snyder, *Myths of Empire*.

[13] This discussion excludes the earlier missions to present-day Greenland and Canada by various Viking explorers, whose missions failed to result in permanent settlements.

[14] For example, on the use of the Columbus analogy during the 1950s' debates about the U.S. response to *Sputnik*, see Walter A. McDougall, . . . *the Heavens and the Earth: A Political History of the Space Age* (New York: Basic, 1985), p. 225.

[15] This account of the Spanish-Portuguese competition draws on the following sources: Roger Bigelow Merriman, *The Rise of the Spanish Empire in the Old World and in the New,* Volume II (New York: Macmillan, 1918); Charles E. Nowell, *A History of Portugal* (Princeton, N.J.: Van Nostrand, 1952); William C. Atkinson, *A History of Spain and Portugal* (Baltimore: Penguin, 1960); H. V. Livermore, *Portugal: A Short History* (Edinburgh, Scotland: Edinburgh University Press, 1973); Daniel J. Boorstin, *The Discoverers* (New York: Vintage [Random House], 1983), pp. 235–54; and J. M. Roberts, *The Pelican History of the World* (New York: Penguin, 1980), pp. 506–8.

al execution of the missions involved costly expeditions relying on the skills of teams of individuals: state leaders, explorers, scientists, and expert technicians. Like Yuri Gagarin and John Glenn in space, the leaders of these voyages—including Christopher Columbus and Amerigo Vespucci—became national heroes. Similarly, the fascinating realms these explorers uncovered created new objects for the popular imagination, as well as opportunities for economic and military advantage.

After Columbus's first voyage in 1492, his sponsor, Spain, made unilateral claims to the new territories, which Pope Alexander VI duly endorsed in April 1493.[16] But King John II of Portugal used his powerful navy to force negotiations with the Spanish crown, yielding a compromise that gave Portugal the right to regions located east of a demarcation line in the south Atlantic.[17] In 1500, Portuguese explorers under Pedro Alvarez Cabral reached Brazil and staked a claim to the rich territory within their zone. It seemed that direct, bilateral negotiations and the formation of a cooperative regime had successfully averted an impending conflict.

But the Spanish-Portuguese entente contained certain fatal flaws. First, it relied on a fragile web of secrecy held together only by the elaborate security precautions taken by the two countries to conceal their maps and specific routes to the New World. Second, it deliberately excluded other European powers in a system characterized by multiple states of relatively equal might. The agreement held for a few decades, but word of the New World's location and its riches eventually leaked out and spread throughout Europe, bringing new challengers and their militaries.[18] Relying on now widely distributed maritime technologies, other European powers soon began to exploit this new route to prospective wealth and colonies. As French King Francis I summed up the views of other European claimants in rejecting the Spanish-Portuguese entente: "The sun shines for me the same as for others: I would like to see that clause in Adam's will that excluded me from the partition of the world."[19] The

---

[16] According to Merriman, the Spanish crown had leverage over the Vatican in this time of trouble because its relative proximity made it the most likely country to send troops in case the Vatican was seized by hostile forces. Merriman writes that "[Pope] Alexander ... was like wax in the hands of Ferdinand and Isabella" (Merriman, *Rise of the Spanish Empire*, p. 202). Nowell (*History of Portugal*, p. 61) also points out that the pontiff was himself a Spaniard.

[17] The agreement was called the Treaty of Tordesillas, signed in June 1494. The accepted dividing line between the respective Spanish and Portuguese claims lay about 370 leagues west of the Cape Verde Islands.

[18] As Boorstin (*Discoverers*) points out, the simultaneous development of the printing press promoted the rapid diffusion of this knowledge as maps to the New World became common.

[19] Reply to the Spanish ambassador in Paris, quoted in Atkinson, *History of Spain and Portugal*, p. 97.

same basic statements about space are now being made by China, India, and other emerging space powers regarding the past history of Russian and U.S. dominance.

The collapse of the Spanish-Portuguese entente and the aggressive activities of France, Holland, and England eventually ruined any chance of managing New World conflicts over colonies and resources.[20] The existence of a system of multiple competing powers in Europe at the time—valuing territory and raw materials as assets of power and seeing nothing to stop their conquests—transformed a peaceful division of spoils (although at the expense of native populations) into a military contest of seek, occupy, and defend.

Notably, the precedent of New World and other multilateral competition on new frontiers has provided a framework for much thinking about space, which assumes a survival-of-the-fittest strategy aimed at edging out the enemy (or suffering similar consequences oneself). Yet such a dire scenario of warfare has not yet emerged in space, contrary to many expectations. One difference may be that, unlike Cold War leaders, reigning kings and queens during the New World struggle viewed war as an acceptable outcome and as fully compatible with the pursuit of expansionist security. But such conflicts were rarely system-destroying, did not involve the elimination of nation-states, and created no crippling environmental damage. One can only imagine, for example, the different outcome in the New World if—as with orbital space debris—all of the arrows and bullets fired in those wars of conquest had continued to speed around the Earth causing damage for decades after they had been fired. These factors (discussed in Chapter Two) make the surrounding context of space security very different. For these reasons, we cannot easily accept arguments about historical inevitability of space conflict and warfare based on the New World analogy.

The Sea and Air Power Analogies

A second common set of analogies used in attempts to explain the dynamics of space security are those of sea and air power. These characterizations of space are most frequently used by military analysts. Common to these studies are references to the great late nineteenth- and early twentieth-century American naval historian Alfred Thayer Mahan. In his time, Mahan played an influential role in rousing the U.S. public, political circles, and the military to abandon its "sluggish attitude" toward maritime competition and join those powers that "cherish . . . aspirations for commercial extension, for colonies,

---

[20] The Protestant Reformation also intervened to reduce the moderating power of Rome in colonial disputes among the European powers.

and for influence in distant regions."[21] Mahan cited the requirement for a canal through Central America and a strong U.S. navy to defend its commercial and strategic interests abroad. Mahan's ideas influenced future President Theodore Roosevelt and other leading officials of the day, resulting in the creation of the Great White Fleet, which toured the world from 1907 to 1909 showing the U.S. flag and America's new naval might.

Supporters of the sea power analogy also emphasize the link between commerce and the military, in that the development of one is viewed as requiring the simultaneous expansion of the other in order to be effective in serving national security. As Air Force Lieutenant Colonel (ret.) David E. Lupton argues, "Space control is very much like past and present concepts of sea control," citing such parallels as lines of communications, cluster points, and the relevance of technological advantages.[22] He compares, for example, Great Britain's ability to dominate access and control of great swathes of the ocean in the nineteenth century to the likely ability of a small number of space-faring states to control upper orbits and keep out adversaries because of their greater technological prowess.[23]

The similar air power analogy has also been studied by analysts seeking to explain and predict the behavior of states in space. What started with civilians Orville and Wilbur Wright in 1903 soon developed into military uses of aircraft during World War I and strategic bombing during World War II. As Steven Lambakis writes: "What Billy Mitchell said about air power . . . is also true of space power and the space environment." In other words, the strong belief is that space will eventually become a dominant field of military endeavor and bring about a revolution in military affairs. U.S. Marine Corps Major Franz Gayl argues in this regard: "As with aviation, access and technology will drive forward to exploit any and all warfighting relevance, application, and advantage from space, quite independent of a nation's will to prevent it."[24] These points lead Major Gayl to conclude that "missions relating to space control, global strike, missile defense, transport, assault support, and such will necessarily follow."[25]

In critiquing these analogies, several points need to be emphasized. In the world that Mahan observed, imperialist states following policies of mercantil-

---

[21] Capt. (USN) A. T. Mahan, *The Interest of America in Sea Power, Present and Future* (Port Washington, N.Y.: Kennikat Press, 1897), pp. 5, 7.

[22] Lt. Col. (USAF, ret.) David E. Lupton, *On Space Warfare: A Space Power Doctrine* (Maxwell Air Force Base, Ala.: Air University Press, June 1998), Chapter 7, p. 5.

[23] Ibid.

[24] Maj. (USMC, ret.) Franz J. Gayl, "Time for a Military Space Service," *Proceedings* (July 2004), p. 44.

[25] Ibid.

ism dominated. In other words, the goal of navies was to protect sea lanes and colonies. Commerce involved trade largely (and sometimes exclusively) with nationally designated monopolies, whose purpose was to funnel resources back to the home economy and provide taxes to the colonial power's government. Today, however, space commerce is becoming more and more international, making it difficult for countries to tell what a company belongs to—it might be registered in Bermuda, operate in the United States, and use technology from Russia. Moreover, as Air Force Lieutenant Colonel Peter Hays observes, "The logic of this 'flag follows trade' argument is clear and has historical precedents, but to date it has not yet prompted any significant calls [by commercial space actors] for better protection."[26] Also, space law currently bans the colonization of the Moon and other celestial bodies, which suggests that maritime analogies centered on securing ownership of and access to colonies may not be relevant to space, at least as long as such treaties remain in force.

Finally, the notion of sea and air power implies the possibility of environmental "control" and the ability of leading countries to exclude other parties. With sea power, littoral seas can be more easily defended than blue water in part because of shore-based support, but there is also the possibility (as in the case of Britain) that a large navy possessing greater speed and firepower than its rivals may dominant the open ocean as well, even to the point of making it inaccessible to weaker navies. In the air, such control has also been realized in actual military contexts, as with the United States over Japan in the latter portion of World War II. But in space the analogy is harder to follow. Unlike in terrestrial, air, or sea environments, it is not clear how orbital space would be controlled. Although analysts cite predominance in space-based platforms and an ability to overwhelm potential adversaries with lasers and kinetic-kill weapons, the viability of "control" remains questionable in a situation where defenses are extremely costly from space, weapons systems are highly transparent (and in predictable orbits), and existing ground systems can be used to attack space assets. As strategist Herman Kahn argued, regarding space, in 1960, "It is very easy to make the obvious Mahan analogy on 'control of the sea' and talk blithely and superficially of 'control of space.' The analogy was never really accurate even for control of the air, and . . . it seems to be completely misleading for space."[27] Similarly, U.S. Navy Commander John Klein observes that "space is a unique

[26] Lt. Col. (USAF) Peter L. Hays, *United States Military Space: Into the Twenty-First Century*, Occasional Paper No. 42 (Colorado Springs: U.S. Air Force Academy, Institute for National Security Studies, September 2002), p. 15.

[27] Herman Kahn, *On Thermonuclear War* (Princeton, N.J.: Princeton University Press, 1960), p. 486.

environment, and any historically based strategic framework—whether naval, air, or maritime—cannot be realistically taken verbatim in its application to space strategy."[28] This is not to say that the sea and air power analogies do not have relevance for space today, but the linkages are not as clear nor as directly relevant as their supporters have argued. Moreover, the experience of at least the first fifty years of space security has shown little evidence of such policies.

Antarctica and Possible Lessons for Space

A third analogy that has been referred to in regard to space is the case of the frozen continent of Antarctica.[29] Although expert Christopher Joyner writes that the Antarctic case has been "largely ignored and undervalued" by experts in international relations,[30] the importance of the continent and its unique governance is now beginning to be recognized because of the "far-reaching impacts exerted by Antarctica on the Earth's climate, atmosphere, and oceans."[31]

Whaling and seal-hunting ships visited Antarctica sporadically beginning in the early 1800s, but the harshness of the Antarctic climate kept the continent from settlement and year-round occupation until after 1945.[32,33] But this did not stop a number of states from making formal claims on Antarctic territory, including Britain (1908), New Zealand (1923), France (1924), Australia (1933), Norway (1939), Chile (1940), and Argentina (1942). In the 1940s, skirmishes among Argentina, Chile, and Britain over territory in the region threatened the militarization of the Antarctic. German and South American troops occupied parts of Antarctica during World War II.

[28] Cmdr. (USN) John J. Klein, *Space Warfare: Strategy, Principles and Policy* (London: Routledge, 2006), p. 20.

[29] See, for example, Phillip Jessup and Howard J. Taubenfeld, *Controls for Outer Space and the Antarctic Analogy* (New York: Columbia University Press, 1959).

[30] Christopher C. Joyner, *Governing the Frozen Commons: The Antarctic Regime and Environmental Protection* (Columbia: University of South Carolina Press, 1998), p. 52.

[31] Ibid.

[32] On the history of human activity on the Antarctic continent, see Jack Child, *Antarctica and South American Geopolitics: Frozen Lebensraum* (New York: Praeger, 1988); M. J. Peterson, *Managing the Frozen South: The Creation and Evolution of the Antarctic Treaty System* (Berkeley: University of California Press, 1988); and Chapter Two in National Academy of Sciences, *Antarctic Treaty System: An Assessment, National Research Council* (Washington, D.C.: National Academy Press, 1986). On the similarities and differences of Antarctica in regard to space, see Jessup and Taubenfeld, *Controls for Outer Space and the Antarctic Analogy.*

[33] Following difficult missions of discovery into the interior of the continent in the early 1900s, a series of more deliberate U.S. expeditions led by Admiral Richard E. Byrd in the 1920s and 1930s set up semipermanent research facilities using modern mechanized equipment and airplanes to survey the continent.

After the war, however, the growing depletion of whaling stocks and evolving public opposition to the excessive killing of wildlife (such as penguins, used for lamp oil) began to dim the perceived advantages of occupying the frozen continent.[34] Instead of military personnel, scientists came to dominate the growing population of visitors after World War II, setting up research stations rather than fortified bases. Still, political and economic claims (particularly over future mining rights—as perhaps on the Moon) continued to be made.[35]

On the initiative of the United States, the twelve countries that established research stations in Antarctica as part of the International Geophysical Year opened negotiations aimed at establishing a cooperative regime to govern international activities there. Within a year, these countries had agreed to the Antarctic Treaty, a sweeping document calling for a suspension of territorial claims, a ban on all military activities, denuclearization, and scientific cooperation.[36] Thus, despite significant international rivalry in the early part of the twentieth century, a strong cooperative arrangement emerged on Antarctica after 1961.

As time wore on, however, Third World nations began to challenge the closed membership of the new Antarctic Treaty system.[37] These outsiders argued that the existing treaty cheated them out of the potential wealth to be had from Antarctic resources. To head off this challenge to the regime, the leading nations of the Antarctic Treaty invited key opponents to visit the region in the late 1960s. Many did, but when they witnessed the extreme harshness of the environment and the obvious near-term impossibility of setting up profitable mining or fishing operations, they dropped their opposition to the restrictive regime.

By the late 1980s, however, developments in technology posed new threats to the Antarctic Treaty, as certain member nations identified commercially viable projects for exploiting such Antarctic resources as oil, minerals, and

[34] Beyond the more widely established sealing and whaling activities of nations, there was an active penguin oil industry run from Australia in the early 1900s in which some 150,000 Antarctic penguins a year were killed and boiled down for their (albeit limited) oil fat. Public outcry in that country, however, finally ended this practice. On these issues related to the industrial killing of Antarctic wildlife, see K. D. Suter, *World Law and the Last Wilderness* (Sydney: Friends of the Earth, 1980).

[35] For a detailed discussion of the competing national claims in the Antarctic, see Child, *Antarctica and South American Geopolitics*; for a formal analysis (using an expected utility model) of these issues, see William E. Westermeyer, *The Politics of Mineral Resource Development in Antarctica: Alternative Regimes for the Future* (Boulder, Colo.: Westview Press, 1984).

[36] The nations agreed to use the right of free inspection of bases as the primary guarantor of this agreement.

[37] It had been limited to those with a commitment to the region, based on the maintenance of a scientific research station.

seafood.[38] In June 1988, representatives from thirty-three countries initialed a mining agreement to begin what many had long expected: the "inevitable" process of commercial development on the Antarctic continent.[39]

These events led a motivated and highly committed group of scientists—headed by French oceanographer Jacques Cousteau—to begin a serious effort to save Antarctica from commercial exploitation and possible environmental disaster. Cousteau's widely recognizable name and worldwide respect for his work on oceans contributed to media coverage and pressure on various governments. Finally, France and Australia—aided by treaty rules requiring unanimity for all changes—blocked the new commercial agreement on the argument that it would cause irreparable environmental damage.[40] By the time the Antarctic Treaty came up for formal renegotiation in 1991, lobbying by scientists and supportive governments in a number of countries had turned the tide against commercial development. After a brief holdout by the United States, treaty members agreed unanimously to prohibit Antarctic mining for a period of fifty years.

In critiquing the relevance of the Antarctic case to space, we can observe similar outcomes for space in the decision of countries to adopt rules of nonterritoriality and to sharply limit military activities. The role of scientific knowledge in altering initial military-led strategies can also be seen. But there are also a number of differences. Space witnessed extensive military testing through 1962, and then sporadic tests of more limited systems for anti-ballistic missile defense and anti-satellite purposes subsequently. Moreover, extensive military support programs played a major role in the U.S. and Soviet space programs throughout much of the Cold War. Similarly, commercial programs made up an increasing portion of space activity, suggesting other priorities besides environmental concerns. Finally, overall spending on space activity dwarfs the amount of money associated with activities in Antarctica, suggesting that coop-

[38] Developments in the areas of oil drilling techniques and shipboard food processing equipment (for Antarctic krill) were particularly important in changing cost-benefit calculations regarding commercialization of the Antarctic. For a detailed description of Antarctic resources and legal issues raised by their possible exploitation, see Gillian D. Triggs, *The Antarctic Treaty Regime: Law, Environment and Resources* (New York: Cambridge University Press, 1987).

[39] The main interest of these states was in oil extraction and the use of Antarctic mineral resources. See Philip Shabecoff, "Development Seen for the Minerals of All Antarctica," *New York Times*, June 8, 1988, p. 1.

[40] A joint statement of the French and Australian and prime ministers argued, "Mining in Antarctica is not compatible with protection of the fragile Antarctic environment." Cited in Malcolm W. Browne, "French and Australians Kill Accord on Antarctic," *New York Times*, September 25, 1989, p. 8.

eration on the frozen continent may be a factor of the relative lack of national attention. For similar outcomes to be realized in space, much greater pulling and hauling, politically, economically, and militarily, would be needed to bring parties around to cooperate and to keep them moving in that direction. Still, some aspects of the Antarctic case, especially in terms of science-based collective learning, may be relevant to explaining the process of conflict limitation seen in space.

What is clear from this discussion is that none of the common analogies applied to space provide a close match to actual dynamics in space. Each, however, helps explain pieces of the puzzle of space security. Analyzing them also contributes to a better grounding in the history of expansionist security and the range of outcomes that are possible. The next task is to move from these general metaphors to more complete explanations and policy perspectives on space that have been developed by analysts over the first fifty years of space security.

Existing Schools of Thought

Despite its growing topicality, the debate on space security has largely been among experts. With the exception of the *Sputnik* period, the years of the Moon race, particularly 1969, and periodic disasters (such as the 2003 *Columbia* accident), public attention to space security has been relatively limited. Nevertheless, among space analysts there has been a wide range of opinion, allowing us to break down the debate between "space defense" versus "space sanctuary" perspectives (mentioned in the Introduction) into four main schools of thought, prediction, and policy representation. Moving from the most conflictual to the most cooperative, they are *space nationalism, technological determinism, social interactionism,* and *global institutionalism.*[41] We will begin with the two poles and also the two largest schools (space nationalism and global institutionalism) and then take on the smaller, but more nuanced technological determinism and social interactionism. Unlike other accounts, this discussion will analyze these schools as dynamic entities (not static perspectives) and track their refinements over time in terms of argument and emphasis as changing space events have stimulated adjustments.

[41] For different categorizations, see Karl Mueller, "Totem and Taboo: Depolarizing the Space Weaponization Debate," in Logsdon and Adams, *Space Weapons;* and Hays, *United States Military Space.* Mueller breaks the debate down into six schools: space racers, space controllers, space hegemonists, sanctuary idealists, sanctuary internationalists, and sanctuary nationalists. Hays employs four categories, although with slightly different criteria than are used here: space hawks, inevitable weaponizers, militarization realists, and space doves.

Space Nationalism

Historian Walter McDougall has argued that space activity—like nuclear weapons development—became possible as a result of large, government-run military programs being conducted by the Soviet Union and the United States after World War II.[42] Without such efforts and the bitter rivalry and fear encapsulated in this competition, human space activity would likely have occurred decades later, on a considerably smaller scale, and with a far slower pace of development. In this context, the school of space nationalism arose. This perspective derives its core dynamics from three sources: (1) the political theory of realism; (2) the competitive history of great power rivalry on prior frontiers; and (3) the context of Cold War hostility. Indeed, given the circumstances of the birth of the space age, in 1957, it is not surprising that most authors of space history have embraced these basic assumptions about space. Such varied analysts as Kash (1967), Harvey and Ciccoritti (1974), Pardoe (1984), Westwood (1984), Gray (1986), Von Bencke (1997), Lambakis (2001), Dolman (2002), and Klein (2006) have all adopted versions of space nationalism, while rejecting the notion that space might somehow be more favorable to cooperation than past great power rivalries in new environments.[43]

The space nationalist perspective on space history is aptly summed up by McDougall's comment that "the international system absorbed space just as it absorbed the atom."[44] But the core of the space nationalist school is rooted in traditions in international relations that go back as far as the work of Thucydides and his observation in the fifth century B.C.E. that "the strong do what they can and the weak suffer what they must."[45] As applied to space, realism makes the assumption that human behavior is essentially static and unchanging, meaning that the prevalence of Machiavellian notions of duplicity, power seeking, and brutality are likely.

---

[42] McDougall, . . . the Heavens and the Earth, pp. 5–6.

[43] See Don Kash, Cooperation in Space (Lafayette, Ind.: Purdue University Press, 1967); Dodd L. Harvey and Linda C. Ciccoritti, U.S.-Soviet Cooperation in Space (Miami, Fla.: University of Miami Press, 1974); Geoffrey K. C. Pardoe, The Future for Space Technology (London: Frances Pinter, 1984); James T. Westwood, "Military Strategy and Space Warfare," Journal of Defense and Diplomacy, Vol. 2, No. 11 (November 1984); Colin S. Gray, "Space Arms Control: A Skeptical View," in America Plans for Space: A Reader Based on the National Defense University Space Symposium (Washington, D.C.: National Defense University Press, 1986); Matthew J. Von Bencke, The Politics of Space: A History of U.S.-Soviet/Russian Competition and Cooperation (Boulder, Colo.: Westview Press, 1997); Lambakis, On the Edge of Earth (2001); Everett C. Dolman, Astropolitik: Classical Geopolitics in the Space Age (London: Frank Cass, 2002); and Klein, Space Warfare.

[44] McDougall, . . . the Heavens and the Earth, p. 405.

[45] Thucydides, "The Melian Dialogue," reprinted in Richard K. Betts, Conflict After the Cold War: Arguments on Causes of War and Peace (New York: Longman, 2008), p. 57.

Regarding space cooperation, Thomas Hobbes's seventeenth-century comment that "the condition of man . . . is a condition of war of everyone against everyone" helps explain these authors' highly skeptical views.[46] Given their assumption about the anarchic nature of the international system, these authors doubt that international agreements to limit competition or ban specific weapons will work, since the states that engage in them can't be trusted *not* to defect from such agreements when they feel threatened. As Colin Gray concludes on this issue, "Much of what has been said and written in favor of various proposals for space arms control amounts . . . to little more than pious nonsense."[47]

According to this perspective, what passed for cooperation during the Cold War was mainly the result of a lack of state interest in pursuing the types of space defenses or other military activities limited by space-related treaties and agreements in the first place. Everett C. Dolman of the U.S. Air Force's School of Advanced Air Power Studies, for example, argues that the 1967 Outer Space Treaty was not a highlight of superpower cooperation and restraint but instead merely a "reaffirmation of Cold War realism and national rivalry, a slick diplomatic maneuver that both bought time for the United States and checked Soviet expansionism."[48]

Space nationalists downplay the significance of U.S.-Soviet weapons restraint after 1963 by pointing to the high costs and technical inadequacy (or redundancy) of early space weapons. Through the late 1980s, they argue, the United States and the Soviet Union cooperated only in areas—such as banning nuclear weapons in space—where they did not intend to deploy such systems anyway. When new, more useful military technologies become available,[49] they predict, states will inevitably deploy them.

Military officers in this school, such as Navy Commander Klein, have outlined a range of desired American preparations for space warfare and the need for eventual establishment of a separate military space service.[50] For such analysts, space weapons are largely a question of defense, a response to the rather dismal global history of ever-escalating military threats whose root cause lies in the anarchic nature of the international system and the inevitable rise of rivals to U.S. interests.

A more extreme form of space nationalism, seeking to move beyond endless space competition, emerged after the end of the Cold War: U.S. space hege-

---

[46] Thomas Hobbes, *Leviathan, Parts One and Two* (original, 1651) (Indianapolis, Ind.: Liberal Arts Press, 1977), p. 110.

[47] Gray, "Space Arms Control."

[48] Dolman, *Astropolitik*, p. 8.

[49] Some of the areas pointed to include kinetic-kill weapons, lasers, and space mines.

[50] See Klein, *Space Warfare*.

mony. Dolman's work is largely responsible for developing this amplification, which combines traditional realism with concepts of social Darwinism popular in the late 1800s and late twentieth-century "democratic peace" theory.[51] From his perspective, space weapons and the competitive instincts they might inspire should be viewed as positive forces for both American security and for space's overall political management and commercial development. Dolman argues that "the United States is the morally superior choice to seize and control space."[52]

These hypernationalist views gained influence among neoconservatives in high positions during the George W. Bush administration. Similarly, according to Joan Johnson-Freese of the Naval War College, documents such as the Air Force's 2004 *Counterspace Operations* moved the United States "down a road leading to near-term weaponization."[53] To provide the hardware for such an effort, a 2006 report by the Independent Working Group (consisting of a number of past Reagan administration officials and longtime missile defense supporters) urged near-term U.S. deployment of one thousand space-based interceptors to defeat any potential adversary and to establish conditions of space dominance.[54] In Congress, after the 2007 Chinese anti-satellite (ASAT) test, Senator Jon Kyl (Rep., Arizona) emerged as a leading proponent of U.S. military space readiness and the need for near-term deployment of similar space-based defenses. While also serving other purposes, the February 2008 U.S. ASAT shot may have been intended in part as a "signal" from administration proponents of such a view to the Chinese military regarding U.S. resolve.

Looking back at the history of new frontiers, one can see a compelling logic behind the space nationalist position. International arms control and other restraint-based regimes have not had a very impressive record, except in recent cases where adequate verification has been possible through new technologies (or onsite inspections) or where groups of states and their militaries have embraced generally shared values and norms.[55] Yet, given the vast developments in technology since the time of the nineteenth-century great powers—which forms the most

[51] On this concept, see Bruce Russett, *Grasping the Democratic Peace: Principles for a Post-Cold War World* (Princeton, N.J.: Princeton University Press, 1993).

[52] Dolman, "Space Power and US Hegemony," p. 39.

[53] Joan Johnson-Freese, *Space as a Strategic Asset* (New York: Columbia University Press, 2007), p. 3.

[54] See the text of the Independent Working Group's "Missile Defense, the Space Relationship, & the Twenty-First Century: 2007 Report," posted on the IWG's Web site at <http://ifpa.org/pdf/IWGreport.pdf> (accessed July 31, 2006).

[55] The exceptions include European restraint regarding chemical weapons use after World War I, which was based on the collective rejection of such weapons by national militaries, largely on

salient point of reference for many space nationalists—one must raise the question of what factors might have changed (or may yet change) due to the influence of enhanced international communications, knowledge about modern weapons and their effects, or conditions related to the specific environment of space.

Consistent with realism, however, space nationalists reject the possible transformative role of emerging actors in space, including transnational corporations, nongovernmental organizations, multistate consortia, venture capitalists, and international organizations. Typical of this line of thinking is James Westwood's prediction that "the historic linkage between commerce and military activities will carry over into space."[56] The end result of this competition concerning space, therefore, is viewed as an increasingly militaristic drive by the leading space powers to secure geostrategic advantages over their rivals, as during the age of sea power.

Global Institutionalism

A second and sharply contrasting perspective, developed around the time of the International Geophysical Year (IGY) organized by scientists worldwide for 1957–58, focused on hopes that space might become a sanctuary from world political conflicts. The IGY had helped bring new attention to space and the desirability of international cooperation in exploring this exciting new environment. The global institutionalist school emphasizes the possible role of new forms of shared human and scientific thinking, supported by international cooperation, treaties, and organizations, in providing space security rather than weapons-based approaches. Its adherents take a far more optimistic view of the lessons of space history and the prospects for future cooperation, seeing space cooperation as a means of transcending conflicts on Earth. As British space writer Arthur C. Clarke wrote in 1959, "Only through space-flight can Mankind find a permanent outlet for its aggressive and pioneering instincts."[57] German-born U.S. space enthusiast Willey Ley similarly hypothesized that "nations might become 'extroverted' to the point where their urge to overcome the unknown would dwarf their historic desires for power, wealth, and recognition—attributes that have so often led to war in the past."[58] Ley noted in this regard the establishment already in 1959 of the U.N. Committee on the Peace-

---

professional grounds. Another example is U.S.-Soviet arms control, which, as will be discussed later in this book, was greatly facilitated by the new technology of space-based reconnaissance.

[56] Westwood, "Military Strategy and Space Warfare."

[57] Arthur C. Clarke, *The Exploration of Space* (New York: Harper, 1959), p. 181.

[58] Willey Ley, *Harnessing Space* (New York: Macmillan, 1963), p. 223.

ful Uses of Outer Space.[59] Another early adherent to the global institutionalist school, physicist Albert R. Hibbs, asked rhetorically in arguing against military-led nationalism in space and instead in support of a human-wide approach to the future manned exploration: "Is it not possible that we will help [in this process] simply because we want a man to stand on Mars?"[60]

Although global institutionalists rarely mentioned political theory, their assumptions expressed concepts going back centuries within so-called idealist approaches to international relations. Seventeenth-century Dutch lawyer Hugo Grotius, for example, observed that man is endowed by his creator with a higher form of reason than animals and argued that "among the traits characteristic of man is an impelling desire for society, that is, for the social life—not of any and every sort, but peaceful and organized according to the measure of his intelligence."[61] A supporting elaboration of these views for space could be traced back to Immanuel Kant's assertion that "perpetual peace" could be achieved by universalist thinking and a federation of nations.[62] As applied to space, analysts used similar concepts to make the case that humans might be able to live peaceably in space through new methods of transnational governance. Indeed, early members of this school saw space as a means of escaping traditional patterns of human conflict, thanks in part to the positive pressures exerted by, on the one hand, international communications and, on the other, a desire to avoid catastrophic war. They depicted cooperation as the *more likely* outcome in space, compared to competition,[63] and argued that as states integrated their economies and national identities began to break down, old notions of state-centric realism could become anachronistic and even fade into history. One especially innovative 1965 book suggested breaking out of superpower military competition via the redirection of defense funding, arguing, "By inviting Soviet cooperation in an intensive program of space exploration . . . we would tend to eliminate warlike preparations."[64] This study concluded that heightened space

---

[59] Ibid.

[60] A. R. Hibbs, "Space Man Versus Space Machine," in Lester M. Hirsh, ed., *Man and Space: A Controlled Research Reader* (New York: Pitman, 1966), p. 87.

[61] Hugo Grotius, "War, Peace, and the Law of Nations," reprinted in Paul R. Viotti and Mark V. Kauppi, *International Relations Theory: Realism, Pluralism, Globalism, and Beyond* (New York: Longman, 1999), p. 411.

[62] Immanuel Kant, "Perpetual Peace: A Philosophical Sketch" (original, 1795), reprinted in John A. Vasquez, *Classics of International Relations* (Upper Saddle River, N.J.: Prentice-Hall, 1996).

[63] More recent concepts of globalization support this case, although few discuss space directly. On globalization, see, for example, Thomas L. Friedman, *The Lexus and the Olive Tree: Understanding Globalization* (New York: Anchor, 2000).

[64] Frank Gibney and George J. Feldman, *The Reluctant Space-Farers: A Study in the Politics of Discovery* (New York: New American Library, 1965), p. 168.

investments would "make further armament expenditures immensely difficult if not impossible."[65]

While some of these more fanciful views did not take hold, evidence to support the global institutionalist case began to emerge early in the space age. The 1963 signing of the Partial Test Ban Treaty, halting space nuclear tests, showed that cooperation between the two rivals had begun and represented a viable alternative to seemingly inevitable space conflict.[66] By the mid-1960s, the two rivals took another major step toward *limiting* the scope of their competition by negotiating the Outer Space Treaty in 1967 and opening it to international membership at the United Nations.[67] This agreement applied existing international law to space, banned all military activities on the Moon and other celestial bodies (on threat of open inspection rights granted to signatory states), and most importantly, removed the Moon and celestial bodies from territorial competition by declaring them to be "the province of all mankind." Soon after, other cooperative efforts followed, including the ABM Treaty and the Apollo-Soyuz joint manned mission. In the commercial area, the Convention on International Liability (1972) and the Convention on Registration of Objects (1974) added further stability and "rules" to space activity.[68] As one analyst observed in 1976, "The USA and USSR have gone further to achieve arms control in space than in any other area."[69] This evidence clearly seems to contradict space nationalist patterns and predictions. Peter Jankowitsch observed in 1976: "In the past [such as with the oceans and the world's airspace], international cooperation was slow to follow new dimensions of human activity."[70] But in space, human activity was "soon followed by the development of new forms of international cooperation, including the rapid formation of a new body of international law."[71]

The global institutionalist school quickly peaked in the early to mid-1970s, when the decline of U.S.-Soviet détente resulted in a sharp decline in civilian space cooperation and yielded to new military space testing in the late 1970s

[65] Ibid.

[66] See Glenn T. Seaborg, *Kennedy, Khrushchev, and the Test Ban* (Berkeley: University of California Press, 1981).

[67] See text of the Outer Space Treaty as passed by the United Nations General Assembly (Resolution 2222) on December 19, 1966, on the State Department Web site at <http://www.state.gov/t/ac/trt/5181.htm> (accessed September 9, 2006).

[68] These agreements are described in greater detail in Chapter Four.

[69] William H. Schauer, *The Politics of Space: A Comparison of the Soviet and American Programs* (New York: Holmes and Meier, 1976), p. 71.

[70] Peter Jankowitsch, "International Cooperation in Outer Space," Occasional Paper No. 11 (Muscatine, Iowa: Stanley Foundation, 1976), p. 3.

[71] Ibid., p. 4.

and early 1980s. By the late 1980s, however, the school had resumed its development. Now somewhat sobered by past disappointments, the global institutionalists had largely abandoned idealist notions for more achievable notions of neoliberalism.[72] In other words, analysts no longer predicted an ultimate philosophical convergence among states in space but instead a form of enlightened self-interest and improved behavior through the benefit of cooperative space treaties, international organizations, and new forms of bilateral and multilateral engagement in space. The rapid growth in U.S.-Russian collaboration in a number of highly sensitive areas of spaceflight after 1991 seemed to confirm their predictions of a coming new era in space. But Bush administration policies after 2001, inspired by concepts of space nationalism, explicitly rejected new treaty-based approaches and additional "rules" for space, thus moving these ideas to the back burner of U.S. policymaking.

Today, a growing *international* pressure for new legal instruments to prevent conflict in space continues to motivate this school of thought, as seen in the nearly unanimous international support at the United Nations for the yearly resolution on the Prevention of an Arms Race in Outer Space. Global institutionalists emphasize the role of international treaties in preserving the benefits of space and the need for expanded efforts to close existing loopholes and create strong prohibitions against the testing and deployment of weapons in space.

Air Force Lieutenant Colonel Bruce DeBlois, for example, rejects the inevitability of space nationalism. He describes the dichotomy of "either defending space assets with weapons or not defending them at all" as a *"false dilemma."*[73] Instead, he argues for broadening the tool kit and abandoning the U.S. "do nothing" diplomatic strategy for space. DeBlois makes the global institutionalist case that a smarter U.S. policy would be one of undertaking "intense diplomatic efforts to convince a world of nations that space as a sanctuary for peaceful and cooperative existence and stability best serves all."[74] As Theresa Hitchens argues, new forms of international cooperation "will be . . . necessary to ensuring the future security of space."[75]

Among European experts, German legal scholar Detlev Wolter has called for

---

[72] Neoliberalism seeks to explain cooperation among states not on the basis of the inherent "goodness" (or perfectibility) of human beings, but instead on the basis of self-interest under conditions of interdependence and the evolution of legal and other rule-based norms.

[73] Lt. Col. (USAF) Bruce DeBlois, "Space Sanctuary: A Viable National Strategy," *Aerospace Power Journal*, Vol. 12, No. 4 (Winter 1998), p. 48.

[74] Ibid., p. 53.

[75] Theresa Hitchens, *Future Security in Space: Charting a Cooperative Course* (Washington, D.C.: Center for Defense Information, September 2004), p. 91.

the negotiation of a Cooperative Security in Outer Space Treaty and the formation of a formal international organization to implement the new agreement.[76]
The treaty would ban destructive weapons from space, including ASATs, space-
strike weapons, and antiballistic missile technologies. It would also set up an
international system for monitoring and verification. Wolter's concept is consistent with treaty proposals at the United Nations offered by China and Russia
in recent years but goes further to institutionalize decision making and implementation at the international level. In the United States, the 2002 proposal
from Congressman Dennis Kucinich (Dem., Ohio) to cut off U.S funding for
space defenses and to negotiate a binding treaty to prevent the weaponization
of space fits into this school as well.[77] Political scientist and former State Department official Nancy Gallagher argues that true space security will "require
formal negotiations, legally binding agreements, and implementing organizations that have both resources and political clout."[78]

Technological Determinism

A third school of thought regarding space security has focused not on political factors but instead on technology and the resulting structural context
of space decision making. This school arose in part out of the technological
optimism that pervaded the United States in the 1950s, when officials predicted
that nuclear power would soon provide safe and virtually free electricity. Space
technology could offer spin-off benefits for life on Earth that would improve
living standards and make work less difficult. But others foresaw a darker evolution: the emergence of military space technologies that would likely lead
to conflict and possibly large-scale destruction in or from space. Such fears
seemed easily predictable given the evidence of the ongoing superpower arms
race. At the same, however, nuclear war had not occurred during the Cold War,
thus giving technological determinists the ability to consider outcomes with
less than cataclysmic consequences and developments that might fall short of
warfare in space.

Early in the Cold War, the optimistic school of technological determinism
emerged in the form of science-based "convergence" theories. Such concepts,

---

[76] Detlev Wolter, *Common Security in Outer Space and International Law* (Geneva: U.N. Institute for Disarmament Research, 2006).

[77] See text of the "Space Preservation Act of 2002" (H.R. 3616), on the Web site of the Federation of the American Scientists at <http://www.fas.org/sgp/congress/2002/hr3616.html> (accessed
February 19, 2007).

[78] Nancy Gallagher, "Towards a Reconsideration of the Rules for Space Security," in John M.
Logsdon and Audrey M. Schaffer, eds., *Perspectives on Space Security* (Washington, D.C.: Space
Policy Institute, George Washington University, December 2005), p. 35.

in fact, linked two very different sets of analysts writing on space security—Soviet and American technologists. Although many of their assumptions differed (such as about the processes of technological change), both subscribed to a fundamentally *materialist* view of history and mankind's space possibilities. This led them to view space activity as a likely driver of new forms of internationalism, thus breaking down existing political barriers between states.

The early U.S. group of technological determinists predicted that cooperation in space would arise out of the objective forces of advanced scientific research and development. Their reasoning was that cost and complexity would eventually drive states to work together in space, as in other areas of high technology. They argued, for example, that the massive, state-funded technological programs developed since World War II would contribute to international stability and create new forms of social engagement by urging caution on their possessors. As Victor Basiuk argued, "Advanced technologies, because of their huge costs, large scale, and, in the case of nuclear weapons, immense destructive power, provide an important impetus to international cooperation."[79] These theorists argued that societies were beginning to converge because of the necessity of performing similar, technologically oriented tasks. Under these conditions, some speculated that competition might itself fade away due to the increasing similarity of erstwhile adversaries, with ideological differences eventually fading into irrelevance.[80] Futurist Neil P. Ruzic foresaw the Moon being settled first by separate teams of Americans and Soviets, who would then begin to cooperate after several decades in the face of shared technological challenges.[81]

Meanwhile, Soviet space analysts writing in the middle of the Cold War enunciated a similar view, although with different political conclusions. Within their communist-inspired framework, Soviet authors portrayed advanced space systems as helping to drive the world beyond its existing conflicts, thus integrating international social forces within the ongoing so-called scientific-technical revolution.[82] Such dynamics, they predicted, would eventually lead to

[79] Victor Basiuk, *Technology, World Politics, and American Policy* (New York: Columbia University Press, 1977), p. 7.

[80] For a recent (and highly detailed) application of broader (economically based) convergence theory, see Harold L. Wilensky, *Rich Democracies: Political Economy, Public Policy, and Performance* (Berkeley: University of California Press, 2002).

[81] Neil P. Ruzic, *Where the Winds Sleep: Man's Future on the Moon, A Projected History* (Garden City, N.Y.: Doubleday, 1970).

[82] For a thorough discussion of this trend among Soviet theorists during the Brezhnev era, see Erik P. Hoffmann and Robbin F. Laird's two exhaustive studies, *The Scientific-Technological Revolution and Soviet Foreign Policy* (New York: Pergamon Press, 1982); and *Technocratic Socialism: The Soviet Union in the Advanced Industrial Era* (Durham, N.C.: Duke University Press, 1985).

the creation of the single world class that Karl Marx had predicted and, therefore, to harmony in space. This Soviet school included such varied authors as Vereshchetin (1977), Lukin (1980), Savitskaya (1985), Zhukov (1985), Gavrilov and Sitnina (1985), and Sagdeev (1986).[83] In the course of social progress, they saw space playing a leading role, since it represented the *most* advanced area of human technology. As Gavrilov and Sitnina argued, "Never before has space played such a role in the business of the transformation of civilization as it does in our time."[84] These analysts saw the so-called atomic space age as a new human era, whose technologies would create the advanced industrial conditions necessary for the emergence of a harmonious, communist society. But censorship and state direction clearly affected this work, as such writings and their scripted unanimity evaporated with the demise of the Communist Party and the emergence of an independent Russian Federation.

On the more pessimistic side of the technological determinist school, a significant group of analysts in the space security debate emerged out of a concern about the U.S. and Soviet military-industrial complexes and the factors driving military and space procurement. It included such authors as Jessup and Taubenfeld (1959), Frutkin (1965), and York (1970).[85] In general, the gloomy predictions of these authors were rooted in their observation that it would be difficult to halt the seemingly inevitable superpower development of more advanced and more destructive military technologies, including those for space. At the domestic politics level, these authors identified several forces at work: natural fears of military leaders about their need to protect the nation, and

[83] See V. S. Vereshchetin, *Mezdunarodone sotrudnichestvo v kosmose* [International cooperation in space] (Moscow: Nauka, 1977); P. I. Lukin, et al., *Kosmos i pravo* [Space and law] (Moscow: Institute of State and Law, 1980); G. P. Zhukov, *Kosmos i mir* [Space and peace] (Moscow: Nauka, 1985); S. Savitskaya, "Gorizonty otkrytogo kosmosa" [Horizons of deep space], *Kommunist*, No. 6 (April 1985); V. M. Gavrilov and M. Yu. Sitnina, "Militarizatsiya kosmosa: novaya global'naya ugroza" [The militarization of space: a new global threat], *Voprosy Istorii*, No. 11 (1985); R. Sagdeev, "Era kosmonavtika—znachit: era cheloveka!" [The era of cosmonautics means the era of mankind], *Kommunist*, No. 5 (March 1986). The relative symmetry of Soviet interpretations was at least in part the result of censorship restrictions favoring pro-cooperation analyses. Below the surface, based on interviews conducted with leading Soviet space experts in Moscow during the late 1980s, the situation was more complicated, involving a significant (though minority) pessimist (or self-described realist) contingent.

[84] Gavrilov and Sitnina, "Militarizatsiya kosmosa," p. 94.

[85] See Jessup and Taubenfeld, *Controls for Outer Space and the Antarctic Analogy*; Arnold W. Frutkin, *International Cooperation in Space* (Englewood Cliffs, N.J.: Prentice-Hall, 1965); and Herbert F. York, *Race to Oblivion* (New York: Simon & Schuster, 1970). See also York's later book with discussion of space developments, *Making Weapons, Talking Peace: A Physicist's Odyssey from Hiroshima to Geneva* (New York: Basic, 1987).

more self-interested motives within industry and the armed services. They described problems such as difficult-to-stop military research programs, patterns of political deference to "expert" advice on technical subjects like space, and simply cagily disguised profit motives sold as the "national interest." Herbert York, former nuclear weapons scientist and defense department research and development director under President Eisenhower, identified interservice rivalries that drove many overly expensive and duplicative missile programs in the 1950s.[86] He quoted Eisenhower on the sometimes harmful effects of collaboration between the "military-industrial complex" and the "scientific-technological elite" in helping to drive these dynamics.[87] York found similar processes at work in the nuclear weapons complex and the related arms race. He wrote in 1976: "In short, the root of the problem has not been maliciousness, but rather a sort of technological exuberance that has overwhelmed the other factors that go into the making of overall national policy."[88] York recognized the risks of scientific fascination with weapons research and the tendency of such trends to lead occasionally to exaggerated perceptions of threat.[89] Similarly, national leaders—whether in Washington, Moscow, New Delhi, or Beijing—face difficult choices in complex national security debates over technology. York, though, believed such harmful deterministic processes could be subverted and short-circuited through collective political action and exposure of these phenomena. But as the space age wore on and as U.S.-Soviet relations moved from détente to renewed hostility and military buildups in the late 1970s and early 1980s, prospects for such developments seemed slim.

At around this time, another variation of technological determinist thinking emerged from European political economy thought regarding space, focusing on the theory of "collective" (or public) goods.[90] This framework viewed space-faring nations as self-interested rational actors making decisions according to shifting economic and strategic calculations. Drawing on Garrett Hardin's famous 1968 article about the "tragedy of the commons,"[91] which had focused

[86] York, *Making Weapons, Talking Peace*, p. 173.

[87] Ibid., p. 125.

[88] Herbert F. York, *The Advisors: Oppenheimer, Teller, and the Superbomb* (Stanford, Calif.: Stanford University Press, 1976), p. xiii.

[89] See York, *Making Weapons, Talking Peace*.

[90] As John Conybeare defines this concept, "A public good is one which has the property of nonappropriability . . . and indivisibility . . . since consumption by one person does not reduce the amount available to anyone else." See John A. C. Conybeare, "International Organization and the Theory of Property Rights," *International Organization*, Vol. 34, No. 3 (Summer 1980), p. 327.

[91] See Garrett Hardin, "The Tragedy of the Commons," *Science*, 162 (1968). Hardin's article used the analogy of cooperative sheep grazing on historical British common grounds and pointed out

attention on the harmful environmental implications of technological change, overcrowding, and failed human management, Swedish analyst Per Magnus Wijkman[92] warned that pressure on states to defect from cooperative space regimes for commercial and military benefit could be expected to grow as the expansion of actors and activities increased the advantages of "enclosure" (or privatization) of space. Unfortunately, as space activity began to move toward multiple actors as the Europeans, Japanese, and others entered space, the prospect for continued cooperation seemed less favorable than during the bilateral U.S.-Soviet space race. On the other hand, as Wijkman argued, conditions of "interdependence" in space (such as involving collision avoidance) gave countries "strong incentives to agree to measures to keep interference at a mutually accepted level."[93] The periodic emergence of U.S.-Soviet restraint in space during the Cold War supported this view, suggesting that in games with repeated "plays," in which states can communicate and adjust their behavior, outcomes might become more favorable. The question after the Cold War was how to extend these lessons into a more advanced technological environment with a greater number of actors and possibly continued military tensions.[94] Likely areas of conflict included critical regions of low–Earth orbit used extensively by the military, locations in geostationary orbit, radio frequencies for satellites, and minerals on the Moon.[95]

Fortunately, trends during the early 1990s seemed to indicate declining national interest in military space weapons technologies. But by the late 1990s, analysts again began to focus greater attention on these issues, as space-based missile defenses returned to active consideration by the U.S. government. The Bush administration's active discussion of space-based kinetic-kill vehicles,

---

the problems that eventually arose from increasing crowding and the presence of even a minority of selfish actors, eventually ruining this cooperation and leading to the grounds' enclosure. Hardin suggested a number of radical policy measures to prevent such international tragedies in the areas of pollution, overpopulation, and nuclear weapons.

[92] See Per Magnus Wijkman, "Managing the Global Commons," *International Organization*, Vol. 36, No. 3 (Summer 1982).

[93] Ibid., p. 535.

[94] As Kenneth Waltz argues, "The likelihood that great powers will try to manage the system is greatest when their number reduces to two." See Kenneth N. Waltz, *Theory of International Politics* (Reading, Mass.: Addison-Wesley, 1979), p. 198. The logic of Waltz's argument extended to space suggests that while meaningful cooperation may have occurred as part of superpower attempts to manage the nuclear world, post–Cold War multipolar conditions (in the presence of multiple states in space) are likely to make cooperative outcomes much more difficult.

[95] On these debates, see Andrew Brearley, "Mining the Moon: Owning the Night Sky?" *Astropolitics*, Vol. 4, No. 1 (Spring 2006).

space-based lasers, and even possible Global Strike weapons renewed interest in the technological determinist school. For most of these analysts, including Lupton (1998), Hays (2002), and O'Hanlon (2004),[96] these dynamics were neither good nor bad, but simply inevitable. The questions then became, what to do about them and, perhaps, how to manage them once they arrived? Although the most recent technological determinists in the space security field do not use a collective goods approach, their arguments are generally consistent with its assumptions and concerns.

Hays foresees a gradual process of weaponization, arguing that "as current political and technological challenges are surmounted . . . it is likely that space . . . will become weaponized and will emerge as an important RMA [Revolution in Military Affairs]."[97] Similarly, the Brookings Institution's Michael O'Hanlon draws implicitly on this notion of gradual, technologically influenced change to make the case for a middle ground in the current space debate, saying, "Extreme positions that would either hasten to weaponize space or permanently rule it out are not consistent with technological realities and U.S. security interests."[98] He and others in this school urge caution on the United States and others to avoid the *aggressive* arming of space, such as that advocated by some space nationalists. This position resonates with many military officers who see space threats on the horizon but also see the desirability of trying to prevent and manage future conflicts, including through possible negotiations with other space powers.[99]

In this context, U.S. Air Force Colonel John Hyten's approach to the challenges of technological pressures in space offers some more optimistic prospects: "If we negotiate openly with the nations of the world; if we allow our industry to exploit space fully and become the unquestioned leader of the information age; and if we develop the means and methods to deal effectively with inevitable conflicts in space, perhaps the new ocean to which President Kennedy referred could remain a 'sea of peace.'"[100]

Thus, according to current technological determinists, management arrangements may be possible but will rely on favorable structural conditions,

---

[96] Lupton, *On Space Warfare*; Hays, *United States Military Space*; and Michael E. O'Hanlon, *Neither Star Wars nor Sanctuary: Constraining the Military Uses of Space* (Washington, D.C.: Brookings Institution, 2004).

[97] Hays, *United States Military Space*, p. 3.

[98] O'Hanlon, *Neither Star Wars nor Sanctuary*, p. 21.

[99] On this perspective, see Col. (USAF) John E. Hyten, "A Sea of Peace or a Theater of War? Dealing with the Inevitable Conflict in Space," in Logsdon and Adams, *Space Weapons*.

[100] Ibid., p. 251.

communications, and political bargaining. To understand these factors and their possible role in space security, we need next to examine the final and most recent school of thought: social interactionism.

Social Interactionism

In the 1980s, with the Cold War beginning to wind down, some political scientists and space analysts began to focus on the possibility of long-term co-operative links in space, even among rival states. The warming of U.S.-Soviet relations and the demise of the Strategic Defense Initiative seemed to indicate a loss of steam for competitive approaches to space security and the diminution of military-led tendencies. New cooperative ventures seemed possible. But the Clinton administration's seeming disinterest in global institutionalist remedies to space insecurities, seen in its failure to propose any new space treaties, suggested that more ad hoc cooperation and management mechanisms might become the norm instead.

Social interactionists rejected the notion of the inevitability of space weapons, given the availability of policy tools among space-faring states to interact with one another, bargain, and prevent the deployment of harmful weapons, which could damage other priorities they have in space. Outcomes, however, were seen as contingent and sometimes imperfect given the nature of political realities. Paul Stares, one of the school's early representatives, observed in 1985 that "ASAT arms control cannot eliminate the threat to space systems, only bound it."[101] But, he continued, "the different approaches to the control of ASAT weapons can work synergistically: the shortcomings of one agreement can to a large extent be remedied by the provisions of another."[102] Stares concluded by suggesting the development of a "rules of the road" approach as one possible remedy.

A number of advocates of this general approach emerged in the mid-1980s among analysts who studied the U.S.-Soviet arms control process. Considerable research up to this time in the international relations field had explored related concepts of cognitive change (or learning) at the individual,[103]

---

[101] Paul B. Stares, *Space and National Security* (Washington, D.C.: Brookings Institution, 1987), p. 172.

[102] Ibid.

[103] For example, see Robert Jervis, *Perception and Misperception in International Politics* (Princeton, N.J.: Princeton University Press, 1976); Richard E. Neustadt and Ernest R. May, *Thinking in Time* (New York: Free Press, 1986); Steven Kull, *Minds at War: Nuclear Reality and the Inner Conflicts of Defense Policymakers* (New York: Basic, 1988); and Deborah Welch Larson, *Origins of Containment* (Princeton, N.J.: Princeton University Press, 1985).

organizational,[104] and state[105] levels. As space analysts had observed, some of these types of behavior had also occurred regarding space, in one form or another. John Lewis Gaddis, for example, described the emergence of a "tacitly agreed upon satellite reconnaissance regime" between the two superpowers as a significant accomplishment made possible by extensive communication, the small number of actors, and the transparency of space.[106] Steven Weber's study of anti-satellite and other arms control attempts in the 1960s and 1970s offered a somewhat more tentative conclusion, pointing out that the two superpowers "did not learn smoothly or in a patterned way."[107] Instead, he identified "lumpy" learning that tended to be concentrated within critical periods when both sides were receptive to cooperative signals from the other. Part of the problem was that space security during the Cold War tended to be dominated by military definitions of security, which restricted the bounds of superpower learning.

Social interactionism requires knowing what will work in existing political and military conditions. For this reason, current representatives of this school do not immediately urge arms control treaties as the best solution, particularly for problems where rules of the road may be easier (and quicker) to obtain among the multiple players in space, some of which are no longer nation-states. At the same time, these analysts caution against current U.S. policies of hyping the space "threat," saying that such statements risk making weaponization a self-fulfilling prophesy. As Joan Johnson-Freese argues: "Relying exclusively on technology for security—in this case, space weapons—does not provide an asymmetric advantage; it creates a strategically unstable environment."[108] These

---

[104] For example, see Graham Allison, *Essence of Decision* (New York: Little, Brown, 1971); John Steinbruner, *The Cybernetic Theory of Decision* (Princeton, N.J.: Princeton University Press, 1974); James G. March and Johan P. Olsen, eds., *Ambiguity and Choice in Organizations* (Oslo, Norway: Universitetsforlaget, 1976); Chris Argyris and Donald A. Schon, *Organizational Learning: A Theory of Action Perspective* (Reading, Mass.: Addison-Wesley, 1978); and Ernst B. Haas, *When Knowledge Is Power: Three Models of Change in International Organizations* (Berkeley: University of California Press, 1990).

[105] For example, see William Zimmerman and Robert Axelrod, "The 'Lessons' of Vietnam and Soviet Foreign Policy," *World Politics*, Vol. 33, No. 1 (October 1981); Lloyd S. Etheredge, *Can Governments Learn? American Foreign Policy and Central American Revolutions* (New York: Pergamon Press, 1985); Joseph S. Nye, Jr., "Nuclear Learning and U.S.-Soviet Security Regimes," *International Organization*, Vol. 41, No. 3 (Summer 1987); and George W. Breslauer and Philip E. Tetlock, eds., *Learning in U.S. and Soviet Foreign Policy* (Boulder, Colo.: Westview Press, 1991).

[106] John Lewis Gaddis, "The Evolution of a Reconnaissance Satellite Regime," in Alexander L. George, Philip J. Farley, and Alexander Dallin, eds., *U.S.-Soviet Security Cooperation: Achievements, Failures, Lessons* (New York: Oxford University Press, 1988), p. 366.

[107] Steve Weber, *Cooperation and Discord in U.S.-Soviet Arms Control* (Princeton, N.J.: Princeton University Press, 1991), p. 288.

[108] Johnson-Freese, *Space As a Strategic Asset*, p. 243.

viewpoints posit a diametrically different understanding of history than that of space nationalists or technological determinists. Rather than portraying the United States as a victim of hostile historical processes, these critics assume that U.S. leaders can instead *influence* military trends in a purposeful manner through their interactions. As Michael Krepon argues, "By virtue of its leadership position in space commerce and military power, the United States now has unprecedented capacity to shape whether space becomes weaponized."[109] Krepon and Michael Katz-Hyman argue that "space, like military activities . . . on Earth, needs a code of conduct to promote responsible activities and to clarify irresponsible ones."[110] In promoting these concepts, Krepon and Katz-Hyman do not assume idealist-inspired value changes among actors, only adaptation to prevent harmful behavior. With increasing crowding in space, such action may be imperative to the continued use of various orbital regions. But if space includes a variety of new state and non-state actors, such cooperation may be difficult to accomplish.

One post–Cold War field of political science that has sought to examine prospects for learning among widely disparate actors is so-called social constructivism.[111] This successor of 1980's learning theory argues that regularized contacts and communication, particularly in an institutionalized setting, can promote common problem-solving and even gradually shared identity formation. In this regard, Alexander Wendt has coined the term "international state" to describe the process of collective-identity formation that takes place among countries under repeated interaction.[112] Such constructivist notions work well in explaining the eventual outcome in the Antarctic, where an "epistemic community"[113] of scientists helped to bring public opinion to bear on a set of negotiations that otherwise was moving toward radical commercialization, thus likely decimating the protective regime governing the Antarctic continent and its surrounding waters. But the applicability of social constructivism (as opposed to more limited social interactionism) to space is limited at present. Instead of witnessing the recent formation of strengthened collective norms and

[109] Michael Krepon (with Christopher Clary), *Space Assurance or Space Dominance? The Case Against Weaponizing Space* (Washington, D.C.: Henry L. Stimson Center, 2003), p. 88.

[110] Michael Krepon and Michael Katz-Hyman, "Irresponsible in Space," *Defense News*, February 5, 2007.

[111] On this literature, see Jeffrey T. Checkel, "The Constructivist Turn in International Relations Theory," *World Politics*, Vol. 50, No. 2 (January 1998).

[112] See Alexander Wendt, "Collective Identity Formation and the International State," *American Political Science Review*, Vol. 88, No. 2 (June 1994).

[113] An "epistemic community" can be defined as a group of experts, analysts, and like-minded officials joined by common beliefs about a body of technical knowledge in an area of public policy.

identities, certain treaties that embodied them have been under challenge in recent years. The U.S. withdrawal from the ABM Treaty in 2002 and China's 2007 ASAT test suggest that norms have not been as powerful in space as they have been in the Antarctic case. One reason may lie in the absence of public lobbying by influential figures—such as former astronauts—against space weapons. Another reason may be the much closer relationship of space to national security. For the United States, international preferences for limits on space defenses have been viewed by recent American officials as unacceptable intrusions on U.S. sovereignty and security. A further reason, however, may be an as-yet poor official understanding of the hazards to the space environment posed by space debris, an understanding that may now be changing given the negative repercussions of China's high-altitude ASAT test.

Toward a New Understanding of Space Security

What emerges from this review of the main conceptual roots of space policy analysis over the past fifty years is a mixed picture. Each of the schools analyzed offers some explanatory strengths, but each also has blind spots and weaknesses. In seeking a better means of structuring our thinking about space security's past and future, we instead return to the discussion of space security that opened this chapter, one that made reference to both man-made and natural threats. In that context, it might be useful to move space security analysis from its traditional focus on states and their militaries to the space environment itself. This shift encourages an emphasis on "softer" tools for achieving space security than military means and refocuses our attention on the "transboundary" environmental problems[114] represented by space radiation and debris.

Viewing space security from the perspective of self-interested actors seeking to protect their access to space in a gradually constricting collective goods environment may offer advantages over tying space security debates to nuclear and other "hard" security issues, which Cold War competition encouraged. Recent recognition of such problems as global warming, the depletion of fisheries, watershed shortages, and deforestation has brought new collective action to address challenges faced by un- or under-protected global commons. To date, space has figured only marginally in these discussions. But growing concerns about orbital debris may be a tipping point in pushing for more attention to such questions in space.

[114] On this category of global challenges, see Mostafa K. Tolba (with Iwona Rummel-Bulska), *Global Environmental Diplomacy: Negotiating Environmental Agreements for the World, 1973–1992* (Cambridge, Mass.: MIT Press, 1998).

Looking back across history for lessons, we can conclude that neither excessive pessimism nor excessive optimism is warranted for space security. The outcomes to date in space have been mixed in regard to cooperation and competition. Yet it is worth observing that surprising levels of restraint emerged during the first fifty years of space activity, despite a global context of political and military hostility. Making sense of these contradictory trends remains a work in progress. Changing the focus of traditional analysis regarding space may be fruitful, as a different lens sometimes brings a new and more accurate perspective to long-studied problems. In the next chapter, then, we consider what might be gained from viewing space security as an environmental management problem.

CHAPTER TWO

Space and Environmental Security

One of the more contentious debates in the space security field has been over what different experts mean by the term "space weapon." Some analysts argue that space was weaponized beginning with the German army's use of V-2 rockets during World War II, because these early ballistic missiles passed through space.[1] However, the V-2 missile made no use of the unique characteristics of space, lacked orbital speed, did not release weapons into the new environment, and was intended to accomplish an existing military mission rather than initiate a fundamentally new, space-driven and space-oriented military objective.[2] Other experts make the case that space has never been weaponized, since no nuclear, kinetic-kill, or laser weapons have ever been stationed in orbit.[3] While this is true, the past testing of nuclear weapons and conventional anti-satellite systems (ASATs) in space and of lasers against space objects seems to stretch such purist definitions beyond reasonable credulity. Still others throw up their hands and say there is no way to define a space weapon because virtually any-

[1] On these points, see Baker Spring, "Slipping the Surly Bonds of the Real World: The Unworkable Effort to Prevent the Weaponization of Space," Heritage Foundation Lecture No. 877, available on the Heritage Foundation Web site at <http://www.heritage.org/Research/NationalSecurity/hl877.cfm> (accessed September 16, 2006).

[2] A perhaps more relevant German effort to develop space-related weapons was the Luftwaffe's research on a suborbital, rocket-powered "skip" bomber for possible use against the United States. But the program was eventually cancelled because of both technical and financial constraints. For descriptions of the German "Sänger bomber" project, see Clayton K. S. Chun, "Expanding the High Frontier: Space Weapons in History," *Astropolitics*, Vol. 2, No. 1 (Spring 2004), pp. 64–67; also Michael J. Neufeld, *The Rocket and Reich: Peenemunde and the Coming of the Ballistic Missile Era* (Cambridge, Mass.: Harvard University Press, 1995), pp. 61–63.

[3] "Space Security or Space Weapons? A Guide to the Issue," Space Security Project, Henry L. Stimson Center, Washington, D.C., 2005, online at <http://stimson.org/space/pdf/issueguide.pdf> (accessed July 20, 2006).

thing in space could be used potentially as a weapon—space debris causes such damage, satellites can run into one another, and the shuttle or Soyuz could be used to "capture" small spacecraft. Some U.S. officials have used this argument to make the case that space arms control is an impossibility.[4] These all-or-nothing definitions may be conceptually satisfying, but they are not useful for promoting a practical understanding of space security.

To overcome this problem, it may be more helpful to adopt a midpoint definition for "space weapon": *any system whose use destroys or damages objects in or from space.* This designation excludes nondevoted, dual-use systems that have the capability to capture and disable satellites, as well as missiles that pass through space on their way to other intended targets without harming space assets. It also excludes systems capable of interrupting the operation of satellites (such as electronic jamming equipment), whose effects are reversible, and unintentional weapons (such as debris fragments or old spacecraft), whose possibly damaging effects are unrelated to their original missions. What it does include are ground-, sea-, or space-based anti-ballistic missile (ABM) and ASAT systems (whether lasers, kinetic-kill vehicles, or explosive systems) whose intent is to destroy objects in space, as well as any other military systems used in space that have damaged spacecraft in the past (such as nuclear weapons tested in space). By this definition, space weapons have existed for many decades and include a variety of U.S.- and Soviet-made systems, some of which are no longer deployed, as well as China's new kinetic-kill and possible laser ASAT systems. Using this definition allows us to understand that the weaponization of space has not been a linear progression. Instead, it has moved in fits and starts, sometimes even reversing itself.[5]

Although space weapons exist today, their impact on the *quality* of space security is influenced by two parallel sets of factors—one technical, one political. On the technical side, security calculations must take into account the quantity of space weapons, their readiness (including test record), their distribution among actors, the nature of their deployment (in storage, on the ground, or in space), and the availability of methods for overcoming or evading them. On the political side, space security is influenced by the breadth, effectiveness, and

---

[4] See, for example, testimony by Amb. Don Mahley, U.S. State Department, presented at the House Committee on Oversight and Government Reform's Subcommittee on National Security and Foreign Affairs hearing, "Weaponizing Space: Is Current U.S. Policy Protecting Our National Security," May 23, 2007, online at <http://nationalsecurity.oversight.house.gov/documents/20070523162935.pdf> (accessed September 20, 2007).

[5] On this point, see Karl Mueller, "Totem and Taboo: Depolarizing the Space Weaponization Debate," in John M. Logsdon and Gordon Adams, eds., *Space Weapons: Are They Needed?* (Washington, D.C.: Space Policy Institute, George Washington University, October 2003), p. 2.

depth of international support for norms, treaties, and other agreements meant to ban, limit, or control such weapons.

But such military calculations alone do not determine space security. Overlaying these specific operational and political factors is an important and often underappreciated set of *environmental* factors that affect security in space. As suggested in Chapter One and analyzed in greater detail in the rest of this chapter, risks from electromagnetic pulse (EMP) radiation, the expanding quantity of orbital space debris, and the increasing population of operational satellites and spacecraft must be understood and taken into account. Failure to do so will affect the accuracy of any predictions about the current or future state of space security.

Jessica Tuchman Matthews wrote presciently in 1989 about an emerging link between scientific understanding among governments and solving security problems in environmental issue-areas involving mixed sovereignty and multiple actors.[6] Although she did not mention space, her prescription that solving environmental security problems will require "far greater technical competence in the natural and planetary sciences among policymakers" is highly appropriate.[7] Notably, she also called for involvement from the private sector in these decisions. The approach, focus, and skill set she invokes are very different from those traditionally applied to military-security problems, which to date have tended to emphasize national responses and have involved relatively limited cooperation with commercial actors. (One notable exception has been in recent efforts at orbital debris mitigation, as will be discussed later.)

Space Weapons and the Nuclear Arms Race

Throughout the Cold War, officials and analysts emphasized the connection between the superpower nuclear arms race and space. Jack Manno argued about the dynamics of nearly three decades of space competition in 1984: "The development of space technology was closely linked to nuclear strategy. Throughout the history of the Space Age they remained inseparable."[8] Similarly, Paul Stares observed in his book *Space and National Security* written near the end of the Cold War, "The superpowers . . . agreed to a variety of . . . measures for regulating the military uses of space, although it is fair to point out that this has not

6 Jessica Tuchman Matthews, "Redefining Security," *Foreign Affairs*, Vol. 68, No. 2 (Spring 1989).

7 Ibid., p. 176.

8 Jack Manno, *Arming the Heavens: The Hidden Military Agenda for Space* (New York: Dodd, Mead, 1984), p. 36.

always been the principal intent of the relevant agreements."[9] In other words, Stares argues that space benefited from superpower arms control. While there is a logic to this perceived connection and some truth in terms of beneficial treaty language, in many cases nuclear arms control benefited from *space* more than the reverse. As Air Force Lieutenant Colonel Peter Hays argues, "Spysats were the essential element in spanning the clear and wide conceptual divide between NTM [National Technical Means] and OSI [on-site inspections]—their promise helped to establish a new type of arms control based on a 'bridge' of trust between the superpowers."[10] This scenario then implies a reversal of the traditional line of causality—with space security now *facilitating* arms control rather than the reverse. Détente-era treaties codified norms of restraint already extant in space but not yet part of nuclear arms control. Thus, the exact nature of this linkage and its validity as a catch-all explanation for military space restraint needs to be reevaluated if we are to better understand the sometimes interrelated but ultimately different causal dynamics of the nuclear and space fields.

The now largely declassified record of presidential and Politburo decision making shows that although there was congruence in the initial competitive motivations behind the nuclear and space programs, the results in terms of deployed systems had begun to diverge significantly even by the early 1960s. Since most historical attention has focused on the nuclear arms race and the highly competitive U.S.-Soviet "Moon race" of the 1960s, few analysts have made much of these countertrends. Despite the Cuban Missile Crisis in 1962, the superpowers continued to race toward much higher numbers of nuclear weapons—eventually reaching peaks of 32,000 on the U.S. side in 1965 and 45,000 on the Soviet side in 1985.[11] By contrast, space dynamics witnessed a sharp *decline* in deployed weapons and the channeling of competition mainly into civilian and military support (and later force enhancement) realms, with devoted weapons research taking place on the margins, but resulting in little testing and almost no deployments. The fact that the two sides decided to set space aside from their continued weapons buildup requires further attention.

In seeking to explain this anomaly, the traditional view is that space wasn't any different from military dynamics in other realms but that space weapons

---

[9] Paul B. Stares, *Space and National Security* (Washington, D.C.: Brookings Institution, 1987), p. 144.

[10] Lt. Col. (USAF) Peter L. Hays, *United States Military Space: Into the Twenty-First Century*, Occasional Paper No. 42 (Colorado Springs: U.S. Air Force Academy, Institute for National Security Studies, September 2002), p. 64.

[11] Robert Norris and Hans Kristensen, "Global Nuclear Stockpiles, 1945–2006," *Bulletin of the Atomic Scientists* (July/August 2006), pp. 64–66.

technologies were redundant or simply weren't mature yet; thus, temporary co-operation emerged. A more sophisticated version of this argument is that space evolved into an exception to the arms race because it was an arena that allowed both sides to spy on each other and thereby stabilize their nuclear relations. But what if we push on these standard arguments?

It seems obvious today that a prerequisite for the use of space for recon-naissance, manned spaceflight, and eventually space-based force enhancements (like the Global Positioning System) was some level of "environmental protec-tion," given the presence of space nuclear testing at the outset of the space race. While no one used this term in the early 1960s, the first steps toward limiting harmful military acts in space took place in 1963 with the ending of orbital nuclear testing. Yet, if the two superpowers in the early 1960s could have tested and deployed a range of space weapons and space reconnaissance systems while also conducting manned spaceflight safely, they certainly would have done so. Instead, they found that nuclear testing proved fundamentally incompatible with the pursuit of other goals in space. These and other weapons threatened to put valuable military and civilian spacecraft at risk. Therefore, specific norms and rules had to be established in order to prevent space's ruination. The main observation here is that military space restraint took place largely for *space*-related reasons, not because of space weapons' immaturity or because of exter-nal pressures of arms control.

To develop these points further, this chapter looks first into the *scientific* and *technical* reasons for space's unique security dynamics. It then seeks to under-stand the *political* processes through which states gradually adapted to them. This two-part model provides the book's main conceptual argument: that *en-vironmental factors have played an influential role in space security over time and provide a useful context for considering its future.*

This is not to say that other factors did not play a role in specific decisions or that actors always knowingly placed "environmental security" at the forefront of their rationales for restraining their behavior. In the early years, other strate-gic calculations often played an important role. Nevertheless, leaders generally acted to protect the space environment in order to preserve their continued use of space for a variety of civilian and military support purposes, even when that meant abandoning competitive strategies for space power projection. The process of environmental learning was partial, sometimes disjointed, and slow. Moreover, the historical record shows that international willingness to engage in military restraint based on environmental factors cannot be viewed as some-how predetermined by space technology or by space itself. Environmental con-straints provided a window of opportunity for state cooperation, but national leaders had to decide to jump through it. For this reason, we must also seek to understand the processes of these cognitive changes, since individual leaderships and the consensus decisions they were able to reach mattered greatly to the achievement of space security at every step on the way.

To highlight the role of environmental factors in space security, however, some discussion is needed, not only on what weapons technologies were developed and deployed in space, but perhaps even more importantly, on what technologies were *not* deployed or were actually *removed* from the competition. As will be argued here, what makes space security dynamics interesting is that they were *not* expressed as a simple linear race—as on past frontiers—but instead as a highly selective race that depended in a fundamental way on the nature of the space environment, international political interactions, and the gradual adoption of shared scientific understandings about how best to proceed. After seeking to understand the critical *technical* factors that created incentives for this restraint, we then need to analyze the *political* processes that allowed it to move forward cooperatively.

A Counterfactual History of Space Security

Contrary to the hopes of the various movements supporting international peace and a nuclear test ban in the 1950s, the weaponization (and "nuclearization") of space occurred almost immediately after the opening of this new environment. As noted earlier, less than a year passed between *Sputnik*'s orbit in 1957 and the U.S. testing of three nuclear weapons in space in the summer of 1958. The Soviet Union soon followed suit. Indeed, from the perspective of military planners at the height of the Cold War in the 1950s, the extension of human activity into orbital space presented unprecedented threats and offensive opportunities, suggesting the need for a range of military responses. The working assumption of military and political organizations on both sides of the Cold War was largely driven by concepts of technological determinism: once space weapons could be developed, they would (and should) be deployed. The Pentagon began planning for Moon-based nuclear weapons aimed at Earth, nuclear ASAT weapons, manned military space stations, and a variety of other offensive and defensive systems. Air Force Brigadier General Homer Boushey foresaw the military space future in a speech given in early 1958:

> The Moon provides a retaliation base of unequaled advantage. If we had a [nuclear-armed] base on the Moon, either the Soviets must launch an overwhelming nuclear attack toward the Moon from Russia two to two-and-one-half days prior to attacking the continental United States (and such launching could not escape detection) or

Russia could attack the continental United States first, only and inevitably to receive, from the Moon—some 48 hours later, sure and massive destruction.[12]

The Soviet military called for similar programs—including plans for an orbital space station, space-to-space weapons, and space-to-Earth weapons—and expected to pursue a vigorous space arms race. Worried by the rapid pace of superpower military preparations, U.S. Secretary of State Dean Rusk cautioned in 1962, "There is an increasing danger that space may become man's newest battlefield."[13] At this juncture, the future seemed already laid out. Surprisingly, events turned out very differently.

To emphasize how sharply subsequent space developments diverged from this trajectory, it is useful to engage in a counterfactual history of space activity; that is, what *could* have happened if the two sides had continued this course and embraced the technological imperative of military space competition that seemed inevitable in the 1950s and early 1960s.

If Washington and Moscow had maintained this direction of unilateral interests and technologically driven military competition in space, nuclear tests in orbit would likely have continued through at least the mid-1960s, while other types of space weapons—including a variety of ASAT and ABM systems using nuclear and conventional warheads—would have entered full-scale deployment to counter the proliferation of each side's missile capabilities and those of other emerging powers (such as China). In all likelihood, some of these space weapons would have been space based, requiring dozens if not hundreds of weapons in orbit to have enough on station at any one time above each adversary's territory. Manned space bombers would have helped to diversify reliance on ground-, sea-, and air-based nuclear delivery systems while also serving as platforms to attack hostile space assets. Such technologies would have been expensive and might have been seen as redundant by some, but they could have been justified (and indeed were at the time in Air Force circles) as "necessary" to national defense, given the Cold War and the obvious threats to national security posed by long-range missiles and ASAT interceptors.[14] However, such

---

[12] Quoted in Erik Bergaust, *The Next Fifty Years in Space* (New York: MacMillan, 1964), p. 99.

[13] Secretary of State Dean Rusk, address of June 16, 1962, *U.S. State Department Bulletin* (July 2, 1962), p. 5.

[14] Many kinds of weapons have been deployed or retained in arsenals over the course of history that have made little sense as military weapons. States often build and field weapons to bolster their credibility, to provide further military insurance, to add to their international (or domestic) prestige, or to serve some domestic interest group. The Polish cavalry in World War II is one example. Another, during the nuclear arms race, was the 2-kilometer-range U.S. Davy Crockett nuclear bazooka (which would likely have incinerated its possessor); still another was the Soviet Union's 100-megaton Tsar Bomba (which would likely have created enough deadly fallout to cripple any attacking Soviet

a militarily and technologically driven process—denying environmental concerns—would have involved important trade-offs.

With continued nuclear testing in space in the 1960s, the U.S. Gemini and Apollo manned programs and the Russian Vostok, Voskhod, and Soyuz missions—regardless of cost—would have been too dangerous to undertake. Radiation stuck in the Van Allen belts from weapons tests would have also made steady development and reliable access to commercial and military reconnaissance satellites impossible. The destruction of any spacecraft, whether in a test program or through offensive action, would have released thousands of metal fragments. By the late 1960s, at least the lower reaches of space would have been cluttered with an enormous amount of orbital debris, making any activity hazardous and extremely costly.

As technologies advanced in the late 1970s in the absence of treaties (or norms) limiting weapons, new types of ballistic missile defenses—orbital magnetic rail guns, lasers, and advanced kinetic-kill weapons—and even space-based, multiple-warhead nuclear weapons might also have been deployed. Thus, by the mid- to late 1980s, space could have been fully weaponized and treated simply as part of the overall U.S.-Soviet offensive and defensive arms race. War might or might not have taken place in space, given general conditions of military deterrence, but space would essentially have been ruined for a number of other productive uses.

Under this scenario, there would have been no prospects for manned spacecraft able to run the gauntlet of debris and automated space weapons to explore the Moon or undertake zero-gravity research. Military space reconnaissance would also have become extremely difficult and costly, except for very short missions and at much higher altitudes than was needed for useful photographic detail. In this environment, accurate intelligence about the respective nuclear arsenals of the two superpowers would have been scarce, thus increasing uncertainty, miscalculation, and the chances of nuclear war. Finally, commercial space services developing for the world economy—from near-Earth remote sensing for agriculture, urban planning, and weather forecasting to geostationary communications satellites—would have been either eliminated or put at severe risk of attack or damage, thus raising costs. Overall, this highly plausible, weapons-driven scenario would have grounded human scientific and commercial aspirations permanently on the Earth, rendering late twentieth-century economic globalization only a dream. For these reasons, it is fortunate that U.S. and Soviet leaders took a different route. Instead, they seemed to follow the

---

army). Thus, the alleged lack of a unique military demand for space weapons need not have prevented their deployment.

advice of U.S. Secretary of Defense Neil H. McElroy, who argued for putting the army's nuclear-tipped Nike Zeus ABM interceptor on hold in 1958: "We should not spend hundreds of millions . . . pending general confirmatory indications that we know what we are doing."[15]

Returning to the actual historical record, notably, we see that none of these technologically feasible systems became part of the deployed space assets of either country. Although competition remained the driving force in space activity and military research programs continued, the two sides limited (through explicit treaties and implicit strategic norms of self-restraint) technologies that would have threatened their use of space for other purposes: civilian manned missions, commercial satellites, and passive military technologies. Although the two sides developed small ASAT programs, these investments can be described as military "hedging" strategies more than full-scale weapons efforts. Despite the potential military risks of such a strategy, this rationale proved sound throughout the Cold War. No incidents of enemy destruction of any satellites took place, even though Moscow (and later, Washington) had the capability to do so. Instead, the two sides decided to enjoy the benefits of mutual restraint. As historian John Lewis Gaddis observes about trends set even at the height of the civilian space race, "The pattern of the early 1960's was one . . . of refraining from developing the full-scale anti-satellite systems that could have been put in place by that time."[16]

Contrary to competing popular images of what some say was a "sanctuary" during the Cold War[17] and what others argue was a period of "essentially unfettered" competition, the reality of the Cold War period in space turned out to be much more irregular, reflecting a messy process of trial and error. Under orders from civilian leaders and with new information about the damage weaponization had already caused to their space assets, the two militaries reversed course in the early 1960s and began to focus the brunt of their space competition on human spaceflight and military support technologies: reconnaissance, communications, early warning, and navigation, a process that changed the whole character of U.S.-Soviet space relations after the 1957–62 period.

General political relations between the two sides did not improve dramatically as a result of these changes. Except for the détente era from 1970–75 and

---

[15] McElroy, quoted in Stephen Schwartz, ed., *Atomic Audit: The Costs and Consequences of U.S. Nuclear Weapons Since 1940* (Washington, D.C.: Brookings Institution Press, 1998), p. 285.

[16] John Lewis Gaddis, *The Long Peace: Inquiries into the History of the Cold War* (New York: Oxford University Press, 1987), p. 202.

[17] On the "sanctuary" argument, see Henry L. Stimson Center, "Space Security or Space Weapons?"

the late Gorbachev-Reagan period from 1986–89, the two sides' space behavior remained fundamentally competitive. What changed were the *tools* employed and the *scope* of the conflict, which became quite restricted in regard to space. However, the underlying reasons cannot be posited either to some deterministic structural theory or to a unidirectional, evolutionary learning concept. Instead, the process was both highly contingent and uneven, reflecting the impact of particular leaders, events, and surrounding political factors. The problem was that knowledge about the space environment first had to be understood and then communicated effectively into national policy on both sides—a hard case even without Cold War rivalry.

Space Physics as an Environmental Constraint

As noted earlier, the first critical obstacle that space physics posed to the U.S.-Soviet military competition in orbit was man-made radiation. This limitation was not recognized at the outset of the U.S.-Soviet space race when nuclear testing in this environment had just begun. But as one technical study describes what we now know about the effects of electromagnetic pulse (EMP) radiation from nuclear explosions in space, "These x-rays, as well as the accompanying gamma rays and high-energy neutrons, strike everything within line of sight, doing severe damage to nearby satellites."[18] Beginning in 1958, the two sides began to test low-yield nuclear weapons in space. They learned that the ionosphere traps EMP radiation, which then dissipates only very gradually. But they did not yet understand the damage it could do.

In the summer of 1962, the United States began testing much higher yield nuclear weapons from Johnston Island in the Pacific to investigate their possible use for antiballistic missile defense against the Soviet Union. The large amounts of EMP generated by these tests caused severe damage to British, U.S., and Soviet communications and reconnaissance satellites then in low–Earth orbit. The tests also affected the power grid on Hawaii and interrupted Pacific region radio and telephone communications for hours.[19] Neither military planners nor civilian leaders could fail to take notice of the risks involved in continued testing. Under the weight of prior planning, however, U.S. tests continued

[18] Daniel G. Dupont, "Nuclear Explosions in Orbit," *Scientific American* (June 2004), p. 104.

[19] See testimony by Dr. George Ullrich, then-deputy director of the U.S. Defense Special Weapons Agency, presented before the House Committee on National Security's Military Research & Development Subcommittee hearing, "Threats Posed by Electromagnetic Pulse to U.S. Military Systems and Civilian Infrastructure, July 16, 1997, available online via the Federation of American Scientists' Web site at <http://www.fas.org/spp/starwars/congress/1997_h/has197010_1.htm > (accessed February 12, 2008).

through the fall of 1962, and the Soviet Union also conducted several high-yield tests in space. Some influential U.S. scientists, like Edward Teller, believed that the requirements of missile defense necessitated continued U.S. testing. Other U.S. officials and representatives opposed an arms control approach because they believed that the Soviet Union might cheat and test nuclear weapons on the dark side of the Moon.[20] This set up a major debate on verification measures, although leaders eventually decided that the known benefits of a space test ban outweighed the small risks of possible cheating. While supporters did not stop pushing nuclear activities in space, the influence of the new technical knowledge about their damaging effects began to weigh more heavily on the minds of U.S. and Soviet decision makers examining the trade-offs involved in space.

Another technical factor influencing space competition is the extremely weak pull of gravity outside the Earth's atmosphere, particularly relative to the high speed of orbital objects. Unlike damaged aircraft in the world's air space and sunken ships in its oceans, orbital space objects have adequate velocity to resist atmospheric and gravitational forces that "cleanse" these other environments by dragging abandoned or destroyed objects down to the ground or to the sea bottom. Instead, even after spacecraft in low–Earth orbit (LEO, or 60 to 1,000 miles above the Earth) are destroyed by weapons, their fragments remain in orbit, traveling at over 17,000 miles per hour (mph),[21] making them arguably even more dangerous, as these pieces are smaller and harder to track. Unfortunately, their orbits decay only gradually over periods of months, years, or decades.[22] Eventually, depending on initial speed and original altitude, they will burn up (if they are small enough) upon re-entering the Earth's atmosphere. In higher orbits, however, virtually no cleansing ever takes place. As physicist Joel R. Primack explains, "Debris in orbits higher than about 880 kilometers (km) above the Earth's surface will be up there for decades, above 1,000 km for centuries, and above 1,500 km effectively forever."[23,24]

[20] Glenn T. Seaborg, *Kennedy, Khrushchev, and the Test Ban* (Berkeley: University of California Press, 1981), pp. 271–73.

[21] Based on figures provided in David Wright, Laura Grego, and Lisbeth Gronlund, *The Physics of Space Security: A Reference Manual* (Cambridge, Mass.: American Academy of Arts and Sciences, 2005), p. 21, for satellites traveling in 200-km orbits. The equivalent speed is 28,000 kilometers per hour.

[22] Natural "solar flares"—occurring every 11 years—heat and expand the Earth's atmosphere, increasing drag and accelerating the process of decay, but only at lower altitudes.

[23] The equivalent distances in miles are 547, 621, and 932, respectively.

[24] Joel R. Primack, "Debris and Future Space Activities," in James Clay Moltz, ed., *Future Security in Space: Commercial, Military, and Arms Control Trade-Offs*, Occasional Paper No. 10 (Monterey, Calif.: Center for Nonproliferation Studies, Monterey Institute of International Studies, July 2002), p. 18.

High-speed orbital debris poses obvious risks to unmanned spacecraft and especially manned vehicles, whose occupants are completely dependent on the maintenance of a pressurized, oxygenated, and temperature-controlled environment. A collision between any sizable piece of orbital debris and a spacecraft would likely cause catastrophic damage and, for manned missions, almost certain death. Currently, some twelve thousand orbital objects longer than two inches are tracked by the military-operated U.S. Space Surveillance Network.[25] Orbital debris must be constantly catalogued, followed, and avoided by spacecraft during the years each item remains in orbit. The reason has to do with simple physics. At orbital speeds, even tiny fragments—such as bolts and shards of metal of about four inches in length—carry a tremendous violent force, equivalent to that of a 10-ton truck traveling at 118 mph.[26] Smaller objects, ranging from metal and plastic released by stage separations to detritus from manned space activities, number well over a million yet are too small to track.[27] Nevertheless, they too can cause significant damage.

An orbital paint fleck that collided with the window of the U.S. space shuttle *Challenger* in 1983 caused serious concern at NASA's mission control for the window's structural integrity during the astronauts' atmospheric re-entry.[28] In 1985, as part of the Strategic Defense Initiative, the United States tested a kinetic-kill ASAT weapon launched from an F-15 aircraft against an aging air force satellite in low–Earth orbit, generating nearly three hundred trackable pieces of debris.[29] As anticipated in arguments made against the test by NASA scientists, these fragments took seventeen years to de-orbit and created dangerous obstacles for the *International Space Station*.[30] China's 2007 ASAT test generated twenty-two hundred pieces of trackable debris, the bulk of which is expected to remain in orbit for approximately forty years.[31] No known technology ex-

[25] A recent figure for traceable objects released by NASA is 11,954. See NASA, *Orbital Debris Quarterly*, Vol. 11, No. 3 (July 2007), p. 10.

[26] Figures based on equivalent metric measures provided in the report coauthored by Simon Collard-Wexler, Jessy Cowan-Sharp, Sarah Estabrooks, Thomas Graham Jr., Robert Lawson, and William Marshall, *Space Security 2004* (Toronto: Northview Press, 2005), p. 3.

[27] Andrew Brearley, "Faster Than a Speeding Bullet: Orbital Debris," *Astropolitics*, Vol. 3, No. 1 (Spring 2005), p. 3.

[28] Incident involving Sally Ride's first spaceflight in 1983, cited in Primack, "Debris and Future Space Activities," pp. 20–21. Fortunately, the window (and the astronauts aboard) survived the re-entry stresses.

[29] David S. F. Portree and Joseph P. Loftus, Jr., *Orbital Debris: A Chronology*, NASA report (NASA/TP-1999-208856) (January 1999), pp. 46–47, available online at <http://ston.jsc.nasa.gov/collections/TRS/_techrep/TP-1999-208856.pdf> (accessed July 20, 2006).

[30] Theresa Hitchens, "CDI Fact Sheet: Space Debris," August 12, 2005, available online on the Center for Defense Information Web site at http://www.cdi.org/program/document.cfm?DocumentID=3106 (accessed July 20, 2006).

[31] *Orbital Debris Quarterly*, Vol. 11, No. 3 (July 2007), p. 1, and remarks by David Wright, Union

ists to clean up orbital debris, given the scope of the problem and the breadth of debris distribution over many thousands of three-dimensional miles in the critical realm of Earth orbital space.

Another challenge that scientists cite in arguing for more urgent efforts to limit space debris is the threat of "cascading" debris. This problem is caused by collisions between individual pieces of debris (and microdebris)—creating even more individual fragments requiring tracking and avoidance—as the constellation of orbital debris increases over time. These interactions threaten to multiply the orbital debris field by a factor of five within this century.[32] What is particularly troubling is that this estimate assumes no increase in the generation of debris. Given the current absence of any rules affecting military space testing, any significant move by states to test and deploy kinetic-kill weapons in space could cause an exponential increase in debris.[33]

A study by the American Physical Society estimates, for example, that to field an effective constellation of space-based interceptors (at an altitude of 186 miles) for U.S. use against a limited missile attack from Northeast Asia or the Middle East would require approximately seven hundred space-based systems for use against liquid-fuel missiles or some sixteen hundred interceptors for use against faster solid-fuel missiles.[34] Other countries' needs might require more or fewer, but likely similar overall numbers. Increased civilian missions, of course, will contribute to this problem as well. Thus, as one study concludes, "Continued annual growth in orbital debris populations represents a clear threat to the sustainability of space security over the longer term."[35]

---

of Concerned Scientists, in a presentation on China's test at the Capitol Building in Washington, D.C., February 9, 2007. Part of the reason for the larger number of particles released by the Chinese test is the greater sensitivity of current-generation space-tracking radars.

[32] "Space Debris: Its Causes and Management," presentation by Nicholas L. Johnson, Chief Scientist and Program Manager, NASA Orbital Debris Program Office, at a seminar, "U.S. Security Interests and the Future Environment in Space: Managing Debris and Radiation," sponsored by the Center for Nonproliferation Studies, Monterey Institute of International Studies, Capitol Hill, Washington, D.C., July 24, 2002.

[33] Because of the greater speed required for objects to remain in orbit at lower altitudes, satellites in LEO do not remain long over specific areas of the Earth. This increases military requirements, since weapons in space need to be over their intended target or in place to intercept missiles launched from any specific site, at all times, in order to be effective. This can be accomplished by increasing the number of weapons deployed, but this raises costs and maintenance requirements significantly compared to air-, sea-, or ground-based systems. As Wright, Grego, and Gronlund observe, "Acquiring the capability to attack a ground target within 45 minutes would be many tens of times more costly if done from space than from the ground." Wright, Grego, and Gronlund, *Physics of Space Security*, p. 95.

[34] American Physical Society, *Report of the American Physical Society Study Group on Boost-Phase Intercept Systems for National Missile Defense* (July 2003), cited in Wright, Grego, and Gronlund, *Physics of Space Security*, p. 98.

[35] Collard-Wexler, Cowan-Sharp, et al., *Space Security 2004*, p. 6.

As seen in the current debate on global warming or earlier debates on acid rain, however, space environmental constraints—like radiation and debris— cannot in and of themselves "determine" space outcomes. Where human decisions are involved, situational, economic, or political factors may encourage cooperation and competitive restraint, but they cannot *require* it. State leaders are sometimes ignorant of potential costs or, on rare occasions, may decide to engage in behaviors that they know will cause harm in order to prove their devotion to a cause, deny some benefit to an adversary, or protect themselves from something they fear even worse (such as economic hardship or nuclear attack). Similarly, inaction by leaders in the face of powerful, self-interested domestic coalitions pushing for particular technologies may lead states to stumble into deploying them, particularly in areas where legislatures find it hard to enact restrictive controls because of countervailing economic incentives, problems of technical complexity, or claims of national security interests. Moreover, as space history shows, decisions to refrain from consideration or actual deployment of particular technologies are not permanent. Leadership changes or shifts in political, economic, or military factors can cause a reassessment of trade-offs, which may result in the breakdown of previous agreements.

For these reasons, Cold War space activity could have moved in a much more negative direction than it did, had different leaders been in power or had situational factors (such as a space-related military crisis) intervened. No single variable is adequate to explain the outcomes that took place, and yet it is possible to clarify the ways in which environmental and political factors *interacted* to cause specific policy changes during this time.

In the process of "bounding" Cold War space competition and channeling it away from space-based weapons into manned spaceflight, military support activities, and weapons "hedging" strategies, environmental factors alone provided only a *situation*.[36] They may have offered the objective possibility of cooperation, but as we have seen, they might not have led to actual agreements because of a lack of trust between the two countries, even in the presence of high objective costs associated with unlimited competition. We need only recall that it was also the bipolar superpower competition that brought about the nuclear arms race in the first place, a rivalry in which the United States spent an estimated $5.5 trillion.[37] For the constraints implied by the structural situation to become active and for the actual limiting of space competition to occur, national leaders had to perceive the desirability of agreement and therefore ·

---

[36] On the concept of "bounding" as a cooperative technique in international relations, see James Clay Moltz, "Managing International Rivalry on High Technology Frontiers: U.S.-Soviet Competition and Cooperation in Space," Ph.D. dissertation, University of California at Berkeley (1989), pp. 44–46.

[37] Schwartz, *Atomic Audit*, p. 3.

alter their viewpoints on the current, unmanaged situation. The existence of environmental interdependence may have provided a necessary condition, but it was not a sufficient one for cooperation to occur. As Steve Weber cautions, "Cooperation is a response to interdependence. But it is not the only possible response."[38]

Interactive Factors and the Process of Limited Learning

To reach agreements regarding space, leaders on both sides had to make two cognitive shifts. First, they both had to recognize that space constituted a particularly *threatening area* in which to engage in an arms race. Such recognition was far from inevitable. If it were, cooperation would have begun much earlier than it did and would not have entailed such heated domestic debates. Moreover, recognition should have been gradual and unidirectional as collective knowledge increased over time. Instead, it proceeded in fits and starts, with reversals in particular areas, especially manned space cooperation, which witnessed a marked decline in the years 1976–84. Supporting a claim of the importance of politics to the recognition of seemingly objective threats is the fact that agreements to ban other destabilizing technologies—such as the sale of advanced arms to the Third World or the deployment of multiple-warhead intercontinental ballistic missiles—were *not* reached by the superpowers in the 1960s and 1970s.

Second, both countries had to desire stability more than superiority; the pursuit of space security via unilateral military means could have continued after 1962. Alternatively, space security could have been sought via looser, tacit mechanisms rather than formal arms control agreements. However, both national leaderships came to value the increased stability offered by explicit treaties. Thus, the physics of the space environment may have provided the *incentive* for military restraint, but leaders had to undertake bold (and strategically risky) initiatives in order to engage in actual self-limitation in regard to a sworn adversary. They then had to use precious political capital to "sell" such agreements to sometimes skeptical audiences at home, including their military leaderships. At times, one side or the other desired an agreement, but a mismatch between the two stood in the way. This occurred with U.S. anti-ABM proposals during the mid-1960s and with Soviet ASAT arms control efforts in the early

---

[38] Steve Weber, *Cooperation and Discord in U.S.-Soviet Arms Control* (Princeton, N.J.: Princeton University Press, 1991), p. 272. Weber continues, "State may choose instead to eschew the potential benefits of cooperation and accept only what they can achieve relying on their own resources."

1980s. How indeed did the two sides reach even minimal consensus about co-operation in such a militarily sensitive field as space?

For this discussion, we can usefully divide space activity into three functional areas: *space science and exploration* (including human spaceflight), *space utilities* (including space commerce), and *military applications*. As a starting point for getting closer to the historical record, we can observe that the level and longevity of cooperation in these three areas of space activity varied considerably. As a means of examining this problem, we can pose the following hypothesis: *If more enduring cooperation occurred in a specific area of space activity, then a higher degree of "environmental interdependence" existed in that area than in others.*

In *space science*, the absence of conditions of environmental interdependence and the major impact of domestic politics hindered cooperation. In other words, each side's deep space scientific missions had no discernable impact on the other's ability to conduct similar activities. Meaningful U.S.-Soviet cooperation in this area proved to be the slowest to develop and, once achieved, the most fragile. Cooperative proposals faced not only domestic obstacles but also bilateral problems, given the difficulty of matching periods of pro-cooperation politics (once achieved) in one country with those in the other: when one side desired cooperation, the other side often did not.[39] Space science cooperation was often held hostage to protest policies of the other side, including areas having nothing to do with space (such as Soviet protests over U.S. action in Vietnam and U.S. disfavor with Soviet involvement in Afghanistan). For this reason, close cooperation in space science between the superpowers did not occur until the 1970–75 détente period, during the Vietnam drawdown and before Afghanistan. Such efforts went into a sharp decline by the late 1970s, when political relations soured. They rose again in the late Gorbachev era and have remained close since then, benefiting from continued, generally positive U.S.-Russian political relations. What is interesting about cooperation in space science is that its very low level of interdependence required a *greater* level of political consensus than the two other types of cooperation.

Cooperation in *space utilities*, by contrast, was encouraged by conditions of environmental interdependence within this field. Transmitting on someone else's radio frequency or trying to occupy the same space in geostationary orbit, for example, ruined communications for both sides. In this light, the evolution of such cooperation by the two superpowers within the context of the International Telecommunications Union was not surprising. Such cognitive change

[39] This can be seen in a series of overtures in the unmanned area rejected by the Soviets in the late 1950s and early 1960s and by the Americans in the late 1970s and early 1980s.

represented adaptation only, as it did not lead to shifts in ultimate U.S. and Soviet values, which remained largely competitive.

In the *military field*, space activities played a central role in the national security organizations of both the United States and the Soviet Union. But each superpower found that its ability to engage in useful defense-related activities in this new environment was directly affected by the actions of the other side. The test of a nuclear weapon even within a few thousand miles of spacecraft in overlapping orbital bands presented very significant (and harmful) effects. Similarly, debris-generating explosions and other types of weapons tests—regardless of their intention—in and of themselves threatened the exercise of national power by others in space. Using Russell Hardin's term, we might call these unique problems related to weapons activity in space *collective bads*.[40] The desire to remain active in such an environment created incentives for cooperation to manage this interdependence, particularly if the collective good of safe access to space for passive military benefits was to be maintained. Over time, the two sides realized that unilateral military means alone could not provide security to their assets in space. This could only be achieved through cooperative *political* means.

So-called structural realists,[41] however, contended during the Cold War that the superpowers were among the *least* interdependent states in history. As Kenneth Waltz observed in 1979, "America and Russia are markedly less interdependent and noticeably less dependent on others than earlier great powers were."[42] But what structural realists meant when they spoke of a lack of interdependence between the superpowers was that these states independently possessed overwhelming power, according to a number of domestic indices: size, national resources, technological skill, and military might. These static measures said nothing, however, about particular dynamics in the *use* of this power in constraining environments such as space. It was here that critical interdependencies arose.

But if environmental interdependence helps explain why the superpowers

[40] As Hardin comments, "Another important class [of international cooperation] includes many of the most important collective actions in contemporary politics: those directed at the elimination of bads rather than the provision of collective goods." See Russell Hardin, *Collective Action* (Baltimore: Johns Hopkins University Press, 1982), p. 4.

[41] Structural (or neo-) realists focus on the unique dynamics of international systems, according to the number of great powers in that system. They argue, in particular, that bipolar systems (such as the Cold War, U.S.-Soviet-dominated world) are unusually stable because of the tendency of states to "manage" the system, their lack of a need to rely on other countries to "balance" the system, and the relatively low cost of communications and bargaining between the two states.

[42] Kenneth Waltz, *Theory of International Politics* (Reading, Mass.: Addison-Wesley, 1979), p. 193.

restrained themselves in military space applications, why did progress toward restraint take more than five years after space's opening in 1957 to begin? The evidence from space activity shows that conditions of interdependence do not dawn on leaders automatically; they have to be learned. Contrary to notions of technological determinism, the existing evidence from actual space history shows that this recognition took place only in the context of the right *political*—not simply technical—conditions. Thus, in order to explain the process of how these policy changes occurred, we must now investigate the second half of our two-part model of environmentally influenced learning: processes of cognitive change.

Cognitive Change in U.S. and Soviet Space Policies

As Joseph Nye argues, states engage in intellectual shifts as a result of one of three developments.[43] The first route is through elections, coups, or changes of leaders by other means. A second way is through normative changes in which the redefinition of moral issues brings about a shift in policy (Nye uses the example of slavery in the United States). Finally, a third way is through cognitive change or learning, as affected by the addition of new information to the decision-making calculus of leaders. We can observe that the emergence of new space norms (such as in late 1962) and major space arms control agreements (in 1963, 1967, and 1972) did *not* hinge on elections or leadership changes in the United States and the Soviet Union. Also, since we know that there was no conclusive moral rapprochement between the U.S. and Soviet leaderships during the Cold War, despite sporadic improvement of U.S.-Soviet relations, we must reject this possibility. Changes in leadership perception during the Cold War, therefore, can only be explained by cognitive change or learning.[44]

Although learning can be viewed as a continuum—ranging from minor changes in operational tactics to wholesale philosophical conversions—most authors in the field of international relations have made a basic division between two types of learning by states. Drawing on Karl Deutsch's 1963 classic, *The Nerves of Government*, Nye makes a basic distinction between "simple" and "complex" learning.[45] Simple learning occurs in situations where actors maintain their original goals and merely change the *tactics* they use to reach them.

[43] See Joseph Nye, "Nuclear Learning and U.S.-Soviet Security Regimes," *International Organization*, Vol. 41, No. 3 (Summer 1987).

[44] Changes after 2001, however, can be ascribed largely to the impact of the 2000 elections in the United States.

[45] Nye, "Nuclear Learning." See also Karl M. Deutsch, *The Nerves of Government: Models of Political Communication and Control* (New York: Free Press, 1963).

In complex learning, by contrast, actors adopt new *goals* out of a reevaluation of the feasibility of certain ends.[46] Ernst Haas's work on learning makes a similar distinction between so-called "adaptation" (or simple learning) and "learning" (or complex learning).[47] Haas defines these two forms of cognitive change within international organizations: "Adaptation is incremental adjustment, muddling through," while learning is reserved for those rare instances when "an organization is induced to question the basic beliefs underlying the selection of ends."[48] Haas's concept is rooted in scientific knowledge, which—in situations of true learning—forms the basis for new consensual *social* goals within an international organization. But how do we approach the more difficult prospect of science-based learning (or adaptation) between rival countries?

Russell Hardin writes of "the fundamental importance of strategic interaction over time in motivating cooperation for collective and therefore for individual benefit."[49] Perhaps the "time" that the hostility of space's physical environment provided to states allowed them to reevaluate the nature of their *inter*-relationships there, especially in terms of security. As Haas has argued, the growth of scientific knowledge increases the possibility that mankind may "learn about its interdependence."[50] In space, interdependence was not a lofty, ideologically motivated goal but a practical concern brought on by environmental factors such as orbits, debris, and mutual vulnerability. How did these factors result in practical policy change to limit space competition?

A number of authors, from Thomas Schelling[51] to Robert Axelrod[52] to Robert Keohane,[53] have presented the case that cooperation is the logical strategy in multiple-play, two-person games such as Prisoner's Dilemma.[54] But simula-

---

[46] Argyris and Schon call these two concepts "single-loop" and "double-loop" learning, with the latter involving a feedback effect that changes organizational goals. See Chris Argyris and Donald A. Schon, *Organizational Learning: A Theory of Action Perspective* (Reading, Mass.: Addison-Wesley, 1978).

[47] See Ernst B. Haas, *When Knowledge Is Power: Three Models of Organizational Change in International Organizations* (Berkeley: University of California Press, 1990).

[48] Ibid., pp. 34, 36.

[49] Hardin, *Collective Action*, p. 4.

[50] Ernst B. Haas, "Is There a Hole in the Whole? Knowledge, Technology, Interdependence, and the Construction of International Regimes," *International Organization*, Vol. 29, No. 3 (Summer 1975), p. 831. Haas's work suggests that technical knowledge often promotes recognition of fundamental interdependencies in particular fields of international activity.

[51] Thomas C. Schelling, *The Strategy of Conflict* (Cambridge, Mass.: Harvard University Press, 1960).

[52] Robert Axelrod, *The Evolution of Cooperation* (New York: Basic Books, 1984).

[53] Robert O. Keohane, *After Hegemony: Cooperation and Discord in the World Political Economy* (Princeton: N.J.: Princeton University Press, 1984), p. 76.

[54] The game of Prisoner's Dilemma is explained by Richard Cornes and Todd Sandler as one "in which two (or more) prisoners are made to turn state's evidence by confessing to a crime that they did

tions show that an iterated game may go through cycles of both mutual cooperation and mutual defection over time.[55] What matters in explaining space restraint is not simply the fact that cooperation becomes more likely over time but instead knowing *how* game rules change (for better or for worse). While structural factors—such as space physics and the prior existence of natural collective goods—may help explain general tendencies, they are not well equipped to tell us how and when states choose to *recognize* these factors. These answers depend largely on cognitive variables: trust, bargaining, and consensual knowledge.[56] In space, new information about conditions of environmental interdependence in the context of military security concerns brought about cooperative agreements in space, which then had a positive feedback effect in limiting future competitive strategies. Over time, as leaders sought to protect the valuable civilian and passive military benefits to be had from space, a greater collective good—safer *access* to space—was created. Unlike in other new frontiers, each side gradually recognized that its own behavior affected (and was affected by) that of the other side. These concepts now appear increasingly relevant in a world beset with fears of global warming and overfishing of the oceans. What this notion does not tell us, however, is under what conditions scientific knowledge about these collective environmental problems will result in meaningful cooperation.

In global environmental politics, scientific experts, nongovernmental organizations, the media, industry personnel, governments, and international organizations have all played central roles in the formation of agreements. Similarly,

---

not commit. To elicit this noncooperative response from the prisoners, the district attorney promises each of them less than the maximum sentence if they both confess. If, however, only one confesses, then the confessor receives a light sentence and the nonconfessor gets the maximum penalty. When neither confesses, they both receive a moderate penalty, greater than the light sentence offered to a lone confessor. Each of the prisoners is interrogated separately, and is thus denied the opportunity to communicate or to cooperate with the other." See their book *The Theory of Externalities, Public Goods, and Club Goods* (New York: Cambridge University Press, 1986), p. 13. Whereas the strategy of defection dominates all others in single-play, Prisoner's Dilemma-type games, in multiple-play games the dominant strategy changes. As Hardin notes, "Among sophisticated players, cooperation in iterated Prisoner's Dilemma may be the norm" (Hardin, *Collective Action*, p. 146).

[55] In space, this can be seen in the rise and fall of different U.S.-Soviet programs over time, based on changing levels of consensual knowledge about space.

[56] By consensual knowledge I mean *the mutually accepted technical and political understandings that form the basis for policymaking in two or more countries at a particular time.* (For another use of this term, see Haas, *When Knowledge Is Power*, p. 74. Haas's definition is more restrictive, requiring the actual formation of consensual social goals.) This logic of "rule" changes in space during the Cold War in space leads us to a number of theoretical questions: How do states move to higher levels of increasingly cooperative games? On what basis does the cumulation of consensual knowledge rest? How can new rules be internalized in the absence of an outside authority?

overriding philosophical concepts such as "global environmental justice" have sometimes captured public opinion and helped to drive negotiations to their successful conclusion, supported in part by powerful, scientifically informed epistemic communities.[57] Such an explanation works well, for example, in regard to the Antarctic Treaty's extension in 1991 (mentioned in Chapter One). In other areas, however, progress has remained elusive due to powerful commercial (and/or national) interests opposing formal agreements or the lack of salience of the issue with the public. In these circumstances, what Pamela Chasek refers to as "turning points" are sometimes necessary to reset the agenda or raise its perceived urgency.[58] Questions of consensus building remain one of the most difficult challenges amidst a multiplicity of actors, who often have equally distinct interests.[59] Conditions of globalization have improved communications, but they have also led to the fragmentation of power, meaning that authority building has become more complex.[60] Power sometimes is needed to promote and enforce environmental regimes, as seen in the security field as well. Finally, the self-interest of actors tends to promote suboptimal decisions and muddling-through, short of complex learning and its embodiment in new institutions, practices, or treaties.

On the face of it, Cold War space restraint may appear to be similar to such largely tactical adaptation and therefore consistent with realist-inspired space nationalist arguments. But realism's assumptions (anarchy, power maximizing behavior, and self-help) are stretched too far to explain space security behavior by the superpowers. First, the restrictive outcomes in space were characterized by *routinized* (tacit and formal) bilateral cooperation, a concept at odds with the notion of international anarchy. Second, military space restraint was based on shared technical and environmental understandings that *altered* the original military understandings regarding space, limiting them largely to nonweapons systems. This knowledge-based cooperation and its cumulation over problems of man-made EMP radiation and then orbital debris showed an evolution to a

---

[57] See Mostafa K. Tolba (with Iwona Rummel-Bulska), *Global Environmental Diplomacy: Negotiating Environmental Agreements for the World, 1973–1992* (Cambridge, Mass.: MIT Press, 1998), p. 183.

[58] See Pamela S. Chasek, *Earth Negotiations: Analyzing Thirty Years of Environmental Diplomacy* (New York: United Nations University Press, 2001).

[59] For a formal theoretical approach to analyzing problems in environmental negotiations, see Scott Barrett, *Environment and Statecraft: The Strategy of Environmental Treaty-Making* (New York: Oxford University Press, 2003).

[60] On this point, see Ronnie D. Lipschutz, *Global Environmental Politics: Power, Perspectives, and Practice* (Washington, D.C.: Congressional Quarterly Press, 2004); also Matthew Paterson, *Understanding Global Environmental Politics: Domination, Accumulation, Resistance* (London: Macmillan Press, 2000).

higher level of scientific-technical understanding rather than static, power-centric thinking. And, third, space restraint involved not just theoretical restraint but actual decisions to *reverse* certain deployed military programs and even to tolerate relative gains by the other side (within the U.S.-Soviet military context) in order to preserve space stability. The U.S. decision to abandon its nuclear-tipped ASAT program in the early 1970s, despite its knowledge of the advancing Soviet conventional ASAT program, is but one example. These factors indicate that other forces—both technical and political—must be considered to better explain patterns in space security relations.

Timing and Policy Change

The last remaining part of our conceptual puzzle regarding space restraint involves timing: how do we explain *when* policy shifts occurred? Theories of cognitive change do not offer much help in this regard, since they are more concerned with process than predicting when a tipping point will be reached.

According to sociologist Robert Nisbet,[61] individuals tend to maintain their current views and approaches to problems as long as existing methods result in reasonable success. This same finding comes out of the international relations literature on national decision making.[62] Sometimes, however, the homeostatic tendencies within areas of activity are shaken up. Environmental analyst Chasek's turning points have their cognates in the strategic and international learning literatures as well.[63] These events, crisis *and* noncrisis, cause breaks in the normal routine of diplomatic discourse and of domestic decision making. Nisbet refers to these cognitive breaks as "the kind of occurrence or happening that has the effect, for however brief a time, of suspending or at least disrupting the normal."[64] In space, such trigger events promoted changes in definitions of national security and—when they were experienced by both the United States and the Soviet Union simultaneously—stimulated the process of bilateral learning.

During the 1957–62 period, the hostility of existing U.S.-Soviet relations and limited understanding about space initially prevented technical knowledge from becoming consensual, even though national leaderships had begun to

[61] Robert Nisbet, *Social Change and History: Aspects of the Western Theory of Development* (New York: Oxford University Press, 1969).

[62] See, for example, John Steinbruner, *The Cybernetic Theory of Decisionmaking: New Dimensions of Political Analysis* (Princeton, N.J.: Princeton University Press, 1974); and Robert Jervis, *Perception and Misperception in International Politics* (Princeton, N.J.: Princeton University Press, 1976).

[63] Chasek, *Earth Negotiations.*

[64] Robert Nisbet, in Robert Nisbet, ed., *Social Change* (Oxford, England: Blackwell, 1972), p. 26.

recognize the need for military space restraint. As will be argued later in this book, it took a trigger event—the U.S. Starfish Prime space nuclear test in July 1962, which put a number of U.S. and Soviet satellites out of service and placed human spaceflight at risk—to coalesce diplomatic efforts on a norm against exo-atmospheric nuclear tests in late 1962, later codified in the 1963 Partial Test Ban Treaty (PTBT). The U.S. *Solwind* ASAT test in 1985 played a similar role in regard to orbital debris, particularly in changing mind-sets within the U.S. military about the severity of this problem. China's 2007 ASAT test, which focused greater international attention on the debris problem, has had a similar effect in helping to rally collective awareness and stimulate transnational cooperation to address the debris threat.[65]

Equally problematic is the reverse phenomenon, or "unlearning."[66] That is, just as events can stimulate learning, other developments can cause past lessons to be forgotten. Specifically, when leaders with little experience in foreign affairs become heads of state, they are more likely to enact policies at odds with established international norms. In these cases, such policies may run into strong opposition from long-serving members of national legislatures or from foreign governments still subscribing to the earlier norms. As will be shown in the coming historical chapters, this resistance frequently succeeded in preventing policies of near-term space weapons deployment from being implemented.

Conclusion

As we look back on the now fifty years of space history, the notion of inevitability is hard to sustain in any particular direction—either cooperative or competitive. A closer analysis of key decision points in the history of U.S.-Soviet space activity shows that much of what seemed predetermined could have gone either way had decision makers failed to receive key information from the other side or had voices calling for renunciation or expansion of space competition been heeded. Supporters of the view that space might somehow be innately cooperative were disappointed, and those who predicted extensive weaponization

---

[65] The February 2008 U.S. ASAT interception took place at approximately 130 miles in altitude, thus posing no significant debris concern because of the rapid de-orbiting of the associated fragments.

[66] For earlier uses of this term, see Franklyn Griffiths, "Attempted Learning: Soviet Policy Toward the United States in the Brezhnev Era," and Peter R. Lavoy, "Learning and the Evolution of Cooperation in U.S. and Soviet Nuclear Nonproliferation Activities," both in George W. Breslauer and Philip E. Tetlock, eds., *Learning in U.S. and Soviet Foreign Policy* (Boulder, Colo.: Westview Press, 1991).

and eventual warfare were also proven wrong. What actually occurred is much more complex and had to do with periodic competitive thrusts and cooperative regrouping, as new space activities revealed unexpected dangers and as U.S. and Soviet leaders sought to minimize risks by establishing norms of unacceptable space behavior.

The causal chain of this environmental security approach applied to the first fifty years of space history can be outlined as follows: (1) following military tests and other experiments, scientists and officials observed the negative security implications that could result from the deployment of certain technologies into this new environment; (2) national leaders gradually recognized that conditions of mutual *interdependence* existed in their conduct of certain harmful space activities; and (3) these new technical and political understandings promoted cooperative restraint in narrowing the scope of space competition in order to protect the use of space as a collective good. The net result was not the elimination of competition in space but its redirection into safer areas, such as manned flight and, increasingly, commercial applications and military support systems. These changes were turned into active policies through a process of interstate bargaining, tacit and formal agreements, and learning; for the most part, these space policies are intact today. As Goldstein and Keohane note, "When institutions intervene, the impact of ideas may be prolonged for decades or even generations."[67] Occasionally, however, "unlearning" took place, showing the contingent nature of such understandings as governments change. The 2000 election victory of Texas Governor George W. Bush, for example, threw the Cold War space framework into disarray, allowing a non–status quo minority in U.S. domestic policymaking to mount a serious challenge to strategic space restraint, which they no longer saw as protecting U.S. national interests. To the dismay of many of these advocates, however, the Bush administration failed to accomplish its goal of establishing U.S. space-based defenses, largely because of the restraining role of Congress. Its use of a sea-based missile defense interceptor to destroy a non-operational U.S. reconnaissance satellite near the end of its decaying orbit in February 2008 represented the high-water mark of this effort to challenge the norms of space security.

China's new military space activities have raised an incipient challenge to U.S. post–Cold War space supremacy. But Beijing has not yet repeated its one destructive, high-altitude ASAT test. Perhaps remarkably, the first fifty years

[67] See Judith Goldstein and Robert O. Keohane, "Ideas and Foreign Policy: An Analytical Framework," in Judith Goldstein and Robert O. Keohane, eds., *Ideas & Foreign Policy: Beliefs, Institutions, and Political Change* (Ithaca, N.Y.: Cornell University Press, 1993).

of space security closed without space war or deployed, space-based weapons. What might be learned from the period from 1957 to the present that would help us both understand past successes and develop new mechanisms for managing the next fifty years of space security, despite the presence of a greater complexity of actors? These are the main challenges addressed in the rest of this book.

PART II

Reassessing Twentieth-Century
Space Security

CHAPTER THREE

Roots of the U.S.-Soviet Space Race

1920s–1962

Whether states join costly technological competitions with other countries depends considerably on their perception of the technology's importance to their national security. In turn, these calculations are influenced by who else is involved in the race, what military advantages the technology might bestow, and finally, its expected expense relative to other budgetary demands. Despite staggering costs, for example, Britain accepted the necessity of an arms race in massive, *Dreadnaught*-type battleships before World War I because of its unwillingness to surrender naval dominance to Germany.[1] British officials saw naval superiority as essential to the maintenance of their empire and therefore critical to national security.

But in regard to missile technology in the 1930s, Britain did not compete. It failed to perceive the immediacy of the threat and did not want to spend precious defense funds on an unproven technology. The United States fell into the same camp, at least initially. It had two large oceans protecting it and decided that missiles—still in their infancy and far from achieving intercontinental range—did not pose serious near-term risks to its security. For both, given the presence of multiple powers in the pre-1945 international system, threats seemed more distributed and less acute, while alliances presented viable options for opposing possible aggressors. Conditions of economic depression also affected the breadth of threats states felt they could afford to deal with.

The exceptions were competing states located in close proximity to their military rivals. This condition held for Germany, France, and the Soviet Union, in particular. But excessively conservative thinking within the French govern-

---

[1] See Holger H. Herwig, "The Battlefleet Revolution, 1885–1914," in MacGregor Knox and Williamson Murray, eds., *The Dynamics of Military Revolution, 1300–2050* (New York: Cambridge University Press, 2001).

ment caused it to ignore the promise (and threat) of rocketry.[2] Paris decided instead to invest in a series of defensive fortifications known as the Maginot Line. Unfortunately, it gambled wrong, and this purely defensive strategy proved to be both shortsighted and very costly. By contrast, Berlin and Moscow were not status quo powers either by ideology or military inclination. Both embraced new technology as a means of overcoming their weaknesses compared to their rivals for continental influence and power. Both sponsored extensive missile research in the early 1930s. But Moscow's self-defeating political paranoia led it to purge many of its top missile scientists and engineers, setting back the program considerably. Drawing on its strong and steady support for rocketry, Germany soon emerged as the unchallenged missile leader. In October 1942, it succeeded in testing the long-range A-4 rocket, whose capabilities—a 200-mile range carrying a 2,200 pound warhead[3]—far exceeded those of any other missile in existence. It would soon be rechristened as the V-2 (or Vengeance Weapon) ballistic missile, which would be used extensively against London, the south of England, Antwerp, and Liege during World War II.[4] Whether the limited military damage it caused (and the 5,400 mostly civilians killed) merited Germany's huge investment in the program remains doubtful today.[5] It also may have stimulated the British to fight harder. The German missile story—replete with its use of Jewish and other European slave labor at Peenemunde and later in the underground Mittelwerk facility—is the subject of other books.[6] What it provided for space's future was a powerful (if tainted) technological base and a cadre of valuable scientists and engineers.

Nazi Germany's defeat in 1945 reconfigured the geography of international missile competition. The country's forcible disarmament, its political division by the four occupying powers, and the U.S.-Soviet seizure and deportation of its top missile personnel removed it from the race entirely. Instead, mirror-

---

[2] On this point, see William E. Burrows, *This New Ocean: The Story of the First Space Age* (New York: Random House, 1998), p. 55. Apparently, after some brief study in the late 1920s, the French concluded that rocketry was a technical field not worthy of attention by serious military officers.

[3] Ibid., pp. 98–99.

[4] The V-1 was a slower and more vulnerable Nazi German cruise missile. For casualty figures, see Burrows, *This New Ocean*, p. 102.

[5] Michael J. Neufeld estimates that the Germans spent the equivalent of about half a billion U.S. dollars on their missile program and concludes (given the comparative size of the German economy), "The Army rocket program imposed a burden on the Third Reich roughly equivalent to that of [the] Manhattan [Project] on the United States." He describes it as a military "boondoggle" and its results as "pathetic" compared to costs of U.S. and British strategic bombing of Germany using conventional aircraft. See his book, *The Rocket and the Reich: Peenemunde and the Coming of the Ballistic Missile Era* (Cambridge, Mass.: Harvard University Press, 1995), p. 273.

[6] See, for example, Neufeld, *Rocket and the Reich*.

ing the emerging bipolar division of the globe, U.S.-Soviet rivalry now drove missile developments. Given Germany's accomplishments with the V-2, neither country could ignore the potential threat posed by future, longer-range ballistic missiles, particularly in a world of nuclear weapons. Although the United States saw fewer reasons to use missiles for political and military purposes, the Soviet Union lacked strategic bomber bases near U.S. soil and had stronger ideological interests in reaching space first: to prove the technological competitiveness of communism to a skeptical world audience. After 1957, the United States would take up this challenge with vigor and commitment. If the incentives for missile development before 1945 had stemmed mainly from national security concerns and broader geopolitical goals, it stood to reason that space would eventually inherit these stimuli as well.

This chapter examines the emergence of space programs within the Soviet Union and the United States from the 1920s to 1962. It shows how missile research evolved from an obscure hobby of enthusiasts to become the focus of one of the largest postwar military research and development programs besides nuclear weapons. Space capability evolved largely as a benefit of missiles, and it became clear early on that the space environment would become contested militarily. By the late 1950s, space had become a central measure of national power, and it seemed unlikely that the two sides would be able to avoid warfare in space. The roots of efforts to prevent such a conflict, however, began to emerge in the early 1960s as a result of harm being done to space by both sides in the form of nuclear testing. But the obstacles to an agreement to prevent further damage proved insurmountable during this period, given the hostility of bilateral politics and the assumption in both capitals that space would be the new high ground of the nuclear arms race. From the perspective of the four schools of thought outlined in Chapter One, space nationalism and technological determinism seemed to have taken over the policy spectrum, with little hope of the type of collective learning required for meaningful change. What passed for attempts at cooperation consisted mostly of fig leaves meant to embarrass or set back the other side's progress.

Roots of the Soviet Space Program

Although Russian rocket and missile research has a long history—dating as far back as the 1600s and leading through the visionary conceptual studies of Konstantin Tsiolkovskiy in the late 1800s and early 1900s[7]—the direct roots

---

[7] On the pre-Soviet history of Russian rocket research, see the following sources: Asif A. Siddiqi, *Sputnik and the Soviet Space Challenge* (Gainesville: University of Florida Press, 2003), pp. 1–3; Peter L.

of the Soviet space program can be traced to the late 1920s. At this time, the Red Army began organizing a centralized rocket research program out of an eclectic group of independent scientific enthusiasts and government research organizations that had existed up until then. The impetus for Soviet interest in rocket research can be traced to two complementing factors: (1) the ideology of Marxist-Leninism, which glorified technology and equated its development with sociopolitical progress; and (2) the emerging trends toward conflict in European politics.

The genealogy of the various Soviet research institutes from the late 1920s onward is complex.[8] It highlights the steadily growing interest of the Soviet government in controlling this key area of defense-related research. In many cases, this meant taking over organizations originally motivated by a fascination with space and propulsion, not weaponry.[9] In most cases, military researchers already working on rockets[10] took over these new organizations. One of the first and most influential of the amalgamated institutes was Leningrad's Gas Dynamics Laboratory (GDL), formed in 1927–28 in the wake of the "war scare" caused by Stalin's fear of an imminent "imperialist"-led attack on the Soviet Union. The Revolutionary Military Council (forerunner of the Ministry of Defense) provided the necessary funding, which soon allowed GDL to expand into the promising field of liquid-fueled rockets. Meanwhile, as the various independent rocket organizations began to disband under government pressure in the late 1920s, new state-led groups gathered up the remaining scientists. Although many of these groups had been founded with peaceful aims, the effect of government support was to turn the direction of research explicitly to

---

Smolders, *Soviets in Space: The Story of Salyut and the Soviet Approach to Present and Future Space Travel*, translated from the original Dutch edition by Marian Powell (London: Lutterworth Press, 1973), pp. 42–53; Michael Stoiko, *Soviet Rocketry: Past, Present, and Future* (New York: Holt, Rinehart and Winston, 1970), pp. 1–16; and Albert Parry, *Russia's Rockets and Missiles* (Garden City, N.Y.: Doubleday, 1960), pp. 69–80. Stoiko (*Soviet Rocketry*, p. 33) points out that these private, Russian scientific and engineering groups interested in rocketry actually *predated* such organizations in Germany. However, it took massive Soviet funding and centralization to produce the first actual rocket weapons in the late 1930s.

[8] On these issues, see the following sources: B. Ye. Chertok, *Rakety i Lyudi* [Rockets and people] (Moscow: Mashinostroenie, 1999), pp. 29–39; B. V. Raushenbakh, ed., *Iz istorii sovetskoy kosmonavtiki (sbornik pamyati akademika S. P. Koroleva)* (Moscow: Nauka, 1983); Leonid Vladimirov, *Sovetskiy kosmicheskiy blef* (Frankfurt, Germany: Possev-Verlag, 1973); Evgeny Riabchikov, *Russians in Space* (Moscow: Novosti Press, 1971); as well as Stoiko, *Soviet Rocketry*, pp. 25–65; Parry, *Russia's Rockets and Missiles*, pp. 81–130; and Smolders, *Soviets in Space*, pp. 42–71.

[9] This includes, for example, the Society for the Study of Interplanetary Communications (*Obshchestvo po izucheniyu mezhplanetnikh soobshcheniy*, or OIMS), formed in 1924.

[10] In 1924, the Soviet government had formed the Central Bureau for the Study of Reaction (Rocket) Problems (*Tsentral'ni buro po izucheniyu reactivnikh problem*, or TsBIRP). Through this organization, which counted many top scientists in its numbers, the government was able to keep track of relevant research and promote its military applications. (See Stoiko, *Soviet Rocketry*, p. 31.)

military applications. Latvian scientist Friedrich Tsander (a younger colleague of Tsiolkovskiy) led those in Moscow pushing to develop operational liquid-fueled rockets by creating the Group for the Study of Reaction Movement (GIRD)[11] in 1931. Within a few years, branches of GIRD formed in many other cities (including Leningrad), eventually bringing the total number of scientists and engineers involved to around a thousand.[12]

By 1933, pressure from abroad had convinced the Soviet military that rocket research needed to be both expedited and centralized even further. The Germans already had formed a military organization for rocket research and seemed bent on finding practical, weapons-related applications for these promising new technologies. In place of LenGIRD, MosGIRD, and the old GDL, a new organization came into being: the Reaction (Rocket) Scientific Research Institute (RNII).[13] Although officially under the auspices of the Heavy Industry Commissariat, the new organization fell under the de facto control of General Mikhail Tukhachevskiy, chief of armaments for the Red Army. Tsander's untimely death in 1933 of typhoid fever left a gap in the leadership of rocket research, with the result that former GDL head Ivan Kleimenov was chosen to run RNII.[14] A young engineer, Sergei Korolev—who had come to Moscow from Kiev in 1929 to complete training in aviation design under the renowned A. N. Tupolev—was appointed deputy director. The move toward greater centralization spurred the development of rocket prototypes as funding flowed in from government sources.

But for Soviet rocket scientists, the benefits of cooperation with the Soviet military soon proved to be short-lived. With Stalin's decision to begin a purge of the top military leadership in 1936, all those with ties to fallen generals soon became suspect as well.[15] Soviet security forces executed many—even in the area of research and development—for allegedly failing in their tasks and thereby committing "treason."[16] By mid-1937, Tukhachevskiy had become a victim of Stalin's paranoia, falling under arrest along with the staffs of his various

---

[11] In Russian, *Gruppa po izucheniyu reaktivnogo dvizheniya.*

[12] Stoiko, *Soviet Rocketry*, p. 65; also Siddiqi, *Sputnik and the Soviet Space Challenge*, pp. 4–5. Martin Caidin's findings put the Germans two years ahead of the Soviets at this time, and the Soviets eight years ahead of the Americans. See his book, *Red Star in Space* (New York: Crowell-Collier Press, 1963), p. 118.

[13] In Russian, *Reaktivniy nauchno-issledovatel'niy institut.* The organization formally opened in 1934.

[14] Chertok, *Rakety i Lyudi*, p. 31.

[15] In the area of aircraft development, for example, the failure of Soviet models to keep up with German planes in the air battles of the Spanish Civil War in 1937 and 1938 led to the arrest of chief designer Tupolev and hundreds of other aircraft engineers.

[16] On the purges in Soviet high-technology industries, see Kendall Bailes, *Technology and Society Under Lenin and Stalin: Origins of the Soviet Technical Intelligentsia, 1917–1941* (Princeton, N.J.: Princeton University Press, 1978).

rocket design bureaus. Within a year, Tukhachevskiy and his top aides were tried on trumped-up charges of spying for the Germans and were executed.[17] The NKVD (or People's Commissariat for Internal Affairs) dispersed the survivors across the burgeoning expanse of the Soviet penal system, sending most to work camps in the far eastern sections of the USSR.

As more and more scientists and engineers of various backgrounds suffered these sentences, however, the Soviet state realized that a terrible waste of valuable skills was going on as these men cut forests, erected dams, and mined precious metals in remote Siberian camps. Many had already been shot or had died from exposure or malnourishment.[18] At about this time, the increasing likelihood of war in Europe caused the Soviet security apparatus to begin organizing a system of military-related research and design bureaus in European Russia to work on critical weapons projects. These gulag facilities of the late 1930s and 1940s (known as *sharashki*) selected skilled prisoners from among the millions in the camps. Perhaps the only benefits of life in the *sharashki* were that they provided relief from the life-threatening physical conditions of the eastern Siberian camps and, spiritually and intellectually, involved support of the Soviet war effort through work in the technical fields for which these engineers and scientists had been trained.

In organizational terms, the state security apparatus soon replaced the military as the dominant force in Soviet rocket research. With the top scientists in the camps, rocket research now followed NKVD chief Lavrenti Beria's own priorities and depended on his delivery of supplies of test materials and fuels.[19] As for the Red Army, while it could promote particular research projects, it now

[17] Vladimirov's account (*Sovetskiy kosmicheskiy blef*, pp. 22–27) points out that Korolev's rapid rise to the top leadership of the later rocket program came about in part because of the virtual clearing out of the senior ranks by executions, a fate which nearly befell Korolev as well. Notably, in 1987, Gorbachev officially pardoned Tukhachevskiy and expressly criticized Stalin for slowing Soviet rocket research through the execution and imprisonment of innocent scientists. On the details surrounding Tukhachevskiy's 1937 execution, see A. Khorev's chapter "Marshal Tukhachevskiy," in *Reabilitirovan Posmertno* [Rehabilitated after death] (Moscow: Uridicheskaya Literatura, 1989), 2nd Edition.

[18] GDL leader Kleimenov and colleague Georgiy Langemak, for example, were executed in early 1938. In March 1938, Korolev received a ten-year sentence for treason and was sent by train across the Soviet Union and then shipped up through the Sea of Okhotsk to the distant Kolyma region, where bitter cold, lack of food, and forced labor in the gold mines created deathly conditions. See James Harford, *Korolev: How One Man Masterminded the Soviet Drive to Beat America to the Moon* (New York: Wiley, 1997), pp. 49–63; and Siddiqi, *Sputnik and the Soviet Space Challenge*, p. 12.

[19] The NKVD eventually picked out Korolev to work in an aircraft design bureau (Central Design Bureau 29) being set up outside of Moscow in 1940 under his former instructor Tupolev. He would eventually spend time designing wing-borne rockets to assist the take-off of bomb laden aircraft. Much of this work fell under Yakovlev and Korolev's former GIRD colleague Valentin Glushko. See Harford, *Korolev*, pp. 56–59; and Siddiqi, *Sputnik and the Soviet Space Challenge*, p. 14.

had to work through Beria to accomplish its goals, for Beria himself answered only to Stalin.[20] Nevertheless, the first successful applications of Soviet rocket technology rose out of these difficult conditions. Wing-borne rockets for aircraft and the deadly, truck-mounted Katyusha canister rockets both saw extensive use during World War II. Most of the scientists spent the duration of the war working on such projects in various camps, lucky to be alive but still not free (even in the very limited sense of Soviet nonprison life under Stalin).[21]

Following the hard-fought Soviet victory at Stalingrad in early 1943, the tide of the war turned and the Red Army moved to the offensive. With this change, scientists gradually began to be released and moved forward to take over German facilities in the Soviet hope of acquiring new technologies. Korolev's work on rocket-assisted jets won him a Soviet decoration and release from prison in late July 1944,[22] although the secret police still tightly controlled his activities and fate. In 1946, the Soviet military shipped Korolev and a small team—under guard—to the Nazi rocket facility at Peenemunde to supervise the salvaging of prototypes and to interview captured scientists and engineers.[23] Within a year of the war's end, the Soviet Union possessed a full-fledged program for the development of long-range ballistic missiles with strong central support. Indeed, the first postwar Soviet Five-Year Plan listed rocketry among its top priorities.[24] Stalin and his top leadership perceived a clear link between state security and possession of this new technology, and therefore pursued it to the utmost despite its drain on Soviet postwar reconstruction.

Accordingly, Soviet Commissar of Armaments Dmitri F. Ustinov[25] began

---

[20] Eminent Soviet historian and legal scholar V. M. Kuritsyn, who studied the first open archives on these matters, confirmed that Beria played a leading role not only in rocket research but also in the Soviet nuclear program during these years. (Remarks by V. M. Kuritsyn, Center for East-West Trade, Investment, and Communications, Duke University, Durham, North Carolina, March 7, 1990.)

[21] Vladimirov's account of "spetzturma" life in Korolev's *sharaga* in Moscow (based on personal interviews with Korolev) is particularly rich in detail. The scientists worked in a fenced-off building on the outskirts of the city for twelve hours a day on weapons-related research and development. For good conduct, individuals were taken to a regular prison in Moscow where they could meet with their families. However, they were forbidden to reveal in conversation their actual work or terms of imprisonment, under threat of an additional eight-year sentence. (See Vladimirov, *Sovetskiy kosmicheskiy blef*, pp. 27–33.)

[22] Siddiqi, *Sputnik and the Soviet Space Challenge*, p. 16. Valentin Glushko was released at the same time.

[23] On the extensive Soviet effort to recover German technology and personnel, see the first-person accounts by Soviet missile scientists and engineers in John Rhea, ed., *Roads to Space: An Oral History of the Soviet Space Program* (New York: Aviation Week Group [McGraw-Hill], 1995).

[24] David Holloway, *Stalin and the Bomb: The Soviet Union and Atomic Energy, 1939–1956* (New Haven, Conn.: Yale University Press, 1994), p. 149.

[25] Ustinov later became a full Politburo member and Leonid Brezhnev's minister of defense.

setting up a new network of military-run (nonprison) missile design bureaus and institutes under the Soviet Council of Ministers' new Scientific Council.[26] The top leadership hoped specifically that a modified German V-2 might eventually be able to make up for the overwhelming U.S. advantage in bombers and its corresponding ability to deliver nuclear weapons (which the Soviet Union was also working feverishly to acquire[27]). The Soviet government commuted the sentences of many of the rocket scientists to accelerate this work.

By 1947, a centralized planning and decision-making body (the Council of Chief Designers) had been set up under the Soviet military, with Korolev at its head.[28] Thus, after a period of NKVD domination from 1938 to 1946, the practical military task of developing a missile for use against Western Europe and, eventually, the United States, transformed the Soviet missile program. It returned formal, operational control to the military, while increasing the direct accountability of the program to high Soviet Politburo leaders, including Stalin himself. Nevertheless, as historian Asif A. Siddiqi writes, even as the camp system wound down, Beria and his deputy in the Ministry of Internal Security, Ivan A. Serov, continued to exercise control over missile research "despite the apparent lack of any formal institutional mechanism."[29]

Korolev now played the leading scientific role in the development of long-range ballistic missiles and, moreover, in the organization of space-related design bureaus, institutes, and production facilities.[30] His Council of Chief Designers soon became the center of rocket technology, handling decision making on all major technical issues.[31] As Keldysh writes of the Council of Chief De-

[26] For more on this period, see David Holloway, "Innovation in the Defense Sector: Battle Tanks and ICBMs," in Ronald Amann and Julian Cooper, eds., Industrial Innovation in the Soviet Union (New Haven, Conn.: Yale University Press, 1982); also Matthew Evangelista, Innovation and the Arms Race (Ithaca, N.Y.: Cornell University Press, 1988), Chapter Five.

[27] For an exhaustive study on this effort, see Holloway, Stalin and the Bomb.

[28] This scientific body appears to have been the functioning organization working under a supervisory State Commission for the Study of the Long-Range Rockets (mentioned by Stoiko, Soviet Rocketry, p. 73), which included Stalin.

[29] Siddiqi, Sputnik and the Soviet Space Challenge, p. 38.

[30] Academician M. V. Keldysh—long credited in the official press with being the head of Soviet space activities in this period—admitted in a remarkably revealing account for the time (1983) that it was in fact Korolev who directed the various areas of military research on rockets. See M. V. Keldysh's chapter "Vospominaniya o S.P. Koroleve," in Raushenbakh, Iz istorii sovetskoy kosmonavtiki.

[31] The council consisted of the project leaders responsible for the development of all rocket-related technologies, including engines, directional systems, and ground-based support equipment. On the role of the council, see the chapters by Islinskiy, Raushenbakh, and Karpov in Raushenbakh, Iz istorii sovetskoy kosmonavtiki. See also discussions in Rhea, Roads to Space. One source points instead to the leading role after World War II of a "Coordinating Council for Space Research" [see Victor Yevsikov, Re-Entry Technology and the Soviet Space Program (Some Personal Observations) (Falls Church, Va.: Monograph Series on the Soviet Union, Delphic Associates, December 1982)]. However, Yevsikov's limited personal

signers, "Thanks to such an organization, the first military rockets were created very quickly."[32] The high-priority status of its projects gave the council direct access to necessary supplies. Notably, both civilian and military research fell under this council,[33] although ultimate decision-making power remained in the Politburo and, more specifically, with Stalin.

Another source of engineering and design skill used by Moscow came from the hundreds of German scientists forcibly removed in 1946 from Peenemunde in the Soviet occupation zone for a seven-year term of work in Soviet rocket design facilities.[34] Although the top designers had gone over to the United States with plans and prototypes, German scientists (led by Helmut Grottrup) did aid Soviet designers in these early postwar years. However, unlike the Americans, the Soviets kept the Germans removed from their most advanced research, literally isolating them from their most sensitive facilities. This deliberate deception caused those German engineers who ultimately left the Soviet Union and reached the West in the mid-1950s to underestimate the actual level of Soviet rocketry, thereby contributing to the shock of *Sputnik I* in 1957.[35]

Under Korolev's direction, work progressed at RNII on development of the first Soviet long-range ballistic missile.[36] Other designers in the program, Mikhail Yangel and Vladimir Chelomey, apparently competed against Korolev in this early stage for the position of chief designer, but failed.[37] Over the course

---

exposure to top-level decision making suggests that the "Coordinating Council" he refers to was, in all likelihood, his own name for the Interdepartmental Commission for Planetary Communications formed within the academy in the 1950s, a body limited to planning and oversight in the space science area only.

[32] Keldysh, in Raushenbakh, *Iz istorii sovetskoy kosmonavtiki*, p. 26.

[33] See Ishlinskiy's chapter "O zhizni i deyatel'nosti akademika Sergei Pavlovicha Koroleva," in Raushenbakh, *Iz istorii sovetskoy kosmonavtiki*.

[34] For a detailed account of the Grottrup group's experience in the Soviet Union, see Parry, *Russia's Rockets and Missiles*, pp. 120–28; also Harford, *Korolev*, pp. 75–90.

[35] The most lasting influence of the Germans probably was not technical. The Germans may have made a more meaningful contribution in ethos and organization. As McDougall writes, "The *managerial techniques* of Peenemunde may have found their way to [the Soviet launch facility at] Tyuratam via the Grottrup team" [italics mine]. See McDougall, . . . *the Heavens and the Earth*, p. 55. Recent interviews conducted by Harford show that the Soviets did not allow Grottrup and his closest advisors to return to Germany until after Stalin's death in 1953, despite the fact that they were only very rarely consulted in their last years in the Soviet Union. See Harford, *Korolev*, pp. 89–90.

[36] Although Korolev had a reputation as a testy manager who could be very hard on subordinates, various Soviet accounts also argue that he was a remarkable organizer who was able to motivate high-quality work from his staff. He successfully meshed the conflicting demands of the numerous military and civilian facilities working on space and rocket research. As the Soviets would find during their problems in the late 1960s after Korolev's death, these managerial skills were a rare individual talent.

[37] For a fuller account of the competition among rocket design bureaus at this time, see Yevsikov, *Re-Entry Technology and the Soviet Space Program*.

of the 1950s, Korolev enlisted the participation of more and more scientists from the Academy of Sciences structure to aid in research, especially on the nature of the upper levels of the atmosphere and on biomedical problems related to spaceflight.[38] Suborbital test flights with dogs (who were first sent into and returned from space in July 1951) contributed a significant amount of biological data useful for later human spaceflight on so-called academic rockets, which remained organizationally separate (and its participants ignorant) of the larger military research program.[39]

The work of nuclear scientists under Academician Igor V. Kurchatov, building the first Soviet nuclear weapon (exploded in 1949),[40] dictated the design specifications facing the rocket scientists. A booster capable of lifting the approximately 2-ton Soviet nuclear bomb became the standard for designers of the intercontinental ballistic missile (ICBM) and therefore demanded a huge rocket.[41] The first successful flight took place in 1948,[42] although this rocket—with a range of only 550 miles—lacked the thrust necessary for use against the United States. This intermediate-range missile began entering the arsenals of the Red Army in 1949.[43] The year 1954 would mark the real turning point in missile development as the Soviet-designated RD-107 and RD-108 engines provided the military with its first boosters capable of intercontinental range.

To cover the burgeoning number of space-related design bureaus and production facilities working on military projects, the Soviet leadership in 1953 established the Ministry of General Machine Building as a "front" organization. Although technically independent, the nature of its work caused it to fall under

[38] Keldysh in Raushenbakh, *Iz istorii sovetskoy kosmonavtiki.*

[39] Ishlinskiy in Raushenbakh, *Iz istorii sovetskoy kosmonavtiki*, p. 18; also, Siddiqi, *Sputnik and the Soviet Space Challenge*, p. 95. Siddiqi notes that Soviet suborbital flights with dogs beat comparable U.S. programs into space by two months.

[40] Evidence from the post-Soviet opening of Communist Party archives and from interviews with surviving Soviet scientists shows that Soviet intelligence acquired on the specifications of the U.S. bomb helped speed the development of these tests considerably. Among other sources, see Serge Schmemann, "1st Soviet A-Bomb Built from U.S. Data, Russian Says," *New York Times*, January 14, 1993, p. A5; Holloway, *Stalin and the Bomb*, pp. 94–108; and Jeffrey T. Richelson, *Spying on the Bomb: American Nuclear Intelligence from Nazi Germany to Iran and North Korea* (New York: Norton, 2006), pp. 65, 67.

[41] Ironically, when the first ICBM was finally developed, its awesome size was a sign of its backwardness, although its design requirements had the benefit of allowing for large space payloads early on in the Soviet program. Smaller U.S. ICBM boosters could not handle such payloads, thereby limiting the early U.S. space program and biasing its work—to the eventual benefit of domestic industry—toward miniaturization of components.

[42] Ishlinskiy (in Raushenbakh, *Iz istorii sovetskoy kosmonavtiki*) and a number of Western sources date this flight in 1948.

[43] McDougall, . . . *the Heavens and the Earth*, p. 55.

de facto military supervision. Korolev's design bureau itself became part of this new ministry, which now occupied the center stage in industrial planning and production for the space program.[44]

Most of the activity under the new ministry took place far from Moscow on the missile testing grounds at Kapustin Yar (near the Caspian Sea) and to a greater extent at the so-called Baikonur (Tyuratam) facility in Kazakhstan.[45] The Soviets kept close guard over the ministry's research, while in the international arena focusing exaggerated attention on the work of the Academy of Sciences. For example, in 1954, negotiations among scientists worldwide planning for the International Geophysical Year (IGY) led to the Soviet designation of a relatively powerless Interdepartmental Commission for Interplanetary Communications (ICIC) under Academician Leonid I. Sedov. This commission sought to give an air of civilian control to the growing Soviet space program.[46]

But the real force behind the Soviet rocket program continued to be the military, whose priority to develop a long-range missile to counter the United States' strategic advantages at the time led to large budgets and a sense of urgency. Tests of Soviet long-range missiles in early 1957 began to suggest that success might not be far off. By August 1957, the Soviets announced their first successful launch of an ICBM, effectively setting the stage for the satellite launch that was soon to follow: *Sputnik I*. The process from initial state-run missile organizations to a space-launch vehicle had taken the Soviets less than thirty years, despite political purges that killed many top scientists and a devastating war that had sapped resources and nearly ended the country's existence.

Roots of the American Space Program

In the United States, missile and space activity developed later and differently than it did in the Soviet Union. In many respects, German and Soviet developments before World War II are much more comparable, being almost exclusively state funded and military oriented. Although the United States military

---

[44] At the "factory" level, the ministry was broken down into a number of design bureaus and production enterprises, which were responsible for developing and building particular systems such as the casing for a rocket, its engines, and its payload. Guidelines for production may have come from design bureaus under the military and, later, certain bureaus under academy control as well.

[45] As early U-2 flights revealed, this facility was actually situated several hundred miles from the village of Baikonur. Therefore, it was often referred to in the West as Tyuratam, the nearby rail spur that marks its actual location. Since the end of the Cold War, the name Baikonur continues to be used by Russia and Kazakhstan, its official owner. Russia now operates the facility under a long-term lease.

[46] The commission's name was later changed to the Commission for the Study and Use of Outer Space. See Congressional Research Service's report prepared for the Senate Commerce Committee, *Soviet Space Programs: 1976–80*, Part I (Washington, D.C.: U.S. Government Printing Office, 1982), p. 322.

had faced British rockets at Fort McHenry in 1814 and used them for signaling in the late 1800s, by the early twentieth century, such small, unguided rockets had "passed into obscurity as weapons,"[47] given comparative improvements in the range and accuracy of long-range, rifled guns.

Nevertheless, on the eve of World War I, individual rocket enthusiasts, including Clark University physics professor Robert Goddard, again began to push the boundaries of propulsion research forward. The U.S. military supported some missile-related research projects during the war, including the Ballistic Institute founded by Goddard's Clark University mentor Dr. Arthur Gordon Webster, but its main aim was to counter Germany's advantage in long-range artillery. The war's end and growing isolationist sentiments in the United States caused federal funding to dry up, typical of past patterns in U.S. military research and development. Many U.S. scientists at the time had little interest in working for the military.[48] But Goddard had no such qualms, and sought whatever support he could find for his research.

Goddard won public acclaim in 1920 for a Smithsonian Institution study he wrote on propulsion and spaceflight. But media attention later yielded to public mockery when journalists and commentators ridiculed his enthusiastic (and seemingly fanciful) statements about the possibility of building rockets to visit the Moon.[49] The scientist felt misunderstood and later bitter when others received attention for discoveries he believed to be his. Goddard finally managed to gain a small navy contract in the early 1920s for work on rocket-propelled depth charges for use against submarines, for which he commuted part-time from Massachusetts to the former World War I Naval Proving Ground at Indian Head, Maryland. The project ended without notable success in 1924, however, in part because of the limits of resources.

Stung by the military's abandonment and by the notoriety given in Europe to Austrian space scientist Hermann Oberth (whom he considered a copycat), Goddard turned inward to focus on his experiments.[50] On a snowy March day in 1926, he put his ideas on liquid-fueled rockets to the test. With the help of a few colleagues, he had put together a strange-looking, A-shaped contraption, with a combustion chamber at the top and separate pressurized chambers of gasoline and liquid oxygen below linked by metal tubing. He and his team

[47] Albert B. Christman, *Sailors, Scientists, and Rockets: Origins of the Navy Rocket Program and of the Naval Ordnance Test Station, Inyokern (History of the Naval Weapons Center, China Lake, California, Volume I)* (Washington, D.C.: U.S. Government Printing Office, 1971), p. 6.

[48] On this point, see Christman, *Sailors, Scientists, and Rockets*, p. 69.

[49] On the media charges and Goddard's hounding by reporters, see Burrows, *This New Ocean*, p. 46.

[50] On this rivalry, see Burrows, *This New Ocean*, pp. 48–53.

heated the fuel chambers to drive the hot gases upward into the rocket chamber itself as they waited for enough fuel to accumulate to reach critical thrust.[51] Suddenly, the rocket moved and shot upward to an altitude of 41 feet and across the field to a distance of 184 feet. This first-of-its-kind test proved that liquid fuels had great potential for rocket propulsion once they could be harnessed more effectively. But a lack of money and materials remained a serious problem for Goddard, and he yearned for a more isolated and climate-friendly test facility. Finally, in 1930, Goddard won a four-year $100,000 grant from the Guggenheim Foundation—thanks to a recommendation from the young, but acclaimed, aviator Charles Lindbergh—allowing him to move his test operations to Roswell, New Mexico.[52] For the next decade, Goddard would continue his work on liquid-fueled rockets, free of the demands of the university and the intrusions of the public. While in New Mexico, he tried again in 1933 to attract military funding for research on anti-aircraft rockets and on rocket-assisted takeoff devices for aircraft, but failed. He continued his research with the help of private funding until the war brought him back into government service.

Meanwhile, a parallel and larger program of missile research had begun to emerge under the academic auspices of the California Institute of Technology (Cal Tech). As early as 1930, the institute's Guggenheim Aeronautical Laboratory was the site of new research on jet propulsion conducted by Frank J. Malina and others under the direction of Hungarian-born mathematician and aerodynamics expert Dr. Theodore von Karman. Attempts made during this time by Cal Tech researchers to pool research information with Goddard in New Mexico were met largely with secrecy and rebuffs from the suspicious inventor. In 1939, the National Academy of Sciences granted Cal Tech funds to develop rocket-assisted takeoff devices, recognizing the university's advantages over Goddard in terms of size, resources, and personnel.

By 1940, the war in Europe was beginning to change the U.S. government's attitude toward missile research. Support for the establishment of a centralized oversight and funding system galvanized as leaders sought to make the most of potential military-related research already going on at U.S. universities. These efforts led to the creation of the National Defense Research Committee. Within weeks of its founding, the committee authorized funding for an "extensive rocket program" based at Cal Tech.[53] Among its key participants were von

---

[51] On the technical details of Goddard's rocket, see "March 16, 1926 Goddard Rocket," on the Web site of the National Air and Space Museum at <http://www.nasm.si.edu/research/dsh/artifacts/RM-RHG1926.htm> (accessed September 22, 2006).

[52] Burrows, *This New Ocean*, p. 83.

[53] Christman, *Sailors, Scientists, and Rockets*, p. 100.

Karman, Malina, and the Chinese-born, U.S.-educated scientist Dr. Tsien Hsue Shen.[54]

For Goddard, the war also brought some military support, although his preference for liquid fuels placed him at a disadvantage, as battlefield conditions favored solid fuels. He received contracts for work on jet-assisted takeoff devices for aircraft from the navy's Bureau of Aeronautics and the Army Air Corps, but Cal Tech's programs soon eclipsed him. As navy historian Albert B. Christman summarizes, "The ordnance structures of the armed forces up to World War II were too inflexible and too limited to make use of Goddard's peculiar form of genius."[55]

Characteristic of the U.S. program right through the war was its lack of central focus and direction. Besides the work at Cal Tech, a significant program had taken shape at the abandoned Indian Head facility in Maryland. Former Goddard associate Clarence Hickman ran the Maryland rocket program with a small military staff.[56] But these efforts could not compare to the work being done in Germany. In the years leading up to World War II, the relative isolation of the United States from European geopolitical tensions and its ideological tradition of laissez-faire capitalism put America almost a decade behind Germany.[57] The U.S. military's "go slow" attitude toward rockets at this time is captured in a December 1941 letter written by Cal Tech's Charles C. Lauritsen[58] to a colleague about his meetings with the official Washington military brass: "Regarding rockets . . . there seemed to be little interest on the part of the armed services."[59]

By mid-1941, however, the Office of Research and Development had given out two major university contracts and several smaller ones to private firms for rocket and propulsion research. On the West Coast, the already well-established Cal Tech facility became the center of activity working largely on navy contracts, while on the East Coast, George Washington University led rocket

[54] On Tsien's background and route to Cal Tech, see Brian Harvey, *China's Space Program: From Conception to Manned Spaceflight* (Chichester, England: Praxis, 2004), pp. 18–19.

[55] Christman, *Sailors, Scientists, and Rockets*, p. 103.

[56] As with all rocket programs at the time, the operation employed relatively few individuals, such that even basic tasks—like the making and packing of primers—were done by the program leaders themselves.

[57] On these funding issues, see McDougall, . . . *the Heavens and the Earth*, pp. 74–77.

[58] Lauritsen eventually left Cal Tech to serve on the National Defense Research Committee, where he promoted expanded rocket research. On his work with the committee, see Christman, *Sailors, Scientists, and Rockets*, pp. 86–114; also Frank J. Malina, "Origins and First Decade of the Jet Propulsion Laboratory," in Eugene M. Emme, ed., *The History of Rocket Technology: Essays on Research Development and Utility* (Detroit: Wayne State University Press, 1964), pp. 63–64.

[59] Lauritsen, quoted in Christman, *Sailors, Scientists, and Rockets*, p. 125.

contracts working at Indian Head, mainly on army contracts. As for results, the Cal Tech effort yielded by the end of World War II an anti-aircraft rocket, an antisubmarine rocket, an air-launched antisubmarine retro-rocket, and naval barrage rockets, all used extensively in the Pacific campaign. Meanwhile, the Indian Head team developed the revolutionary "bazooka" and an air-to-air rocket,[60] both of which played important roles in the U.S. war effort. But the fate of missile efforts remained unclear.

In strategic terms, the end of World War II found the United States in a fundamentally different position from the one it occupied in 1941. It was now the unquestioned—if reluctant—leader of the noncommunist world and the sole guarantor of democracy in Western Europe and beyond. The potential of rocketry, however, threatened to erase U.S. military superiority over its primary adversary, the Soviet Union. If the Soviets perfected a long-range rocket before the United States, they could terrorize the continent and America itself as the Germans had been unable to do with the much shorter-range V-2s. But a more important weapon was the atomic bomb. For this reason, the United States had proposed the so-called Baruch Plan to the United Nations in 1946.[61] The program called for all states to agree to a nonnuclear status and to submit their facilities to inspection as a step toward Washington's subsequent destruction of its own arsenal. Not surprisingly, Stalin had no faith that the United States would dismantle its weapons once Moscow had opened up its facilities. He was also unwilling to submit to second-class nuclear status. Soviet representatives rejected the offer. Still, the surfeit of bombers in the U.S. arsenal, the availability of bases in Europe, and the lack of even an intermediate-range Soviet missile at least temporarily ameliorated the threat faced by the United States. Under these conditions, developing long-range missiles was not a number one priority, as it was in the Soviet Union.[62]

Nevertheless, an initial move to expand U.S. rocket research began in 1946. The U.S. Army now had access to the ideas and know-how of the numerous German scientists who had intentionally surrendered to the American side in 1945, bringing prototypes, plans, and personnel they had removed from their Peenemunde launch facility. Budget expenditures for fiscal year 1946 for missile research in the army air force climbed from $3.7 million in 1945 to $28.8

[60] Although it seems unbelievable today, Christman's account highlights just how limited American missile research was at this time by pointing out what a "first" this program represented. "On July 6, 1942, an Army fighter aircraft fired a series of 4.5 inch rockets at Aberdeen [Maryland] marking the first forward firing of an American rocket from a plane in flight" (Christman, *Sailors, Scientists, and Rockets*, p. 145).

[61] On the Baruch Plan and Soviet reactions, see Holloway, *Stalin and the Bomb*, pp. 163–65.

[62] On this point, see McDougall, . . . *the Heavens and the Earth*, p. 98.

million.[63] However, the enthusiasm of the rocketeers did not match the mood of the president and Congress, whose goal was to return to some degree of budgetary "normalcy" after the war's vast expenditures. By 1947, as the Soviet Union continued to push ahead, Truman had cut the missile budget back to $22 million.[64]

The American effort to use the technological expertise developed in the German V-1 and V-2 projects had been code-named "Project Paper Clip."[65] The military sent a number of missile experts from Cal Tech—including Malina and Tsien—to Germany to assist in the process.[66] Beginning in the fall of 1945, some 120 German scientists and engineers had arrived at the White Sands facility in New Mexico to work on perfecting their V-2 design. The U.S. government whitewashed the military, Nazi, and even SS backgrounds of a number of missile experts—including Wernher von Braun—to make them more palatable to the American public.[67] Assisted by the Germans, the General Electric Company took charge of launch operations under an army contract that saw forty test flights of instrumented V-2 rockets in the course of the next five years.[68] Even more notorious Nazi rocketeers would later join the U.S. Army effort, including V-2 program head (and German Army General) Walter Dornberger and Mittelwerk operations director (and longtime Nazi Party member) Arthur Rudolph.[69]

Meanwhile, the navy had begun to organize a new program of research in conjunction with the Applied Physics Laboratory at Johns Hopkins University. Other military-funded rocket research at this time included a project between the Jet Propulsion Laboratory (at Cal Tech) and Aerojet General Corporation,

[63] Ibid., p. 87.

[64] Ibid.

[65] Richard Rhodes, *Dark Sun: The Making of the Hydrogen Bomb* (New York: Touchstone, 1996), p. 228.

[66] Harvey, *China's Space Program*, p. 19.

[67] Von Braun, for example, had joined both the Nazi Party and the SS to further his missile career. On U.S. motivations, see Burrows, *This New Ocean*, p. 122.

[68] For more on this German-aided test program, see Constance McLaughlin Green and Milton Lomask, *Vanguard: A History* (Washington, D.C.: Smithsonian Institution Press, 1971). During this period, the United States also began experimenting with suborbital animal flights. Unlike the Russians, who preferred dogs, the Americans used monkeys and mice to test the effects of acceleration and high altitude on metabolism. In most cases, the animals died, not because of the physical effects of spaceflight itself but, instead, faulty rockets or recovery systems. On these V-2 animal flights, see Neil McAleer, *The Omni Space Almanac* (New York: World Almanac, 1987), pp. 14–15.

[69] Dornberger first had to serve two years in a British detainment facility for his responsibility for the V-2 attacks on London. Rudolph, an unrepentant Nazi, made it to the United States in 1958 and worked on the Saturn launcher before finally leaving the country in the 1980s to escape charges of having been a war criminal. See Neufeld, *Rocket and the Reich*, pp. 226–30, 267; and Burrows, *This New Ocean*, p. 527.

working for the Navy Bureau of Aeronautics, and at the Glenn Martin Company, working on contracts (beginning in 1946) from the Naval Research Laboratory. Despite the active interest and past experience of the navy in missile and propulsion research, the military's missile specialists still continued to have difficulty convincing interservice committees and civilian technical overseers that new rocket programs had valuable enough missions to justify their costs.

Not to be left out of the growing U.S. military competition over space, the army air force commissioned a study by Project RAND—then part of the Douglas Aircraft Company—to examine the potential of Earth-orbiting satellites.[70] Entitled "Preliminary Design of an Experimental World Circling Spaceship," this 1946 report discussed the advantages of satellite reconnaissance in providing invulnerable access to information about sites in enemy territory, although largely to inform bombing decisions. The report also noted the likely value of satellites for communications uses.

By 1948, analysts perusing the first "Secretary of Defense Annual Report" discovered the heretofore hidden fact that all three branches of service were conducting *competing* missile research programs. The results proved perhaps predictable: public outcry and some reductions in funding across the board. Still, the multiplicity of programs went on.[71]

Without the budgetary fervor of a crash program, progress in the U.S. missile sector proved slow through the end of the decade. But by 1950, public concern and congressional reaction to apparent Soviet progress in missile development led President Harry Truman to appoint Kaufman T. Keller—then head of the Chrysler Corporation—to conduct a study of the state of the various U.S. rocket research programs and to provide advice on areas deserving more attention and greater funding.[72] The next year, Keller's support helped begin a modest U.S. ICBM program (Atlas) under the air force. Yet overly stringent design specifications and the lack of a small enough nuclear weapon to use as a possible warhead kept the program in a "back burner" mode. Ironically, with the McCarthy trials in full swing, the government imprisoned Cal Tech missile expert Dr. Tsien in 1951 for alleged communist ties, and later deported him.[73]

Meanwhile, in the area of scientific research, the Naval Research Laboratory

[70] On this report, see Paul Stares, *The Militarization of Space: U.S. Policy, 1945–1984* (Ithaca, N.Y.: Cornell University Press, 1985), pp. 24–26.

[71] The army continued work on an intermediate-range ballistic missile, while the NRL worked on a completely separate research program—the Viking series—aimed at exploring the upper atmosphere. On these points, see Green and Lomask, *Vanguard.*

[72] On Keller's work, see Rip Bulkeley, *The Sputniks Crisis and Early United States Space Policy: A Critique of the Historiography of Space* (Bloomington: Indiana University Press, 1991), pp. 74–77.

[73] Harvey, *China's Space Program*, pp. 20–21. Tsien reluctantly returned to China in 1956 and was eventually recruited by the Chinese government to begin a national missile program.

began developing the Viking series of upper atmospheric rockets in 1949–54. Over the five-year period, however, its cumulative funding remained modest, at less than $6 million. With the outbreak of the Korean War, interest grew in missile research but only in work directly related to immediate military applications. As Green and Lomask comment, "In these years the Department of Defense was unwilling to spend more than token sums on research that appeared to have only [a] remote connection with fighting equipment."[74]

Analysts conducting research for the air force within the RAND corporation, however, recognized the emerging military importance of space. In secret reports issued in November 1950 and April 1951, RAND experts predicted the value of Earth satellites for military reconnaissance.[75] Work on this subject continued under the rubric of Project Feedback.

The election of General Dwight D. Eisenhower as president in 1952 brought major changes to the U.S. Department of Defense, turning over an organization that had experienced no top-to-bottom shake-up since Roosevelt's election in 1932. In addition, the development of the first hydrogen bomb in 1953 and adoption of the administration's "New Look" strategy promoted the expansion of the U.S. nuclear arsenal for more credible deterrence ranging from massive retaliation to regional war scenarios.[76] Finally, the promotion of Brigadier General Bernard A. Schriever as head of the air force's Western Development Division of the Air Research and Development Command (ARDC) put a dynamic and committed project manager behind the sluggish Atlas program.

As much as these factors promoted work on developing a long-range missile in the United States, they still could not override the organizational problems inherent in the interservice rivalries of the time. The multiplicity of programs divided rather than concentrated the national effort to develop an ICBM, splitting up key personnel into different services and causing each program to accept unnecessary budget limitations in order that each might continue to get a piece of the federal research pie.

Ironically, funding proved somewhat easier in the nuclear sector. In tune with the "atomic enthusiasm" of the time, the United States began a classified program to develop a nuclear-powered rocket, Project Rover. Initial studies of the concept had been completed in 1947 by North American Aviation and in 1953 by Oak Ridge National Laboratory, but no actual hardware had resulted.[77] Finally, in 1954, the Pentagon requested funds for the following year to con-

---

[74] Green and Lomask, *Vanguard*, p. 12.

[75] Stares, *Militarization of Space*, p. 31.

[76] On the New Look strategy, see Lawrence Freedman, *The Evolution of Nuclear Strategy*, 3rd ed. (London: Palgrave, 2003), pp. 76–85.

[77] On these studies, see Astronautics.com, at <http://www.astronautix.com/project/nerva.htm> (accessed August 23, 2006).

duct joint air force–Atomic Energy Commission research at the Los Alamos Scientific Laboratory on reactor designs for lofting ICBMs and for conducting manned space exploration.[78]

Results from Project Feedback began to flesh out parameters for a U.S. satellite reconnaissance program, calling the matter a "vital strategic interest."[79] Air force officials accepted the recommendations of hundreds of RAND and external experts working on the problem and began to move forward with a multiyear development project in late 1954. The program aimed at ultimate acquisition of photo-intelligence on Soviet airfields and missile sites, as well as electronic intelligence and weather forecasting data.[80] But funding for this air force program, which eventually came to be known as Sentry (and later the Satellite and Missile Observation System, or SAMOS) remained modest.

By 1955, the National Security Council (NSC) began to suspect that the Soviet Union would launch a military satellite within the next few years, causing potential damage to the worldwide reputation of the United States as a technological leader.[81] Therefore, its members reasoned, the United States should seek not only to beat them, but to beat them with an "open" satellite program. After reviewing classified documents regarding the Soviet ICBM program, the NSC decided in a secret May 20 report (NSC 5520, entitled "Draft Statement of Policy on U.S. Scientific Satellite Program") that the United States should set up a parallel "civilian" satellite program—albeit technically under the navy—and use it to beat the Soviet military into space.[82] The report also mentioned two devoted military satellites under development and asserted to the president U.S. rights to overfly the Soviet Union without Moscow's prior consent. By way of further justification for the civilian satellite, the document mentioned that it too would provide information "clearly relevant to missile and anti-missile research" and "military communications."[83] As a result of this decision, President Eisenhower announced on July 28 that America would launch a satellite as part of the International Geophysical Year (IGY) planned for 1957–58. The goal of this research year was to further scientific knowledge about the Earth's geophysical environment by a worldwide exchange of results from sixty-seven

[78] John W. Simpson, *Nuclear Power from Underseas to Outer Space* (La Grange Park, Ill.: American Nuclear Society, 1995), p. 122.

[79] Stares, *Militarization of Space*, p. 30.

[80] Ibid.

[81] According to declassified documents, the United States government was relatively well informed about the progress of Soviet missile research. (See Bulkeley, *Sputniks Crisis*, p. 68.) The American public, however, remained sorely ill informed, thus contributing to the great shock of 1957.

[82] NSC 5520, "Draft Statement of Policy on U.S. Scientific Satellite Program," printed in John M. Logsdon, ed., *Exploring the Unknown: Selected Documents in the History of the U.S. Civil Space Program*, Volume 1: *Organizing for Exploration* (Washington, D.C.: NASA, 1995), pp. 308–10.

[83] Ibid., p. 311.

national programs, involving some thirty thousand scientists. In both the United States and the USSR, the IGY encompassed a range of scientific activities studying subjects from Antarctica to the seabed, most of which had no relation to space. In fact, until the U.S. announcement, no space activities at all had been planned for the IGY. But political leaders soon recognized the advantages of being first into orbit and of appearing to be "peace loving," even if they had to use military technologies to do it.[84] Eisenhower's statement stressed that this program was merely a continuation of the ongoing program of U.S. meteorological research on the upper atmosphere.[85]

Yet problems continued to plague the American satellite program. From the start, Project Vanguard—a hybrid program of the National Academy of Sciences and the navy[86]—had been intentionally underfunded (due to other military priorities), and its rocket underpowered.[87] It had beaten out a more developed, army-sponsored program, Jupiter, largely because of the technical superiority of its instrumentation package and the favorable impression this made on the scientist-dominated committee. But the choice came with a cost, since the Jupiter C missile had a track record of accomplishment. In September 1956, it had flown 3,300 miles and could likely have placed a payload into orbit had one been ready and had it been given such a mission.[88] However, there were also other considerations. Former Nazi scientist von Braun headed the Jupiter team, and U.S. government sentiments strongly opposed turning over the first American satellite project to a German who had fathered the V-2 missile scarcely a decade earlier. In addition, the priority for the nation was missile research aimed at the construction of an ICBM for the air force, not scientific research aimed at space achievements. As Herbert York notes, "We had specifically ordered that the scientific IGY satellite program not be allowed to interfere with any of our military [launch] programs."[89]

Given the recognized U.S. lag in missile programs, it is not surprising that

---

[84] As Manno writes, "The president believed that military space objectives should receive the highest priority. He also believed, however, that the United States should *appear* to be interested in outer space for strictly peaceful purposes." See Jack Manno, *Arming the Heavens: The Hidden Military Agenda for Space* (New York: Dodd, Mead, 1984), p. 30.

[85] Dodd L. Harvey and Linda C. Ciccoritti, *U.S.-Soviet Cooperation in Space* (Miami, Fla.: University of Miami Press, 1974), p. 6.

[86] The division of responsibilities gave the Academy of Sciences' IGY National Committee control over instrumentation, tracking, and construction of the satellite's shell.

[87] Behind the scenes, official guidelines for the program stipulated that Vanguard should not impede higher priority defense work on intermediate- and inter-continental-range ballistic missiles.

[88] On this point, see M. S. Hunt, "History of Astronautics" (Chapter 1), in N. H. Langton, ed., *The Space Environment* (New York: American Elsevier, 1969), p. 11.

[89] York, *Race to Oblivion*, p. 110. In support of this account, see also U.S. National Security Council Report, "U.S. Scientific Satellite Program" (cited in Bulkeley, *Sputniks Crisis*, p. 157).

American proposals for limiting international military access to space began coming from Washington. Eisenhower realized that intercontinental rocket technology, once developed, would revolutionize warfare by dramatically shortening the lead time before an attack and by rendering defense virtually impossible. Thus, it was clearly to U.S. advantage to halt the suspected Soviet move toward ICBMs, since the Americans already had bases in Turkey and Western Europe from which they could reach the Soviet Union with bombers and shorter-range missiles. At the root of Eisenhower's attempts, therefore, was a U.S. desire to cement its existing strategic advantages over the Soviet Union.

On January 12, 1957, the United States argued before the United Nations that space should be covered by an international arms control regime that would ban all military systems from entering space and mandate its use only for peaceful, scientific purposes. The proposal noted:

> No one now can predict with certainty what will develop from man's excursion into this new field. But it is clear that if this advance into the unknown is to be a blessing rather than a curse the efforts of all nations in this field need to be brought within the purview of a reliable armaments control system.
>
> The United States proposes that the first step toward the objective of assuring that future developments in outer space would be devoted exclusively to peaceful and scientific purposes would be to bring the testing of such objects [satellites, missiles, and other possible space weapons and platforms] under international inspection and participation.[90]

The Soviet Union, not surprisingly, saw the situation in space quite differently and did not respond favorably.

Soviet strategy, as implied earlier, included even fewer elements of cooperation. Under the leadership of Communist Party First Secretary Nikita Khrushchev, the Moscow leadership knew that its military was far weaker than that of the United States and justifiably feared Western tactical advantages. This led Moscow to hold to a policy of talking peace internationally but keeping its real program away from foreign view.

To stem the growth of the U.S. nuclear arsenal, to divert attention from its own growing arsenal, and to win propaganda points internationally, the Soviet Union had announced a policy on July 21, 1955, of seeking to achieve a complete ban on nuclear testing.[91] However, the proposal lacked a plan for verification

[90] United States Memorandum Submitted to the United Nations General Assembly, January 12, 1957, reprinted in Harvey and Ciccoritti, *U.S.-Soviet Cooperation in Space*, p. 1. Later, on July 22, Secretary of State John Foster Dulles reiterated the main points of this proposal in a nationally televised speech.

[91] See Thanos P. Dokos, *Negotiations for a CTBT 1958–1994: Analysis and Evaluation of American Policy* (Lanham, Md.: University Press of America, 1995), p. 3.

and met with U.S. opposition. World public opinion had begun to be affected by the harmful health and environmental effects of atmospheric nuclear testing, and various public organizations began to lobby for the test ban. In the United States, Democratic presidential candidate Adlai Stevenson ran on a platform in 1956 of ending large atmospheric tests. His defeat in November placed the onus back on Eisenhower.[92]

Of course, Khrushchev was not about to trade away the technological potential he saw in ICBMs for overcoming an otherwise immutable situation of overwhelming U.S. strategic superiority in deliverable nuclear weapons. While it would be an exaggeration to blame Moscow wholly for the lack of progress during this period, Khrushchev admits in his memoirs that it was official Soviet policy to hold back on cooperation with the West until Moscow was sure it had completed the design of an effective ICBM: "We felt we needed time to test, perfect, produce, and install the booster by ourselves. Once we got our feet planted firmly on the ground and provided for the defense of our country, *then* we could begin space cooperation with the United States."[93]

Sputnik and Soviet Space "Superiority"

Following plans to include space in the IGY, the Soviet press mentioned in April 1956 Moscow's intention to launch a "scientific" satellite during the planned eighteen-month program.[94] In planning meetings for the IGY, however, the Soviets provided few details on their intended research.[95] Unlike the Americans, they simply went about quietly moving forward with the building blocks to carry out their claim. After three failures, Korolev's team succeeded in launching the world's first ICBM (the R-7, or U.S. designated SS-6) on August 21, 1957.[96] Not only did it mark a military milestone, but for First Secretary

[92] George Bunn, *Arms Control by Committee: Managing Negotiations with the Russians* (Stanford, Calif.: Stanford University Press, 1992), p. 18.

[93] See Strobe Talbott, trans. and ed., *Khrushchev Remembers: The Last Testament* (New York: Little, Brown, 1974), p. 55. In support of this perspective, Dr. V. S. Vereshchetin, then-vice-chairman of the Soviet space organization Interkosmos, admitted in a remarkably candid interview with the author in Moscow in April 1988 that the USSR *was* largely to blame for the lack of progress on space cooperation during the early 1960s.

[94] On the arcane politics and planning surrounding the IGY in space, see Burrows, *This New Ocean*, pp. 166–78.

[95] The United States, by contrast, provided considerable information to the world organization regarding its plans for *Vanguard I*.

[96] See M. V. Keldysh's report of the launch in Raushenbakh, *Iz istorii sovetskoy kosmonavtiki*, p. 28. On

Khrushchev it proved to be a brilliant propaganda coup. His provocative claims to the West about the R-7's capabilities and of Soviet stockpiles of the missile led many in the West to believe that the Soviet Union had developed a whole arsenal of effective, nuclear-tipped ICBMs. In fact, the original design of the R-7 never went into large-scale production as a military missile because of its long launch preparation time.[97] Khrushchev's bluff was designed to counter the by now insurmountable U.S. lead in medium-range missiles and bombers deployed around the Soviet periphery.

For the Soviet space program, however, the ICBM booster proved to be a *real* asset. Within two months of the test of the R-7, Korolev had organized plans for the launch of *Sputnik I*, providing frequency information for its signal and orbital parameters for its planned movement. To protect themselves from possible failure, the Soviets announced the actual launch time only after the satellite had already successfully achieved orbit on October 4, 1957. Despite the prior knowledge and expectation of the scientific and military communities, *Sputnik I* had an unexpected political effect.

Americans wondered aloud how the "red" Soviet Union, long seen as a technological backwater, could have beaten the United States into space. As recently as the early 1920s, Herbert Hoover and U.S. relief teams had helped Russians and Ukrainians survive the deadly famine that struck their country in the aftermath of World War I and civil war. In the 1930s, Americans companies had exported some of the basic technologies necessary to bring Soviet manufacturing into the twentieth century. During World War II, according to then popular U.S. perceptions, it was *American* trucks, planes, and food that had provided the necessary edge to Soviet forces in their desperate struggle against the German *Wehrmacht* through the Lend-Lease program. For all of these reasons, *Sputnik* shattered American perceptions of their own technological superiority as a nation.[98] Governmental officials, right up to President Eisenhower, engaged in hand-wringing and tried unsuccessfully to downplay *Sputnik*'s significance.

---

the prior failures, see Burrows, *This New Ocean*, pp. 165–66. The key to Russian success was the clustering of engines in the critical first stage.

[97] The liquid-fueled R-7 rocket took 17.5 hours to prepare for launch. However, it became the model for the so-called Vostok space-launch booster, which became the main Soviet space rocket for three decades.

[98] As renowned scientist and Truman presidential advisor Vannevar Bush testified before Congress in 1954, "In any field where the Russians can copy a method used elsewhere, they can do so promptly and effectively. . . . But whether they can strike out into new fields and do original things, I rather doubt. I don't think they are that good" (testimony before the House Government Operations Committee, quoted in Bulkeley, *Sputniks Crisis*, p. 68).

Senate Majority Leader Lyndon B. Johnson launched a congressional investigation into the administration's failure[99] and called for massive U.S. investment in
space, which he called the "key" to world power.[100]

In the highly symbolic world of the Cold War, however, international reac-
tion to *Sputnik* was immediate and, for the United States, extremely negative.
The result of public opinion polls in Western Europe following the Soviets'
Sputnik flight showed that by a large margin the French, Italian, German, and
even British populations had become convinced of Soviet superiority over the
United States in scientific development.[101] Worse though, at least as far as the
White House was concerned, was the fact that all now rated the Soviet Union
ahead in overall military power, albeit by a lesser margin. Thus, although President Eisenhower personally opposed the idea of racing with the Soviet Union
in space and rejected the need for additional U.S. military expenditures,[102] any
cooperation with the Soviets now seemed to mean admitting to a second-place
status. The American public and Congress demanded to be the leader in space,
and soon. These events and the military-political pressures that surrounded
them put down the roots of what would soon become a hostile U.S.-Soviet
space relationship.

While Americans waited for their own satellites, the Eisenhower administration played for time. As Eisenhower reflected on this period, "In the weeks and
months after Sputnik many Americans seemed to be seized not only with a
sudden worry that our defenses had crumbled, but also with an equally unjusti-
fied alarm that our entire educational system was defective."[103] On October 10,
on the heels of the *Sputnik I* launch, Eisenhower's Representative to the United
Nations Henry Cabot Lodge proposed the immediate negotiation of an agreement for internationally controlled prelaunch inspection of all space rockets
to ensure their peaceful intentions.[104] Lodge's impassioned speech to the U.N.

[99] For an insightful analysis of Johnson's thinking and the political context that surrounded
it, see John M. Logsdon, *The Decision to Go to the Moon: Project Apollo and the National Interest*
(Cambridge, Mass.: MIT Press, 1970), pp. 21–22. Although Logsdon rates Johnson's concern for U.S.
national security and reputation as the key factors, another motivation was Johnson's hope of run-
ning for the presidency in 1960.

[100] Hugo Young, Bryan Silcock, and Peter Dunn, *Journey to Tranquility: The History of Man's
Assault on the Moon* (London: Jonathan Cape, 1969), p. 98.

[101] See polls cited by Harvey and Ciccoritti, *U.S.-Soviet Cooperation in Space*, pp. 49–50; also by
McDougall, . . . *the Heavens and the Earth*, pp. 240–41.

[102] See Dwight D. Eisenhower, *The White House Years: Waging Peace (1956–61)* (New York: Doubleday, 1965), p. 217.

[103] Ibid., p. 216.

[104] Leadership-to-leadership proposals for cooperation were not the only source of space arms control measures discussed during this period, however. One of the more interesting proposals for U.S.-

delegates referred to the previously lost chance at nuclear arms control under the Baruch Plan. Unfortunately, after deployment of nuclear fission weapons and then hydrogen bombs by the United States, it now proved even more difficult for Washington to convince the world (and especially the Soviet Union) that space could be free of weapons. Still, Ambassador Lodge argued:

> Mr. Chairman, in 1946 when the United States alone had nuclear weapons, it proposed to the United Nations . . . a plan to ensure the peaceful use of the new and tremendous force of atomic energy by putting it under international control. We made that proposal. The world knows now that a decade of anxiety and trouble could have been avoided if that plan had been accepted. We now have a similar opportunity to harness for peace man's new pioneering efforts in outer space. We must not miss *this* chance. We have therefore proposed that a technical committee be set up to work out an inspection system which will assure the use of outer space for exclusively peaceful and scientific purposes.[105]

Given its new advantage, Moscow expressed no interest in the U.S. proposal, unless linked to an overall arms control deal that would take away existing U.S. advantages in bombers and missiles based close to the Soviet border. Chairman of the Council of Ministers Nikolai Bulganin outlined Moscow's official policy in a letter to Eisenhower in February 1958: "[The Soviet Union] is . . . prepared to discuss the question of intercontinental missiles, provided the Western powers are prepared to agree on the prohibition of nuclear and hydrogen weapons, the cessation of test of such weapons, and the liquidation of foreign military bases in the territories of other states."[106]

In the verbiage of the Eisenhower administration, the Soviets were once again selfishly blocking a truly revolutionary arms control agreement.[107] For the Soviets, however, this was merely another U.S. attempt to cement an existing lead in military power by keeping the Soviet Union out of the game. But

---

Soviet cooperation came from Representative Robert Hale of Maine in U.S. Congress on January 23, 1958. Hale suggested a plan which he called "satellites for peace" (drawing on Eisenhower's "atoms for peace"). According to this proposal, the two countries would agree immediately that all satellites would be devoted to peaceful purposes while working separately toward agreement on the more difficult problem of banning missiles in space. Other members apparently were not impressed with the proposal, as it never made it to the floor for a vote. (On Hale's proposal, see Harvey and Ciccoritti, *U.S.-Soviet Cooperation in Space*, pp. 27–29.)

[105] Harvey and Ciccoritti, *U.S.-Soviet Cooperation in Space*, p. 13.

[106] Bulganin, quoted in Harvey and Ciccoritti, *U.S.-Soviet Cooperation in Space*, p. 16.

[107] Although an agreement with the Soviet Union proved unattainable, the U.S. Congress did pass Concurrent Resolution 332 on July 23, 1958, declaring the "sense of the Congress" that the exploration of space should be by and for peaceful means and, if possible, pursued by the United States through cooperation with other nations. Thus, a majority of the American representatives felt uncomfortable with the prospects of unrestricted space military competition.

Moscow now held the trump card, since it had already developed intercontinental-range missile technology and done it *before* the United States.

The heavy-lift capability of this clumsy ICBM allowed the rapid progression of the *Sputnik* series of scientific launches, which included the orbital flight of the dog Laika on *Sputnik II* in November 1957 and the heavily instrumented orbital probe *Sputnik III* early the next year.[108]

The possibility of manned spaceflight became the next challenge facing Korolev and the space program. Beginning in 1958, the Soviets began to train a cosmonaut team in a former aircraft factory outside of Moscow. Within two years, a sprawling complex known as Star City[109] had sprung up in a nearby forest, isolated from the attention of foreign journalists based in Moscow but close enough for access to the central scientific community. The Soviet air force assumed responsibility for the selection and training of prospective cosmonauts, choosing from among active-duty pilots on the basis of age, stamina, adaptability, and small size.[110]

The U.S. Reaction

Although Eisenhower opposed the concept of racing in space, officials in the administration could no longer deny that something needed to be done to address this new crisis of credibility in regard to U.S. space capability, particularly in the highly visible civilian area. In late 1957, Eisenhower charged the President's Science Advisory Committee under James R. Killian to reassess national programs in science and technology. Within the military, similar reforms took place, including the establishment of the Advanced Research Projects Agency (ARPA) and the position of director of defense research and engineering (DDRE), both part of the Defense Reorganization Act of 1958. Eisenhower named Lawrence Livermore Laboratory Director York as the first head of DDRE. York would comment later on the reasons behind the U.S. space failure:

[108] The Soviets had no plans to return Laika from orbit. But she died earlier than was planned—likely on the fourth day of her flight—due to the overheating of her capsule. Her experience proved cautionary for Soviet scientists, who subsequently had to focus considerable attention on developing a more reliable cooling system before human spaceflight could begin. On Laika's flight and these preparations, see the chapter by Abram Moiseyevich Ghenin (doctor of biological sciences) in Rhea, *Roads to Space*, pp. 42–50; see also the account of Laika's flight in Siddiqi, *Sputnik and the Soviet Space Challenge*, p. 174. Her capsule eventually re-entered the atmosphere on April 14, 1958.

[109] Technically, *Zvezdniy gorodok* translates as "Star Town," but it has become known in the West as "Star City."

[110] For a detailed account of the early cosmonaut program, see Riabchikov, *Russians in Space*.

"Unnecessary duplication was rife, and vicious interservice struggles over roles and missions were creating confusion."[111]

Within Congress, the series of special committees looking into the space problem eventually led to the creation of the House Science and Astronautics Committee and a similar committee in the Senate under Democratic Majority Leader Johnson.[112] Other changes in the areas of education, energy, and science policy followed.[113] Perhaps most significantly, however, *Sputnik* set into motion the process of transforming the weak National Advisory Committee on Aeronautics (NACA) into the National Aeronautics and Space Administration (NASA), formalized with Eisenhower's signing of the Space Act on July 29, 1958.[114]

In the military realm, leading defense officials began to call for a rapid expansion of programs. As Brigadier General Schriever had stated back in February, "Our safety as a nation may depend upon our achieving 'space superiority.' Several decades from now the important battles may not be sea battles or air battles, but space battles."[115] In November, the air force quickly accelerated work on a planned missile-launched, manned space bomber, the so-called Dyna-Soar (or X-20). Possible missions for the new, reusable vehicle included anti-satellite (ASAT) attacks, space-Earth bombing, and reconnaissance.[116]

In the critical area of results off the launch pad, early U.S. attempts to match the Soviet feat of *Sputnik I* met with humiliating failure. The first Vanguard satellite–Viking rocket combination exploded on the launch pad in December 1957. The president turned to Science Advisor Killian to study the options at hand. Killian's scientific panel came back with a report just before the end of the year urging the president to shift responsibility for the first launch to

[111] Herbert F. York, *Making Weapons, Talking Peace: A Physicist's Odyssey from Hiroshima to Geneva* (New York: Basic, 1987), p. 135. York continues, "The Army diverted large sums from programs designed to improve its ability to fight conventional wars into missile and space programs whose purpose was to preempt Air Force activities in those areas. The Air Force, in turn, distorted its own programs in order to head off the Army in those same areas, again neglecting some other responsibilities in doing so."

[112] Recommendations for similar changes had been made a year earlier by the Symington committee, but it took the political furor caused by *Sputnik* to get them adopted.

[113] To cite a few examples, the Joint Committee on Atomic Energy formed a subcommittee on space propulsion and backed the formation of a Division of Outer Space Development in the U.S. Atomic Energy Commission. Congress also passed the 1958 National Defense Education Act, which provided $903 million for improved science, engineering, and foreign language training in U.S. schools and universities.

[114] Logsdon, *Decision to Go to the Moon*, p. 30. See also Herbert York, *Race to Oblivion* (New York: Simon & Schuster, 1970), p. 145.

[115] Schriever quoted in Stares, *Militarization of Space*, p. 48.

[116] Burrows, *This New Ocean*, p. 252.

the Jupiter C missile, despite the stigma of von Braun's Nazi service.[117] Given the circumstances, Eisenhower accepted their recommendation during the first week of January.

American pride began to be restored on January 31, 1958, when the Huntsville (Alabama) army team succeeded in launching the scientific satellite *Explorer I* aboard a modified Redstone (Jupiter C) rocket.[118] In March, the navy also managed to overcome its two prior failures with the orbiting of *Vanguard I*. But American humiliations were far from over. From the launch of *Vanguard I* in March 1958 to the end of 1959, four Explorer spacecraft and six Vanguard satellites failed to achieve orbit.

Meanwhile, the United States scrambled to complete development on a range of reconnaissance satellites. The Central Intelligence Agency hoped soon to orbit a secret Corona satellite under the civilian cover of the Discoverer series. The air force continued to move forward with both SAMOS and with an early-warning satellite program known as the Missile Detection Alarm System (MIDAS).[119] Others looked further afield for possible military advantage, including to the Moon. Air Force Lieutenant General Donald L. Putt testified before the House Armed Services Committee in March that "the Moon appears to be of such significance that we should not let another nation establish a military capability there ahead of us."[120] General Putt went on to predict that a U.S. military base on the Moon "is only the first step toward stations on planets far more distant . . . from which control over the Moon might then be exercised." But the air force's enthusiasm for this potential new military environment had gotten slightly ahead of itself. The main attentions of Congress and administration focused on the more pressing issue of the military significance of *Sputnik I*'s launcher.

As the U.S. weapons laboratories sought to gauge the seriousness of Khrushchev's missile threats and what to do about them, work began on a series of tests in the spring and summer of 1958 to determine the feasibility of using nuclear explosions for missile defense purposes. The first three tests fell under

[117] On this report, see James R. Killian Jr., *Sputnik, Scientists, and Eisenhower: A Memoir of the First Special Assistant to the President for Science and Technology* (Cambridge, Mass.: MIT Press, 1977), pp. 120–21.

[118] On the Redstone and Jupiter C modifications, see Loyd S. Swenson Jr., James M. Grimwood, and Charles C. Alexander, *This New Ocean: A History of Project Mercury* (Washington, D.C.: NASA, 1966), pp. 21, 29. This rocket also became known as Juno I.

[119] Norman Friedman, *Seapower and Space: From the Dawn of the Missile Age to Net-Centric Warfare* (Annapolis, Md.: Naval Institute Press, 2000), p. 92.

[120] Quoted in Dwayne A. Day, "Take off and nuke the site from orbit (it's the only way to be sure . . .)," *The Space Review*, June 4, 2007, online at <http://www.thespacereview.com/article/882/1> (accessed September 30, 2007).

Operation Hardtack I, a series of thirty-five tests conducted in the Pacific for a variety of purposes.[121] The Yucca shot involved a small 1.7 kiloton bomb detonated from a balloon at an altitude of 16 miles. Then, two additional tests—Teak on August 1 and Orange on August 12—used much higher-yield bombs of 1.9 megatons at altitudes of 48 miles and 27 miles, respectively. From these tests, it became clear that while such blasts could destroy incoming missiles, communications would be severely affected by electromagnetic pulse (EMP) radiation. The blasts charged the night sky above Johnston Island in the South Pacific with phosphorescent colors and blacked out all radio transmissions for hours. Results of the test program also showed that any pilot flying above 50,000 feet during a high-altitude nuclear blast would receive a lethal dose of radiation and die before reaching his target.[122]

In late August and early September 1958, the United States conducted its first nuclear weapons tests above the stratosphere (60 miles in altitude) from a ship in the South Atlantic under the rubric of Project Argus. These three secret tests each exploded a 1.7 kiloton warhead from missiles shot to altitudes of 100, 182, and 466 miles, respectively, to analyze the effects of EMP radiation above the Earth's atmosphere. The finding that the Earth's magnetic field traps charged particles confirmed the predictions of eclectic inventor and periodic U.S. government advisor Nicholas Christofilos.[123] However, the tests also showed that radiation would behave differently in space than in the stratosphere, where fallout dissipated in intensity and migrated from the location of the test relatively quickly. By contrast, radiation spread and persisted much longer in space. This would later have serious implications for both manned spaceflight and for satellites carrying electronics.

Confrontation on the International Scene

The United States' disappointment with the International Atomic Energy Agency (established in 1957) had led the Eisenhower administration to doubt

[121] On Operation Hardtack I, see the Web site of the Radiochemistry Society, at <http://www.radiochemistry.org/history/nuke_tests/hardtack1/index.html> (accessed September 18, 2006).

[122] Manno, *Arming the Heavens*, pp. 56–59.

[123] Christofilos was a Greek immigrant and elevator engineer with an eccentric and yet highly innovative mind. Despite his lack of a security clearance, he provided valuable insights into scientific phenomena in a variety of areas. Often, his speculations proved outlandish and simply incorrect. But on this occasion, he proved correct, where others with much more training had failed to predict these dynamics. (Author's interview with Wolfgang Panofsky, a frequent interlocutor of Christofilos's, at his office at the Stanford Linear Accelerator, Stanford, California, March 9, 2006.)

the possibility for an international space agency. But its desire to force the Soviet Union's hand and to isolate it from the world community led to a proposal
by U.S. Ambassador Lodge in September 1958 at the United Nations for an
international program of space cooperation. U.S. efforts at establishing an international space organization achieved at least political, if not yet operational,
success in late 1958. On December 13, the member nations approved a resolution establishing an Ad Hoc Committee on the Peaceful Uses of Outer Space
(COPUOS).[124] After all of the effort expended by U.S. negotiators, however,
this victory proved to be short lived. The Soviet Union, Poland, and Czechoslovakia immediately undertook a boycott of the new organization, claiming
that the voting membership had been deliberately skewed to favor the West.[125]
Although it was true that the committee membership included a preponderant
number of Western, developed countries, this was merely a reflection of those
nations possessing the technical prerequisites of space activity rather than an
intentional political move.

Outside events did not encourage the United States to accommodate Soviet
demands regarding space. Khrushchev's Berlin Ultimatum on November 27,
1958, stated that East Germany would in six months assume control over the
new "free city" of Berlin, which was still jointly occupied, setting up an inevitable confrontation. In addition, Khrushchev's frequent (and false) claims of Soviet possession of a whole arsenal of nuclear-tipped ICBMs to back his threats
cast an increasingly militaristic shadow over the space frontier and made Soviet
policy appear all the more menacing to American leaders.

American strategy during this period—in terms of cooperation—was one
of non-engagement. Beyond a fundamentally competitive core in the area of
military missiles, there was a certain desire to pursue multilateral (as opposed
to bilateral) cooperation in space science, but there was not a forthcoming attitude of reaching out to the Soviet Union, largely for security reasons. Eisenhower wanted neither to race in space nor to entangle the United States in
intrusive cooperative programs.[126] He used space science proposals primarily

---

[124] U.N. General Assembly Resolution 1348, "Question of the Peaceful Uses of Outer Space," December 13, 1958.

[125] India and the United Arab Republic (UAR), also members of the new committee, subsequently
refused to participate, arguing that nothing could be accomplished unless both the Soviets and the
Americans played a part in the organization.

[126] Eisenhower commented unapologetically on the U.S. "loss" to the Soviet Union's *Sputnik*: "Since
no obvious requirement for a crash satellite program was apparent, there was no reason for interfering
with the [Vanguard] scientists and their projected time schedule" (see Eisenhower, *White House Years*, p.
209). He also added that the United States could have beaten *Sputnik* by a year had it chosen to use the
army's Redstone rocket, but—in the context of the IGY—that would have meant the release of sensitive
military information on its developmental stages to Soviet scientists, as members of the International

as a political tool, aimed not at substantive cooperation but rather at exposing the military character of the Soviet program.

Facing mounting pressure on the nuclear test-ban issue as well, Eisenhower tasked a National Security Council ad hoc working group in 1958 (under physicist Hans Bethe) to study questions related to its possible verifiability. The experts on another panel (under physicist Wolfgang Panofsky) considered specific problems of verifying nuclear testing in space. Among other issues, they examined Edward Teller's concerns that the Soviet Union might cheat by trying to test nuclear weapons behind the Moon. The committee's classified report in March 1958 concluded that cheating in space by sending weapons and detectors behind the Moon—while not impossible—"was simply too expensive to make it worth it."[127] However, the group recognized that testing above a certain altitude, and certainly behind the Moon, was not detectable. Still, the Bethe group's overall conclusion was that it should be possible to establish a broadly effective verification regime through fixed stations and on-site inspections, subject to adequate political clearance on both sides to conduct such work.[128]

Armed with this information, Eisenhower proposed that an international conference of experts be convened in the summer of 1958 in Geneva to discuss possible verification mechanisms for a comprehensive nuclear test-ban treaty. Their report in August 1958 outlined a proposed international system of land- and sea-based detectors, although it recognized the existence of certain gaps. One such gap had to do with the absence of provisions for monitoring nuclear tests above 50 kilometers in altitude.[129] As Robert Gilpin's 1962 study on the test-ban negotiations explains:

> The success of the [Hardtack I] and Argus shots centered American and world attention on the fact that the Geneva System would not detect nuclear explosions at very high altitudes. Further, the American tests made it obvious that it was no longer appropriate to "brush aside" this type of testing as unimportant, as the Conference had done. High altitude testing was not only feasible, but it had now been proved that it could be accomplished at a reasonable cost.[130]

Committee of Scientific Unions. (As it happened, the Redstone *did* become the first U.S. satellite booster, although by the time of its use in 1958 its development was no longer secret.)

[127] Author's interview with study leader Wolfgang Panofsky, Stanford, California, March 9, 2006.

[128] Despite his participation in this unanimous report, Teller later opposed the Partial Test Ban Treaty and testified against it in Senate hearings in 1963 (interview with Panofsky, March 9, 2006). On Teller's testimony, see Glenn T. Seaborg, *Kennedy, Khrushchev, and the Test Ban* (Berkeley: University of California Press, 1981), p. 272.

[129] Dokos, *Negotiations for a CTBT 1958–1994*, p. 7.

[130] Robert Gilpin, *American Scientists and Nuclear Weapons Policy* (Princeton, N.J.: Princeton University Press, 1962), p. 225.

Despite these drawbacks, international political pressures not to give up on the process and growing international health concerns resulted in an agreement by the three nuclear states at the time (the United States, the Soviet Union, and Britain) to begin actual negotiations toward a treaty. The Soviet Union's vehement opposition to on-site inspection slowed negotiations, while the U.S. government's own position remained tentative due to the widespread belief among military leaders and weapons laboratory personnel that continued nuclear testing was necessary for reasons of national security.[131] In 1959, a status report from Technical Working Group I of the negotiations (examining space testing) reaffirmed that possible shielding of an explosion could seriously degrade the viability of verification of tests conducted farther than 200,000 miles from Earth, such as behind the Moon.[132] Thus, the specter of cheating in space—despite the obvious monetary costs—continued to be raised. Closer to Earth, testban critics at the time argued that detectors for space tests in the higher realms of low–Earth orbit "would not be possible for many years" and, even when deployed, "would be extremely limited in capability."[133] The assumption that space would be weaponized remained the dominant belief among policymakers. In October 1959, the U.S. Air Force carried out a test of Project Bold Orion, launching a missile interceptor from a B-47 aircraft to within four miles of the Explorer 6 satellite.[134] Bold Orion's intended nuclear warhead meant that the United States, ironically, had beaten the Soviets in carrying out the world's first successful ASAT test.[135]

In August 1960, the defense department sought to strengthen its ASAT options by approving the air force's Satellite Inspector (SAINT) program: a coorbital satellite "inspection" system that would double as a space weapon. The classified program involved use of an Atlas first stage, an Agena B second stage, and the satellite interceptor (with a small propulsion unit) as the third stage.[136] The interceptor would be launched into space and maneuver into a nearby location. Then, it would either destroy the Soviet target satellite or (in one later

[131] On these internal U.S. government discussions on verification and on questions of the reliability of U.S. nuclear weapons, see George B. Kistiakowsky, A Scientist at the White House: The Private Diary of President Eisenhower's Special Assistant for Science and Technology (Cambridge, Mass.: Harvard University Press, 1976), pp. 5–9.

[132] Gilpin, American Scientists and Nuclear Weapons Policy, p. 233.

[133] Ibid., p. 276.

[134] Stares, Militarization of Space, p. 109.

[135] Steven Lambakis, On the Edge of the Earth: The Future of American Space Power (Lexington: University Press of Kentucky, 2001), p. 101.

[136] Stares, Militarization of Space, pp. 114–15.

option) disable it by releasing a cloud of paint to cover and blind its sensors.[137] The program included plans for nineteen test vehicles. When word of the basic parameters of SAINT leaked out, however, some Christian groups protested the sacrilegious name given to a project that would move the U.S. space program beyond its supposedly peaceful parameters.[138]

Despite growing U.S. and Soviet preparations for space warfare, officials and scientists continued to push ahead toward a possible nuclear test-ban treaty over the course of 1958–60 during ongoing meetings in Geneva.[139] One reason was that the official U.S. position had begun to change. Eisenhower now believed that the Soviet Union, being behind in the nuclear race, had more to lose than the United States.[140] Yet continued U.S. efforts to gather intelligence on the Soviet nuclear arsenal via U-2 high-altitude surveillance planes—in the absence of satellites—brought an end to chances for an agreement until 1962. Despite U.S. denials of conducting reconnaissance overflights, Soviet success in shooting down an intelligence-gathering U-2 flown by Gary Powers on May 1, 1960, exposed the U.S. deception and led to increased bilateral hostility.

Not surprisingly, it was in the nongovernmental area that some progress toward U.S.-Soviet space cooperation first began to be made, far removed from other, ideologically charged political fora. In mid-1959, the respective U.S. and Soviet academies of science signed several agreements to begin exchanges of scientific data from their space research during the IGY.[141] The agreements dealt with the exchange of recorded tracking data from the eight IGY satellites launched (three Soviet, five U.S.).[142] By July, the American scientists had already provided to Moscow approximately forty-six tape recordings of the *Sputnik I, II,* and *III* flights, as received on U.S. soil.[143] After several months the Soviets

---

[137] William E. Burrows, *Deep Black: Space Espionage and National Security* (New York: Random House, 1986), p. 145.

[138] Stares, *Militarization of Space*, pp. 115–16.

[139] Bunn, *Arms Control by Committee*, p. 24.

[140] As Bunn explains, "By 1959 Eisenhower favored agreement to end all testing." Bunn, *Arms Control by Committee*, p. 26.

[141] Office of Technology Assessment, *U.S.-Soviet Cooperation in Space* (Washington, D.C.: U.S. Congress, Office of Technology Assessment, OTA-TM-STI-27, July 1985), p. 9.

[142] For a list of the satellites, see Office of Technology Assessment, *U.S.-Soviet Cooperation in Space*, p. 17.

[143] On this exchange, see comments by NASA Chief Administrator Glennan, quoted by Harvey and Ciccoritti, *U.S.-Soviet Cooperation in Space*, p. 44. Arnold W. Frutkin estimates the number of tapes at between thirty and forty. See Arnold W. Frutkin, *International Cooperation in Space* (New York: Prentice-Hall, 1965), p. 20.

still had not responded to the agreement, thus showing their ultimate dependence—unlike the American scientists—on official political controls.

Evidence of a slight change in the Soviet position came in September with the notification given to the IGY World Data Center for Rockets and Satellites in Ft. Belvoir, Virginia, of the weight and radio frequency of the Soviet *Luna II* spacecraft. Some U.S. scientists claimed that this move marked a definitive decision by Moscow to increase the cooperative aspects of its program with greater exchanges of data.[144] Others doubted the significance of this limited overture, since the real meaning of the *Luna II* flight was purely competitive: to become the first country to strike the Moon's surface and then to distribute metallic "CCCP" (USSR) emblems on the lunar soil. *Luna I* had missed the Moon in January 1959, but *Luna II*'s use of an advanced spatial guidance system (which allowed it to adjust its speed at different points in the flight instead of following a simple ballistic trajectory) helped it to reach its target.[145] The event occurred fortuitously on the eve of Khrushchev's first visit to the United States. In response to U.S. questioning, however, the Soviets denied that the mission was any attempt to claim the Moon as Soviet territory.[146]

In the commercial area, the United States and Soviet Union both took part in the proceedings of the International Telecommunications Union (ITU), whose members agreed in late October to schedule a meeting for 1963 to allocate radio frequencies to the various national programs.[147] Though important, this was an expressly multilateral agreement and involved prospective (rather than immediate) cooperation. It did little in the way of providing a precedent for improved U.S.-Soviet ties in the short run.

Meanwhile, at the United Nations, the Soviet Union had made a slight concession in its demand for equal communist bloc representation in COPUOS. With the consent of the United States, the new agreement added Albania, Bulgaria, Hungary, and Romania—none of which had space programs—to the organization. The body now totaled twenty-four member nations (12 Western, 7 communist, 5 neutral). U.N. Resolution 1348, passed on December 13, 1959, formalized this compromise solution. However, as they had after the 1958 agreement, the Soviets promptly raised new objections over practical issues

---

[144] See *New York Times*, September 15, 1959, p. 20.

[145] For a detailed analysis of the *Luna II* flight, see Albert Ducrocq, *The Conquest of Space: Moon Probes and Artificial Satellites: Their Impact on Human Destiny* (London: Putnam, 1961), Chapter X, pp. 195–213.

[146] *New York Times*, September 15, 1959, p. 1.

[147] *New York Times*, November 1, 1959, p. 135.

in committee voting (insisting on veto power) and on the principle of not cooperating with the West in space without prior agreement on limiting U.S. military activities in other areas. Clearly, Cold War sentiments ran deeper than a desire to cooperate in space. COPUOS would not hold its first meeting until 1961.

NASA officials, eager to develop the cooperative aspects of their organizational charter and hoping to steal some glory from the so-far vastly more successful Soviet program, began a new round of bilateral space science proposals in late 1959 and in 1960. In November, NASA Deputy Chief Administrator Dr. Hugh Dryden took the occasion of a meeting of the American Rocket Society to escort visiting Soviet Academicians Leonid I. Sedov (president of the Soviet Commission for Interplanetary Communications) and Anatoliy A. Blagonravov (vice president of the USSR Academy of Sciences) on an excursion to NASA's Langley, Virginia, center. There, he showed the Soviet scientists a variety of current NASA projects, including a prototype of the *Mercury* space capsule (which would not be launched for another two years). After this rather spectacular visit, however, the Soviets failed to respond with a reciprocal invitation of their own, either to NASA officials or to members of the National Academy of Sciences. Again, the problem seemed to be the Soviet scientists' own lack of power and access to classified space facilities, which remained under the control of the Soviet military.

Within the Soviet armed forces, the need to push the development of new booster technology for space and for the delivery of nuclear weapons had brought about a major organizational innovation in late 1959—the creation of the Strategic Rocket Forces (SRF). As part of Khrushchev's new emphasis on the revolutionary significance of Soviet ICBMs (and concomitant demotion of ground forces),[148] this move formalized the military's already existing control of missile research and testing. The SRF's primary responsibility was defined as the launching of strategic nuclear missiles, but it had the related, secondary task of handling all space launches as well.

The U.S. government hoped to use international pressure and obligations to either open up the Soviet space program or, through Soviet rejections of cooperation, to expose its military character to the world. In the latter task, it proved fairly successful. In late 1959, NASA Chief Administrator T. Keith Glennan (working through the National Academy of Sciences) offered the Soviet Academy of Sciences use of the U.S. space tracking network to make up for

---

[148] On Soviet nuclear doctrinal changes, see Freedman, *Evolution of Nuclear Strategy*, pp. 143–45.

"blind" periods when communications from Soviet spacecraft could not reach Soviet receiving stations because of the curvature of the Earth.[149] Again, officials from the Soviet academy noted their appreciation for the offer but failed to follow up on it.[150]

Keeping up the pressure on the Soviets to open up their program, NASA Deputy Chief Administrator Dryden conveyed a broader plan of U.S.-Soviet cooperative proposals in a letter to Soviet Academician Sedov in late January. Dryden listed several areas of possible cooperation: exchange of data from the U.S. *Explorer VII* satellite, a proposal to place an American experiment on a Soviet spacecraft, and joint activity using the future U.S. comsat *Echo*. Like many before them, these proposals languished through the rest of 1960 without favorable response from the Soviet academy.

Meanwhile, the U.S. military had begun to make significant advances in conducting a variety of intelligence-collection and remote-sensing activities from space. These programs had begun in 1956, when the air force had granted contracts to Lockheed, Eastman-Kodak, and CBS Laboratories. Capsules sent into suborbital space took pictures of cities and other objects below and then ejected small film canisters that were retrieved by specially equipped air force planes. Others tested equipment to intercept communications, gather weather data, or detect missile launches. These programs worked on the assumption of the coming of U.S. orbital flights over the territory of the Soviet Union and the Eastern European communist bloc states.

With the emerging capabilities of photo-reconnaissance technology, the intelligence community had assumed responsibility for the secret Corona program. After many attempts, the CIA finally succeeded in August 1960—over two years into the U.S. orbital flight experience—in conducting the first U.S. film-return reconnaissance mission over the Soviet Union with the so-called *Discoverer 14* mission.[151] William Burrows describes the impact of this revolution in intelligence collection when the images were rolled out across the floor in the Oval Office: "The president was stunned by what he saw: . . . photographs that cov-

[149] On this offer, see William H. Schauer, *The Politics of Space: A Comparison of the Soviet and American Programs* (New York: Holmes and Meier, 1976), p. 223.

[150] As Schauer notes, U.S. motives in the proposal were not completely altruistic: "Use of American tracking facilities would require prelaunch announcement of Soviet flights, which not only would depart from the Soviet policy of secrecy but would have the effect of guaranteeing easy detection and wide publicity of delays and failures of Soviet launches. As such, NASA could expect the Soviet response, and this expectation probably entered into the decision to offer the use of American facilities" (Schauer, *Politics of Space*, p. 223). Notably, Eisenhower failed to weigh in on Khrushchev to push these projects.

[151] On these early efforts, see Jeffrey T. Richelson, *America's Space Sentinels: DSP Satellites and National Security* (Lawrence: University of Kansas Press, 1999), pp. 8–9.

ered 1.5 million square miles of Soviet and Eastern European territory and that turned up sixty-four airfields, twenty-six new surface-to-air missile sites, and a third major rocket-launch facility at Plesetsk. Discover 14's 'take' was nothing short of phenomenal."[152] But Eisenhower and his advisors decided to keep the program secret, lest they reveal to the Soviet Union their ability to spy on its assets from space.[153] Thus, the technological impact of this secret milestone was lost on an American public still in awe of ongoing Soviet space accomplishments with much bigger rockets and far heavier payloads.[154] The importance of U.S. launches seemed like child's play in comparison, even though these missions now contained more sophisticated and militarily valuable payloads.

Corona's success in overcoming a long history of problems in U.S. satellite reconnaissance and the CIA's leadership in developing new hardware helped spur the development of the top-secret National Reconnaissance Office in August 1960 to coordinate U.S. intelligence gathering from space.[155] The organization's very existence would not be admitted until after the end of the Cold War, as the administration hoped to put both the American media and U.S. adversaries off the scent. But the Soviet Union already knew of U.S. space reconnaissance plans. It also knew of the first successful orbiting of a MIDAS infrared, early-warning test satellite in May 1960, which unlike Corona had received front-page coverage in major U.S. newspapers.[156] What Moscow did not know about the growing list of U.S. military spacecraft was how to stop them, at least short of war.

Space Politics and the 1960 U.S. Presidential Race

The problems of the U.S. civilian space program became a major issue in the American presidential campaign between Democratic Senator John F. Kennedy and Republican Vice President Richard M. Nixon in 1960. In a reversal of the political spectrum of the later 1980 and 2000 elections, it was the Democrat

---

[152] Burrows, *This New Ocean*, p. 234.

[153] For more on the Corona program, see Dwayne A. Day, John M. Logsdon, and Brian Latell, eds., *Eye in the Sky: The Story of the Corona Spy Satellites* (Washington, D.C.: Smithsonian Institution, 1998).

[154] As Burrows writes about the U.S. loss of the public relations battle, "The day Discoverer 14 was snatched by an Air Force Flying Boxcar over the Pacific, the Soviet Union sent up Korabl Sputnik 2 with two more mutts and other creatures." Burrows, *This New Ocean*, p. 307.

[155] Stares, *Militarization of Space*, p. 44.

[156] Richelson, *America's Space Sentinels*, p. 15. The satellite functioned briefly but stopped transmitting only two days into its mission. The MIDAS program would later struggle with other technical problems associated with its infrared detectors.

Kennedy who stressed space's importance to national defense, railed at the past administration for allowing a "missile gap" to develop,[157] and urged competitive spending by the United States to "catch up" in space. Nixon knew that these charges and even some U.S. government statements about Soviet capabilities were false. However, intelligence limitations kept him from revealing that the Soviets had not in fact developed any sizable missile stockpile. But this *perceived* imbalance in U.S.-Soviet capabilities affected the American political scene greatly and provided Kennedy with a cause célèbre. Nixon attempted to argue that despite operational setbacks, the United States held a lead in space technology. Given the secrecy of the Corona program, however, his line of reasoning proved hard to sell to a now doubtful American public. Ironically, despite Nixon's aggressive reputation in regard to the communist threat as a former leading ally of Senator Joseph McCarthy, he argued Eisenhower's more moderate line in stating that it made no sense to run blindly down a competitive path that could touch off a damaging military space race with Moscow. However, as in 1980 and 2000, the American people backed the candidate of change, competition, and military vigilance in a closely contested election.[158]

With Kennedy's victory, a cautious military-led program that had rejected racing was replaced by a daring, expensive, and highly *competitive* program with the unabashed goal of establishing the United States as the leading space power. Not only did the young president stress a more active policy than did Eisenhower, but he also sought a more interactive role as well, especially in regard to the Soviet Union. This meant a tactical shift from a multilateral to a bilateral approach.[159] The Eisenhower administration wanted to use cooperative overtures to expose the hypocrisy of the military-dominated Soviet program, where all defense-related launches and weapons plans were kept secret. In essence, Kennedy had decided to start playing the space game by Soviet rules. Accordingly, one of his first directives, in January 1961, was to prohibit release of prelaunch information on all U.S. military flights, not just photo-reconnaissance missions.[160] While the civilian side of the program would increase under Ken-

[157] Of course, as U.S. intelligence had already revealed, there was no missile gap at all. In fact, these overflights, still highly classified at the time, showed that the United States had a commanding lead in deployed nuclear-tipped missiles. Once he found out in early 1961, however, Kennedy felt pressed by his own rhetoric to push the arms race forward with a 25 percent increase in defense spending for fiscal 1962.

[158] As if to underline the difficulties of the U.S. space program, the first Mercury-Atlas qualification test rocket, launched on election day, exploded two-and-a-half minutes after takeoff from Wallops Island, Virginia.

[159] See Harvey and Ciccoritti, *U.S.-Soviet Cooperation in Space*, p. 64.

[160] Stares, *Militarization of Space*, pp. 63–64.

nedy, the military would also continue to expand, but even more quietly now. Although flight announcements and the names of the missions dropped from public view, within the next year launches of Corona satellites would double from thirteen to twenty-six.[161] Meanwhile, the defense department's space budget would triple from Eisenhower's 1960 budget of $561 million to Kennedy's 1963 budget of $1.55 billion.[162] Already in March, the services began to put new ideas for military activities before the Congress. One was "Early Spring," a navy effort to join the space rush by using a Polaris sea-launched missile to destroy Soviet satellites by releasing hundreds of metal pellets into their path.[163] Again, there seemed to be no shortage of possible threats and conceivable responses, but little coordination within the Pentagon.

Meanwhile, Moscow's smaller, but more focused, space program forged ahead by grabbing the headlines with remarkable new nonmilitary capabilities. On April 12, 1961, the Soviets succeeded in putting the first human into Earth orbit. Cosmonaut Yuri Gagarin's flight in *Vostok I* led to public criticism of President Kennedy for allegedly dragging his feet on space.[164] For Kennedy, things would get worse before they would get better. Within weeks, the disastrous Bay of Pigs operation in Cuba—a failed CIA effort to overthrow Fidel Castro with a small expatriate army—put him deeper into trouble with the electorate. His now obvious inexperience with the consequences of executive authority contrasted sharply with his bold preelection pledges to reinvigorate what he had described as lagging U.S. military power. The tremendous political impact of the Gagarin flight gave Khrushchev every reason he needed to make the most of a brilliant propaganda victory. As remembered by Soviet affairs analyst Arnold Horelick, "No event since the death of Stalin had been so widely publicized by the USSR as the April 12, 1961 flight of Major Yuri Gagarin."[165] In his own flamboyant political style, Khrushchev crowed to the international media about the flight's significance in proving "socialist superiority" in science, technology, and military power.

[161] Ibid., Appendix II, Table 4, p. 266.

[162] Ibid., Appendix II, Table 1, p. 255.

[163] Bob Aldridge, "Anti-Satellite Warfare: Little Heard of and Never Seen," Pacific Life Research Center report, Santa Clara, Calif., August 29, 2000; see also Stares, *Militarization of Space*, pp. 109–10.

[164] By this time, following a series of failures using two types of boosters, the United States had successfully tested both its Mercury-Redstone (January 31) and Mercury-Atlas (February 21) rockets. The first flight had been especially significant because it involved the safe launch and return of the chimpanzee Enos. Although Enos had to be rescued from his nearly submerged Mercury capsule when it landed downrange and had begun taking on water, his flight set the stage for the manned, suborbital flights that were to follow in the spring.

[165] See Horelick, quoted by Harvey and Ciccoritti, *U.S.-Soviet Cooperation in Space*, p. 74.

President Kennedy knew that he could not sit idly by. After receiving from Vice President Johnson a detailed plan on how to put the political building blocks in place to rally Congressional support,[166] Kennedy tasked Vice President Johnson on April 20 to review U.S. space activities and to identify areas where the United States could *beat* the Soviets.[167] From his prior work on the Senate Science Committee, Johnson was well versed in the plans generated under the Eisenhower administration, which included a possible Moon landing, although costs had seemed prohibitive.[168] Meanwhile, in a classified report submitted to Johnson on May 8, NASA Chief Administrator James E. Webb and Secretary of Defense Robert S. McNamara identified the nonmilitary field as a top American priority, stating, "It is vital to establish specific missions aimed mainly at national prestige."[169] The report explained that while these missions would affect U.S. military strength "only indirectly, if at all," they had tremendous symbolic value: "Our attainments are a major element in the international competition between the Soviet system and our own."[170] In a direct shift from the defense-oriented rhetoric of the previous fall's campaign, the Webb-McNamara report argued that such "non-military, non-commercial, non-scientific but 'civilian' projects such as lunar and planetary exploration" must be a key focus if the United States were to turn the tide in what they called "the battle along the fluid front of the cold war."[171] With this report in hand, Johnson, already a known space backer and expert, briefed the president. After further deliberations, Kennedy took his action plan to the Congress on May 25, 1961.[172] Entitled "Urgent National Needs," Kennedy's speech proposed to get the United States out of its space crisis with a dynamic mission, saying, "I believe that this nation should commit itself to achieving the goal, before the decade is out, of landing a man

[166] On their meeting on April 19, see Logsdon, *Decision to Go to the Moon*, p. 109.

[167] See Kennedy's Memorandum to Vice President Lyndon Johnson. Kennedy writes in exasperation, "Are we working 24 hours a day on existing programs? If not, why not?" See the memorandum, reprinted in Smolders, *Soviets in Space*, p. 381.

[168] On these points, see Logsdon, *Decision to Go to the Moon*, pp. 34–35.

[169] "Recommendations for Our National Space Program: Changes, Policies, and Goals," May 8, 1961 (declassified) report to Vice President Lyndon Johnson, by NASA Administrator James E. Webb and Secretary of Defense Robert S. McNamara, p. 8. (Personal papers of Dr. Herbert York, University of California, San Diego, Institute on Global Conflict and Cooperation.)

[170] Ibid.

[171] Ibid.

[172] For all of his advisors' concern about the urgency of the space competition, the report gave a very positive net assessment of U.S. space capabilities in everything but the area of prestige missions. The report made the following comparisons to the Soviet program: "Scientifically and militarily we are ahead. We consider our potential in the commercial/civilian area to be superior" (Webb and McNamara, "Recommendations," p. 7).

on the moon and returning him safely to earth."[173] Although Kennedy also mentioned in the speech the possibility of cooperating with the Soviets on such a mission, that message was clearly a sidebar to the main competitive thrust of the address. As presidential advisor Ted Sorensen writes on Kennedy's perspective, "The President was more convinced than any of his advisors that a second-rate, second-place space effort was inconsistent with this country's security, with its role as world leader and with the New Frontier spirit of discovery."[174]

At the Vienna Summit with Soviet leader Khrushchev in June 1961, the two sides discussed space cooperation but reached no agreements. The failure of the meeting to achieve any results in this area only fanned the sparks of the U.S. competitive drive in space. While Kennedy raised with Khrushchev the possibilities of a joint Moon mission and greater coordination on space launches, the Soviet leader had no intention of exposing to world scrutiny what he viewed as his political ace-in-the-hole, the tremendous symbolic power of the as yet unchallenged Soviet space program. Why allow foreigners to meddle in state secrets, especially when they would reveal the depths of his own bluffs with regard to Soviet missile power?[175] Khrushchev light-heartedly and without any commitment "accepted" Kennedy's proposal regarding the Moon, successfully deflecting media attention from the U.S. initiative.[176] Later, however, he downplayed the possibility of meaningful U.S.-Soviet cooperation anywhere in the world, saying to the forty-four-year-old president: "If I were your age, I would devote more energy to our [communism's] cause. Nevertheless, even at sixty-seven, I am not renouncing the competition."[177]

Ironically, a small step toward symbolic space cooperation came following the peak of the Berlin crisis in the summer of 1961. Khrushchev's blustering strategy to force the West out of Berlin had failed, leaving the stop-gap measure of the Berlin Wall as the only alternative to halt the huge and embarrassing outflow of East Germans to the West. With Sino-Soviet problems escalating at the

---

[173] President Kennedy, quoted in Harvey and Ciccoritti, *U.S.-Soviet Cooperation in Space*, p. 77. Interestingly, this civilian program announced by Kennedy was not the first U.S. proposal on a Moon landing. Schauer (*Politics of Space*, p. 164) writes that in 1957 the U.S. Air Force's Ballistic Missile Division had plans for a man to land on the Moon by the end of 1965. According to Manno (*Arming the Heavens*, p. 50), by 1958 the defense department had even awarded contracts to U.S. companies for feasibility studies on prospects for a Moon base under so-called Project Lunex. Kennedy's speech in 1961, therefore, marked a civilian seizure of a previously military-controlled project.

[174] See Theodore C. Sorensen, *Kennedy* (New York: Harper & Row, 1965), p. 525.

[175] As Sorensen reports, "At Vienna Khrushchev . . . said [space] cooperation was impossible . . . because he did not want his rockets observed." Sorensen, *Kennedy*, p. 529.

[176] See account of the Vienna summit by Special Counsel to the President Theodore Sorensen (Sorensen, *Kennedy*, pp. 543–50).

[177] Soviet First Secretary Nikita Khrushchev, quoted in Sorensen, *Kennedy*, p. 544.

same time in the shape of a growing ideological schism and practical concerns over the cavalier Chinese attitude toward nuclear weapons, Khrushchev moved to ease U.S.-Soviet tensions. By the fall of 1961, Soviet negotiators in the United Nations dropped their previous insistence on maintaining veto power in COPUOS and agreed to simple majority rule. In addition, they promoted progress on legal questions relating to space. After negotiations within the United Nations, the Soviets agreed to support the U.S.-sponsored General Assembly Res. 1721 on "International Cooperation in the Peaceful Use of Outer Space." This nonbinding resolution, passed on December 20, 1961, recommended that states consider in their space activities a number of principles: the application of international law to space, the nonappropriation by states of the celestial bodies, and international registration of all vehicles launched into space. In addition, the resolution called for "the urgent need to strengthen international cooperation."[178] However, the resolution lacked the status of a formal agreement and included no specific requirements on states or enforcement mechanisms.

As the measure had gone through the United Nations during the course of the fall, the sincerity of the Soviet position and the willingness of Khrushchev to follow up on this space initiative seemed dubious. In October 1961, for example, the Soviets had exploded two huge hydrogen bombs from air-dropped balloons above their northern test site: one with a 30-megaton yield and another of over 50 megatons (the largest nuclear explosion ever). Afterward, Khrushchev boasted to the West, "We placed Gagarin and Titov in space, and we can replace them with other loads that can be directed to any place on Earth."[179] If cooperation was an actual second track in Soviet strategy, it clearly suffered from a lack of follow-through.

Most U.S. military planners still believed that nuclear war with the Soviet Union was inevitable and that space would play a critical role. For this reason, the United States launched a special experiment called Project West Ford on October 21, 1961. Over the protests of a small group of space scientists, who feared the possible impact of orbital space debris, the United States intended to distribute 480 million copper filaments into an Earth orbit of 1,970 miles in altitude as a means of testing U.S. ability to maintain military communications in a nuclear-deteriorated, wartime environment.[180] The goal was to bounce communications

[178] See Office of Technology Assessment, *U.S.-Soviet Cooperation in Space*, quoting G.A. Res. 1721 (XVI), p. 18.

[179] Cited in Manno, *Arming the Heavens*, p. 81.

[180] On Project West Ford, see online chapter from Donald Martin's book *Communications Satellites*, 4th ed., posted on the Web site of the Aerospace Corporation at <http://www.aero.org/publications/martin/martin-4.html> (accessed September 18, 2006). This detailed study cites a figure of

off of the expected "5-mile wide and 24-mile long belt" that project scientists expected would be formed.[181] But the dispensing mechanism malfunctioned, leaving the satellite in an otherwise useless, long-duration orbit.[182]

Not surprisingly, fiscal year 1961 marked a new watershed in U.S. spending for space activities.[183] The overall budget (military and civilian) for space programs jumped from $1.1 billion in 1960 to $1.8 billion.[184] Included in the NASA budget was support for further development and testing of the propulsion reactors developed by the military's ROVER nuclear reactor program, which had become nearly dormant due to the Pentagon's decision in the late 1950s to focus on conventional propellants. The new goal was to develop a long-duration manned spaceflight rocket for NASA's Space Nuclear Propulsion Office. The new, so-called NERVA (nuclear engine for rocket vehicle application) Project foresaw an optimistic future using civilian nuclear power for deep-space missions.[185] A large, long-term contract to build and test prototype reactors in the NERVA program soon went to the Astronuclear Laboratory at Westinghouse. Funding accelerated thanks to President Kennedy's direct support. Indeed, Kennedy's decision to mention the program in his May 1961 speech meant that it formed a central part of his new competitive drive—using superior technology to beat the Soviets.[186]

Behind the scenes, the Kennedy administration had greatly increased space weapons programs as well. However, unlike in the civilian area, he continued to prefer that they be kept from public view. The president was outraged when Lieutenant General Schriever—head of the new Air Force Systems Command— had called publicly for growth in the "military oriented space program" in a

---

480 million filaments and offers other technical information on the project. However, NASA's Web site uses a figure of 350 million filaments. See "Aeronautical and Astronautical Events of July–September 1961" on the NASA Web site at <http://history.nasa.gov/Timeline/1961-3.html> (accessed September 17, 2006).

[181] Ibid.

[182] The West Ford satellite remains in orbit today as a piece of useless (but dangerous) debris. Given its altitude, it will likely stay there for centuries, requiring continuance tracking and avoidance by other spacecraft passing through that orbital plane.

[183] Stares, *Militarization of Space*, Appendix 1, Table 1, p. 255.

[184] The defense department held a 45 percent share, and the rest was parceled out to a variety of lesser actors, including the Department of Energy.

[185] Simpson, *Nuclear Power from Underseas to Outer Space*, pp. 133–35.

[186] By 1963, the staff for the NERVA project had grown to 1,940 people (Simpson, *Nuclear Power from Underseas to Outer Space*, pp. 133–35). According to a 1962 article in *Time* magazine, funding more than doubled for the project from 1961 to 1962, and Kennedy himself visited the test site at Jackass Flats in Nevada later in 1962 to check on the progress of the work. See "The Care & Feeding of Rover," *Time*, December 21, 1962, available online at <http://www.time.com/time/magazine/article/0,9171,940130-2,00.html> (accessed August 23, 2006).

speech in the fall of 1961.[187] In March 1962, the Kennedy administration moved to strengthen the formal "blackout" directive to classify the names of a variety of payloads and prohibit prior announcement and press coverage.[188] Kennedy wanted to keep the military "on message" in order to maintain space's civilian face in front of the public, the media, and international observers.

After a thorough review of military weapons programs, Secretary of Defense Robert McNamara had begun to move forward on a number of fronts. In the ASAT area, the military had a range of programs, owing to interservice rivalry and a perception that political winds were blowing in this direction. A navy ASAT effort, called appropriately Hi-Ho, conducted two nondestructive tests of a satellite interceptor launched from a Phantom F-4B fighter in April and July 1962.[189] By this time, high-energy laser research had also begun as a possible option for both ASAT and eventual anti-ballistic missile (ABM) use. Already, officials like General Curtis LeMay worried that the Soviets could develop these technologies and seek to "deny space to us."[190] But the technology remained too immature to deploy. Meanwhile, the co-orbital SAINT interceptor program began to face budget cuts due to delays and cost overruns.[191] It would not recover. Instead, McNamara had already begun to move ahead with yet additional new programs. He and his advisors had become convinced that Khrushchev meant what he said about the Soviets' ability to station nuclear weapons in space and wanted the ability to respond. But, as one air force study of U.S. ASAT programs concluded, threat inflation may have played a significant role in these decisions: "The U.S. military frequently had attributed to the Soviets strategic motivations, objectives, and capabilities that were of concern within the walls of the Pentagon. Thus, military leaders often jumped to conclusions not supported by the facts."[192] Indeed, the CIA argued that no current threats merited such a costly and possibly provocative program.[193] But President Kennedy and Secretary McNamara distrusted the Soviets and stuck with the Pentagon's recommendations.

For ASAT purposes, it still seemed that nuclear weapons provided the only

[187] Stares, *Militarization of Space*, pp. 64–65.

[188] Ibid., p. 65.

[189] Burrows, *Deep Black*, p. 146; also Stares, *Militarization of Space*, p. 111.

[190] LeMay quoted in Stares, *Militarization of Space*, p. 111.

[191] See the NASA History Division's report "On the Shoulders of Titans" on the NASA Web site at <http://history.nasa.gov/SP-4203/ch6-2.htm> (accessed September 23, 2006).

[192] Lt. Col. (USAF) Clayton K. S. Chun, *Shooting Down a "Star": Program 437, the US Nuclear ASAT System and Present-Day Copycat Killers*, Cadre Paper No. 6, Institute of National Security Studies, U.S. Air Force (Maxwell Air Force Base, Ala.: Air University Press, April 2000), p. 7.

[193] Ibid., p. 6.

certainty of taking out the target. A previously existing army program—Project 505 (also known as Mudflap)—planned to put a Nike-Zeus ABM interceptor into service as an ASAT using a 400-kiloton weapon. But its payload capacity could not accommodate larger nuclear weapons believed to be necessary for the mission, and its radar system could be easily overwhelmed in a barrage attack. Finally, it could not destroy satellites beyond a 250-mile range.[194] In an effort both to edge out the army from a critical mission area and to create more "bang for the buck," air force planners proposed a system (later known as Project 437) using a larger Thor booster, which had the advantage of being able to carry a larger nuclear weapon to greater range. Unlike Project 505, Project 437 used a less expensive, all-military launch crew and could employ spare Thor missiles from the air force's already extant intermediate-range ballistic missile program.[195] Given these advantages, McNamara gave the program a green light in November 1962.

Although it continued to gain congressional support, the air force's manned space bomber Dyna-Soar faced increasing scrutiny from Secretary McNamara because of its unproven technology and unclear mission. In the face of this pressure, the service and its congressional supporters came up with two bold ideas in February 1962: an air force space station called the Manned Orbital Development System and a military space ferry known as Blue Gemini.[196] While NASA strongly resisted this seeming air force poaching on its territory, the effort really consisted of a Pentagon effort to buy a stable of transport vehicles—which would be developed by NASA—on the cheap. Congress and the air force set to work putting funding for the projects into the 1963 federal budget. But none of these programs helped Kennedy in the real race for U.S. and international public opinion. For this effort, he needed to be more active in pushing NASA to the forefront and making actual progress.

Another False Dawn for U.S.-Soviet Civilian Cooperation

As the administration planned for fiscal year 1963, it succeeded in winning congressional support for increasing NASA's budget to $3.67 billion.[197] The already pro-space Johnson, who as vice president headed the National Space Council, now became the administration's leading advocate for the civilian

[194] Ibid., pp. 8–9.

[195] Ibid., p. 10.

[196] NASA History Division, "On the Shoulders of Titans."

[197] NASA, "Managing NASA in the Apollo Era," SP-4102, Chapter 7, on the NASA Web site at <http://history.nasa.gov/SP-4102/ch7.htm > (accessed September 23, 2006). Part of the increases in NASA expenditures included orders for hardware from the military.

space program. A priority for achieving high-prestige accomplishments was the development of a heavy-lift booster, one area where the Soviets still led the Americans. Under Kennedy, this effort was now tied explicitly to the Moon mission and the development of U.S. manned spacecraft.

These massive budgetary injections helped NASA expand rapidly, consistent with the competitive thrust of the Kennedy administration. The space organization expanded its base of operations into huge new facilities, including the Houston-based Manned Spaceflight Center under Robert Gilruth. Vice President Johnson, in alliance with fellow Texan and Chairman of the House Appropriations Committee Congressman Albert Thomas, had wangled this center away from its existing Virginia-based headquarters. NASA also acquired 80,000 acres at Merritt Island, Florida, for construction of a permanent main launch facility at Cape Canaveral.

In addition, NASA officials took major steps through the summer of 1962 to decide on the much needed plan and booster to be employed in the Apollo program. Gilruth fought and won out over von Braun with a flight design calling for a separate Lunar Excursion Module (LEM) rather than a direct-ascent approach, which would have used the huge NOVA booster already under development. Gilruth's design brought forth the somewhat smaller Saturn C-V, a rocket still unequaled in power, whose massive engines could lift 45 tons to the Moon (or the equivalent of sending 120 Mercury capsules simultaneously into Earth orbit).[198] By October 1962, the Grumman Corporation had been awarded the contract for the lunar module, and von Braun quickly shifted gears to take on the task of constructing the Saturn V at Huntsville.

In the still-nascent area of commercial space policy, the United States made a decision to stress the development of the private sector rather than maintain government control over all aspects of space activity.[199] According to the much debated Communications Satellite Act passed by Congress in August 1962, however, the government retained an interest in this area of space operations by sponsoring the development of a U.S. monopoly corporation (COMSAT) that would purchase rockets from NASA, mate them with privately built satellites, and market a worldwide communications network to both the developed and underdeveloped world, thus guaranteeing U.S. dominance in the field. The

[198] On these comparisons, see Joseph J. Trento, *Prescription for Disaster: From the Glory of Apollo to the Betrayal of the Shuttle* (New York: Crown, 1987), p. 49.

[199] As Schauer summarizes American aims, "It was the intention of the President and Congress to secure a wider role in the space program for nonaerospace firms, to substitute the market mechanism for political decision-making in allocation of satellite communications channels, and to provide a flexible mechanism to encourage *international* use of an American satellite communications system" (Schauer, *Politics of Space*, p. 34).

idea was another less touted brainchild of Kennedy's May 1961 speech before Congress. It aimed at a paternalistic U.S. technological alliance to shepherd the rest of the world into space and the modern age of satellite telecommunications, while keeping the Soviets out.[200]

Nevertheless, on February 21, 1962, in a congratulatory letter to Kennedy on the success of the first U.S. orbital flight by John Glenn, Khrushchev expressed Soviet interest in a "pooling" of U.S.-Soviet space efforts. For the first time, the Soviet leader did not specifically mention the prerequisite of a general disarmament agreement.[201] It seemed that some hint of shared interests had begun to dawn on the two leaderships regarding space.

Inspired by the Khrushchev letter and eager to raise U.S. prestige with a high-profile cooperative program—as well as to learn more about the Soviet space program—Kennedy sent back a list of concrete proposals. Notably, this new list focused exclusively on bilateral programs, although it left out the previous spring's offer for a joint Moon program. Instead, it emphasized a series of smaller missions: cooperation in weather satellites, remote sensing, and space tracking. Continuing this first-ever round of direct leader-to-leader proposals, Khrushchev responded by adding a few further possibilities for discussion: cooperation in the return of spacecraft, the rescue of cosmonauts, and a joint research program (presumably unmanned) for exploration of the Moon, Mars, and Venus.

Practical negotiations began in March between NASA's Dryden and Soviet Academician Blagonravov in New York. Due to Soviet complaints about the active U.S. military space program, however, the atmosphere was far from cordial. Blagonravov spent much of his time not in serious negotiation but in criticizing the United States for its space nuclear testing, Project West Ford, and overflights of the Soviet Union by U.S. spy satellites (to date, the Soviets had failed to develop their own reconnaissance capabilities). In the course of the talks, the Soviet preference for largely symbolic multilateral international cooperation—paralleling the Eisenhower approach—became clear. Such programs would involve less apparent acceptance of U.S. policies in other areas and more free publicity around the world, while not exposing to the Western media weaknesses in the Soviet space and missile program. After the spring, these differences led both Kennedy and Khrushchev to lose interest in the civilian space negotiations.

---

[200] For more on the development of COMSAT, see McDougall, . . . *the Heavens and the Earth*, pp. 352–60.

[201] See Frutkin, *International Cooperation in Space*, pp. 92–94; and Harvey and Cicorritti, *U.S.-Soviet Cooperation in Space*, p. 86.

Yet NASA continued to pursue an agreement (without direct White House participation or congressional support) in a second round of talks, held in Geneva from May 29 to June 7, 1962. Scientists on the two sides reached agreement on the pursuit of three areas of discussion: space communications, space meteorology, and geomagnetic surveys.[202] But because of the lack of top-level political backing, the range of these programs became progressively narrower as the negotiations continued, and the actual fulfillment of the agreements was not scheduled to take place, in most instances, for several years. The Soviets—still fearful of opening their program—expressed interest only in limited, post-mission exchanges of data, not in the integration of research agendas, personnel, and programs desired by the U.S. scientists.[203] The Memorandum of Understanding signed between the two sides elicited little interest from either Kennedy or Khrushchev. As events would happen, the Soviet Union would not even approve the agreement until August 1963, when subsequent changes had watered the cooperative elements down even further. The Khrushchev initiative of early 1962, in the end, yielded few results.

Soviet Military Visions for Space

The personnel, resource-related, and security demands of the newly expanded space program—from its more limited ICBM mission—required enormous organizational initiatives. On June 23, 1960, the Central Committee of the Communist Party had issued a secret decree entitled "On the Creation of Powerful Carrier-Rockets, Satellites, Space Ships, and the Mastery of Space."[204] The plan called for the development of a range of heavy-lift boosters, the deployment of a military space station for the purpose of engaging in space battles, and smaller space-based weapons capable of conducting reconnaissance and firing against targets in space and on the ground. This ambitious program was designed to support a number of independent and ongoing research efforts under a coherent master plan.[205] By 1961, the Soviet military had begun to centralize a

[202] The Soviets rejected U.S. proposals to cooperate in space tracking by the establishment of facilities on one another's territory.

[203] For example, see NASA official Frutkin's book, *International Cooperation in Space*. Frutkin argues that the Soviets consistently failed to follow through on agreements and balked at even the most minor exchanges of technical information. Thus, he says, while the United States at this time had already begun numerous cooperative projects with foreign scientists, the Soviet Union had no significant cooperation with any nation.

[204] Siddiqi, *Sputnik and the Soviet Space Challenge*, pp. 239–40.

[205] Less clear was where the funding would come from for the costly elements required to make the plan a reality, given simultaneous goals of expanding the Soviet Union's nuclear complex and weapons arsenal.

number of specialized institutes that had developed within the Academy of Sciences over the course of the 1940s and 1950s to help streamline its efforts. Many of the institutes moved into military-controlled ministries as line production of boosters replaced research tasks, while others fell under the direct control of the Ministry of Defense.[206] In another organizational move affecting space research, the Soviet Council of Ministers announced in 1961 the formation of a special committee to plan state activities in the areas of science and technology. The evidence suggests, however, that this body—like many others formed in the late Khrushchev period—floundered due to internal contradictions.

In response to the U.S. Hardtack I and Argus programs, the Soviets began a series of high-altitude nuclear tests in 1961 and 1962 to explore EMP effects and the possible use of space nuclear explosions for ABM purposes. On September 6, 1961, the Soviet military launched nuclear-tipped missiles from its Kapustin Yar missile test range to altitudes of 14 and 26 miles, exploding 10.5-kiloton and 40-kiloton weapons, respectively. On October 27, 1961, it extended the test program into space by carrying out two 1.2-kiloton blasts at altitudes of 93 and 186 miles. The tests mirrored closely the earlier Hardtack and Argus tests in altitude, purpose, and yield. Clearly, the Soviet military believed it needed to match American programs and capabilities. It seemed that there would be no backing down from the military competition.

The 1957–62 period had witnessed remarkable Soviet firsts in spaceflight, from *Sputnik I* to Gagarin. But the reality behind Khrushchev's claims of technological superiority over the United States told another story, despite the Soviets' undeniable achievements.[207] Soviet comparative weakness in computers, miniaturization, and advanced electronics put their program several years behind that of the less publicized U.S. military program, leading Khrushchev to pursue a policy that focused on high-publicity civilian firsts. Part of the pressure for haste had led to a disastrous explosion on October 24, 1960, at Baikonur, which killed the new head of the Strategic Rocket Forces, Field Marshal Mitrovan I. Nedelin, and 126 scientists, engineers, and soldiers working at the site, setting back the Soviet space program considerably.[208]

[206] For a description of these reforms, see David Holloway's "Innovation in the Defense Sector," in Ronald Amann and Julian Cooper, eds., *Industrial Innovation in the Soviet Union* (New Haven, Conn.: Yale University Press, 1982).

[207] The Soviets had failed to report on a number of attempted space coups, which had turned into expensive mistakes instead. For example, two probes launched toward Mars during Khrushchev's trip to the United Nations in October 1960 failed even to achieve Earth orbit.

[208] According to several accounts, the new heavy-lift, R-16 booster failed to ignite during a test. Nedelin (under instructions from the impatient Khrushchev to complete the test as soon as possible) ordered an immediate investigation, overruling the scientists on hand who argued that the propellant first

In the military sector, Moscow had become increasingly frustrated with U.S. advances in photo-reconnaissance satellites, which had effectively replaced the vulnerable U.S. U-2 aircraft and now provided a unilateral strategic advantage over the USSR in intelligence-gathering capabilities. While struggling to develop its own spy satellites, the Soviet Union complained vehemently in international fora of American "violations" of international law and sovereignty of airspace.[209] U.S. launch rates by now far exceeded those of the Soviet Union, despite the opening of Kapustin Yar as a second space launch facility in 1962. Part of the Soviet fear was that the United States would discover the truth behind its "missile bluff," and Khrushchev shunned space cooperation for the same reason: fear of intrusion and discovery.

Rising Tensions and Expanded U.S. Nuclear Testing in Space

By mid-1962, the United States had decided to move ahead with its controversial exo-atmospheric nuclear testing program (Project Fishbowl), despite complaints from scientists worldwide who feared damage would be caused to the Van Allen radiation belts, possibly affecting the Earth's climate. It seemed that the Pentagon and the nuclear laboratories had begun to run at cross purposes with the state department. Indeed, Secretary of State Dean Rusk cautioned in a speech in June, "Steps must be taken at this early stage to keep outer space from being seeded with vehicles carrying weapons of mass destruction."[210] Privately, however, an interagency memorandum from Deputy Under Secretary of State Alexis Johnson outlining a draft U.S. diplomatic strategy for the president showed that U.S. officials worried that a declaratory ban would not be enforceable.[211] In fact, Johnson's top-secret memo suggested that the United States should "dissuade Canada from pressing for consideration of such a ban"

---

be drained from the rocket engines and fuel lines. Minutes after the scientists and engineers left their protective bunkers and began examining the missile, the second stage of the rocket suddenly ignited, causing a tremendous explosion. The heat of the fireball literally vaporized some members of the team, while blowing others to their deaths in a 200-ft. pit below the launch pad. On this disaster, see Siddiqi, *Sputnik and the Soviet Space Challenge*, pp. 256–58.

[209] The United States knew, for example, that only in 1961 did the Soviets even begin to engage in full-scale deployment of ICBMs with the new SS-7 booster (the R-16, by Soviet designation).

[210] Secretary of State Dean Rusk, address of June 16, 1962, *U.S. Department of State Bulletin*, July 2, 1962, p. 5. (Quoted in Harvey and Ciccoritti, *U.S.-Soviet Cooperation in Space*, p. 163.)

[211] Deputy Under Secretary of State U. Alexis Johnson, "Memorandum for NASA—Mr. Webb, Defense—Mr. Nitze, Air Force—Dr. Charyk, ACDA—Mr. Fisher, CIA—Dr. Scoville, W.H.—Mr. Kaysen" (Top Secret), July 10, 1962, National Archives (Ref: NSAM-156).

as U.S. policy would have to "make clear its firm opposition to any declaratory ban."[212]

Despite American opposition to foreign testing and deployment of nuclear weapons in space, the U.S. test series moved forward, exposing the United States to considerable criticism both outside and within the government. These experiments proved far from successful, resulting in damaging environmental consequences both in space and on the ground. A number of the tests in Operation Dominic had sought to determine the effectiveness of large nuclear explosions in space for ABM purposes as well as establish the survivability of military communications satellites in an all-out war. Indeed, the expectation of the military was that it would have to operate in such an environment within the near future, given the course of relations with the Soviet Union. However, the Thor missiles being used at Johnston Island to launch the tests experienced a number of troubling problems leading to failures of attempted tests on June 5 and June 20. Finally, on July 9, U.S. scientists conducted the Starfish Prime test, which exploded a huge, 1.4-megaton hydrogen bomb at an altitude of 248 miles, lighting up the night sky with bright colors visible thousands of miles away. The blast disrupted radio transmissions as far away as California and Australia for several hours.[213] As Atomic Energy Commission Chairman Glenn Seaborg noted in his memoirs, "To our great surprise and dismay, it developed that STARFISH added significantly to the electrons in the Van Allen belts. This result contravened all of our predictions."[214] The test proved embarrassing and costly, particularly as British and U.S. scientists had cautioned against its likely effects. The EMP radiation it generated eventually disabled at least six satellites, including the British *Ariel I*, the U.S. *Traac, Transit 4B, Injun I, Telstar I*, and the Soviet *Kosmos 5*.[215] *Telstar I* had only recently been launched, providing unprecedented live television coverage from the United States to Europe and back again before it went silent later in the year.[216]

---

[212] Ibid., under heading "Recommended U.S. Position on a Separate Ban on Outer Space Weapons of Mass Destruction," p. 8.

[213] For more on the tests in Project Fishbowl, see Stares, *Militarization of Space*, p. 108; and Manno, *Arming the Heavens*, pp. 82–84.

[214] Seaborg, *Kennedy, Khrushchev, and the Test Ban*, p. 156.

[215] On this satellite damage, see Herman Hoerlin, "United States High-Altitude Test Experience: A Review Emphasizing the Impact on the Environment," Los Alamos Scientific Laboratory Monograph, Los Alamos, New Mexico, October 1976, pp. 25–26, posted on the Web site of the Federation of American Scientists at <http://www.fas.org/sgp/othergov/doe/lanl/docs1/00322994.pdf> (accessed September 18, 2006). Some sources indicate that a seventh satellite may also have been damaged.

[216] Burrows, *This New Ocean*, p. 269.

But the United States followed up with its third attempt to carry out the so-called Bluegill test (now dubbed Bluegill Triple Prime) on July 25. The result proved to be a catastrophic failure. Because of problems with the missile, the range safety officer had to blow up the rocket just a few feet off the launch pad with its warhead still aboard, spewing fire and plutonium across the site. The disaster forced the United States to shut down the facility for weeks in order to conduct a clean-up operation.[217] Despite these problems, the United States had already made plans for a fall series of tests, including the very high-altitude shot, Urraca.

Beginning in the spring of 1962 when the test series had been announced, Urraca had raised objections from scientists internationally, who feared lasting damage to the Van Allen belts from this planned, large nuclear explosion at an altitude of over 800 miles.[218] But this proto-epistemic community of space scientists remained too weak during this time to have much influence on policy. The sudden unavailability of the launch site, however, delayed plans for the test. Notably, this pause revealed the emergence of high-level concerns about the growing impact of the nuclear test program on humans in space, given a planned Soviet launch in early August. As Seaborg reports, "On August 11 [1962] we received word that the Soviet Union was anxious about the safety of its astronaut, Nikolayev, whose orbit around the earth was expected to last for several more days."[219]

The Soviet military, for its part, had also begun to worry about the expanding U.S. nuclear program and possible vulnerability to attack. By 1962, the Soviets deployed a system of some thirty nuclear-tipped surface-to-air missiles for anti-aircraft and possible ABM defense near the coast at Leningrad.[220] Although they lacked sophisticated radars and would have been of dubious value against ICBMs, the effort showed that Moscow too believed that war was becoming increasingly likely. They also had begun construction of a dedicated, nuclear-tipped A-35 ABM system (termed "Galosh" in the West) to defend the government in Moscow and the region's many strategic facilities.[221]

Overall, by midsummer 1962, U.S.-Soviet space relations seemed bent on ignoring obvious information that might shake them from what seemed to be a collision course, or at least the full-scale weaponization of space. Although

---

[217] Seaborg, *Kennedy, Khrushchev, and the Test Ban*, p. 156.

[218] Ibid., p. 153.

[219] Ibid., p. 156.

[220] Donald R. Baucom, *The Origins of SDI: 1944–1983* (Lawrence: University of Kansas Press, 1992), p. 30.

[221] Pavel Podvig, ed., *Russian Strategic Nuclear Forces* (Cambridge, Mass.: MIT Press, 2001), p. 413.

scientific knowledge had begun to accumulate for some collective action to protect space and prevent such an outcome, the political mechanisms and willpower did not yet exist. As U.N. General Secretary U Thant commented, the U.S. space nuclear test program was "a manifestation of a very dangerous psychosis which is in evidence today."[222] After the Berlin crisis of the previous summer (and the Soviet-ordered construction of the Berlin Wall), the tide of U.S.-Soviet relations seemed to be worsening. Stung by U.S. satellite overflights of Soviet launch facilities—which exposed the myth of the Soviet ICBM force—Khrushchev attempted to compensate for the lack of a militarily useful, long-range missile through a short-cut road to military parity: stationing Soviet intermediate-range missiles on the newly communist island of Cuba. Unfortunately, this effort would set back emerging back-channel discussions on EMP radiation in space and push out the timetable for a possible agreement.

Conclusion

Given the context of the Cold War, the dominance of policies of narrow space nationalism in these early years of spaceflight is not surprising. The brief U.S.-Soviet wartime alliance had receded quickly into memory as Moscow solidified control over Eastern Europe and the United States moved to consolidate its defensive relations with Western Europe through the 1949 North Atlantic Treaty Organization. The press in both Washington and Moscow emphasized daily the fundamental political disputes that divided the two sides and had established them as rival hegemons fighting over an unclaimed realm in space. Although the horrors of World War II induced some caution on both leaderships, the two sides had already faced off in a costly war over Korea, with the Chinese doing the Soviets' fighting against the U.S.-led United Nations forces. The secrecy of the Soviet side and evidence of its active espionage activities in the United States only added to the psychological tension. Space represented a new opportunity for both states to prove their merit and extend their military capabilities for what some expected to be an ultimate, deciding battle. Diplomats pushing for cooperative outcomes still held little sway. Similarly, some leading scientists had begun to doubt the sagacity of ambitious weapons efforts in space,[223] but they faced opposition from nonexperts with more political clout.

[222] Cited in Manno, *Arming the Heavens*, p. 84.

[223] See, for example, the skeptical analysis of space bombers and military Moon bases at the end of the report provided by the President's Science Advisory Committee to the president in March 1958. "Introduction to Outer Space," Appendix 4, in Killian, *Sputnik, Scientists, and Eisenhower*, p. 297 (paragraph 2).

For military officials, one problem caused by the close relationship of space activities to national defense capabilities was that it raised serious doubts about the "safety" of cooperation.[224] Letting down one's guard could mean allowing the other side to gain a sudden and decisive advantage. Even in human space-flight, cooperation could provide the other side with possibly critical information that might be used to provide data on questions relating to the launch, detection, and tracking of one's ICBMs.[225]

Politically, cooperation posed other problems. The Soviet Union feared that cooperating even in space science with the West would not only send the wrong signal to its allies—especially the Chinese—regarding capitalist countries but also allow American scientists to expose to the world certain glaring weaknesses in Soviet space technologies. The sharp division within the Soviet program between the academy officials appointed to negotiate cooperative arrangements with the United States and the military officials actually in charge of the Soviet space program left no room for the normal give-and-take of negotiations. This limited contact seemed to rule out the formation of consensual norms or even agreement on scientific truths about dangers, such as EMP radiation. Although some individual Soviet scientists may have had the desire to cooperate and build such contacts, under the Soviet system they lacked not only means but also crucial information on what was going on even in their own space program.[226]

In the momentous years from 1957 to 1962, the U.S. space effort had shifted from a modest military space program with a small civilian component under Eisenhower into a massive, civilian-led but militarily potent program under Kennedy. The country's concern over Russian space accomplishments had proven much deeper than imagined by Eisenhower. Americans had lost faith in their own self-image as a world technological leader. President Kennedy was caught between two competing conceptions of reality. In one, he recognized the need to race with the Soviet Union in space for military advantage. In the other, he realized that restoring Americans' belief in themselves (and therefore his presidency) was perhaps even more important. But it was becoming increasingly clear that this could not be done through military means alone. Kennedy's decision to emphasize NASA would provide a major opening for other options

---

[224] As one analyst argues, "Until 1962 Soviet-American space cooperation was largely limited to the nongovernmental and multilateral cooperation of scientists." See Schauer, *Politics of Space*, p. 222. Notably, this would no longer be true after 1963.

[225] In terms of limiting cooperation, this factor proved particularly damaging in the field of manned spaceflight, as in other areas that would require actual exchanges of personnel or equipment.

[226] Author's interview with then-Interkosmos official Dr. V. S. Vereshchetin in Moscow, April 1988. For a detailed record of U.S. contacts with Soviet scientists, which confirms Vereshchetin's views, see also Frutkin, *International Cooperation in Space*.

and have a profound effect on the next twenty years of the American space program. Yet, in the early Kennedy years, a secondary but not inconsequential factor emerging in the American reevaluation of space competition had to do with newly appreciated technical risks of implementing the original aims laid out by military space planners in the 1950s. The Kennedy administration—and especially Defense Secretary McNamara—had begun to worry about the political, environmental, and strategic implications of a number of proposed U.S. (and Soviet) space weapons. As had been learned from Project Fishbowl, the space environment could not adequately contain nuclear radiation, which rapidly dispersed along the Van Allen belts, rendering human spaceflight dangerous and multimillion-dollar communications and reconnaissance satellites inoperable. For the United States, this meant the possible loss of the one area of space where it now led. In addition, the defense department worried about possible developments within the Soviet weapons program (if left unconstrained by arms control), increasing pressure from the international scientific community, and even more severe consequences for emerging U.S. commercial space interests as a result of dangerous EMP radiation from the nuclear test program.

The Soviets too had begun to worry about space radiation creating problems for manned spaceflight, an area where they had established a clear lead and had won tremendous international acclaim. By 1962, the two sides had reached a crossroads. Their existing path pointed in an easily identifiable direction: the increasing dominance of the respective military programs, the continued development of a range of space weapons, and the eventual reduction or elimination of competing civilian and commercial space programs. The alternative route of interaction and cooperative rule building for space had curves and potholes but might eventually lead to a better place: namely, eliminating a number of weapons programs, restraining others, and creating agreements to protect other emerging space activities. These choices had now become increasingly clear-cut.

CHAPTER FOUR

The Emergence of Cooperative Restraint

1962–1975

Scientific information to support a cooperative agreement to limit electromagnetic pulse (EMP) radiation in space, either in the context of general nuclear test-ban negotiations or separately, had emerged by the late summer of 1962. Other factors seemed to support the notion that putting weapons of mass destruction into space and fighting over bases on the Moon would be dangerous new steps in the arms race. Yet space cooperation and military restraint remained blocked by a number of factors. First, no arms control between the two sides had ever taken place. Second, strong military-technological forces continued to push a seemingly inevitable process of testing and development of both nuclear and other systems for space. And, third, the exact mechanisms through which this new knowledge of EMP's harmful effects in space might be put into action remained murky. Instead, traditional, military-led patterns of "business as usual" continued even during the Cuban Missile Crisis, when both sides tested nuclear weapons in orbit and at high altitudes, risking nuclear war. Given this situation, it seemed unlikely that anything good might follow this very dangerous event. As physicist Sidney Drell, who participated in U.S. military space efforts at the time, and Steven Weber observe regarding this period, "The United States and the USSR appeared to be poised on the brink of an arms race in space."[1] In fact, for some supporters of a *greater* military role in space, this arms race was not happening fast enough and the United States was in danger of losing it.[2]

[1] See the chapter by Weber and Drell in Alexander L. George, Philip J. Farley, and Alexander Dallin, eds., *U.S.-Soviet Security Cooperation: Achievements, Failures, Lessons* (New York: Oxford University Press, 1988), p. 381.

[2] As author and space enthusiast Erik Bergaust wrote in a scathing critique of U.S. space activities under Kennedy and Johnson's supervision: "The Kennedy Administration's failure to build a

By the spring of 1963, however, the main sticking points in the test-ban talks finally broke, with the United States abandoning hopes of a breakthrough in on-site testing and instead opting to take the most readily achievable formula for a ban: everywhere except underground. Although this proved a major disappointment to nuclear disarmament hopes, it marked a great accomplishment for securing the future of space activity. Indeed, whereas the nuclear arms race continued unabated, more concrete progress followed regarding space.

Satellite developments in the Soviet Union also brought a change in Moscow's attitude toward reconnaissance, which now shifted from opposition to support, thus opening new opportunities for space's regulation. The two sides agreed in October 1963 on a resolution at the United Nations to ban a host of activities that might lead to U.S.-Soviet conflict in space. This framework smoothed the path toward a new period of increasingly civilian-influenced activities in space, the development of additional types of military support satellites, and later on, talks toward a formal space treaty. While space weapons continued to have advocates arguing for their unique capabilities, national leaderships consistently overruled or sidelined these enthusiasts into back-burner programs because of the feared impact of these systems on other, higher-priority uses of space, namely, civilian programs and military support activities. Bilateral competition continued, but billions of dollars flowed instead toward a surrogate, civilian competition: sending spacecraft to far planets, carrying out increasingly complex manned missions, and finally, putting people on the Moon. By 1972, patterns of social interaction had replaced narrow self-interest and purely technologically determined competition, with elements of global institutionalism becoming firmly entrenched in the policies of the Richard M. Nixon administration.

The space talks of the early 1960s established some of the critical linkages—bilaterally and within the United Nations—that had led to the negotiation and signing of the Outer Space Treaty in 1967. This agreement marked a singular departure from unilateralist policies on both sides and the first clear effort to build an institutionalized, multilateral framework for space security. Similarly, a gradual norm of noninterference with each other's passive military satellites also began to emerge by the late 1960s. This initially tacit cooperation became institutionalized in the Anti-Ballistic Missile (ABM) Treaty and the first Strategic Arms Limitation Agreement in 1972. Indeed, space restraint had now been formally recognized as having a major "spin-off" value. As one participant in

---

strong military space capability could prove to be disastrous." See Erik Bergaust, *The Next Fifty Years in Space* (New York: MacMillan, 1964), p. 104.

the secret U.S. Corona program notes on these satellites' impact on the détente era, "It was imagery that set the stage for the arms limitation talks."[3] The
reason was simple: "With imagery, we could go to a numbers-based strategic
arms limitation negotiation with the high confidence that we didn't need any
help from the other side to verify it."[4] In short, the protection of space had now
become recognized as crucial to the safeguarding of key assets for national security. Contrary to predictions in the late 1950s, leaders had begun to recognize
that space was now too valuable to be used for war.

Agreements on space cooperation appeared in other areas as well. In the
broadcasting area, interdependence in using radio frequencies and potential
conflicts over geostationary orbital slots stimulated agreement over a coordinating mechanism. Similarly, agreements on the registration of objects and the
new regulations for compensating parties harmed by a foreign country's space
objects began to institutionalize an arena that had once seemed ripe for a military free-for-all. Finally, in the civilian and space science arena, the latter part
of this period saw dramatic new forms of cooperation, some of which began to
suggest that the two sides might give up their space competition altogether. At
least on the American side, the impact of Project West Ford and the gradual accumulation of new scientific data began a process by the mid-1960s of NASA's
consideration of a second major environmental threat to space: orbital debris.
By the mid-1970s, this would emerge as an incipient international issue, as explosive Soviet ASAT tests began to expand the previously manageable debris
field. But consensual knowledge in this field would not coalesce for international action yet. The data were limited, space was not yet that crowded with
long-lived assets, and a political consensus did not yet exist.

This chapter analyzes the steps that gradually moved the space race away
from its initial focus on nuclear weapons and the deployment of near-term
space defenses—essentially, military-led strategies of achieving space security.
Instead, by the mid-1960s, the two capitals had agreed on the need to preserve
the space environment for other purposes—largely civilian programs, space
science, and military support systems. Doing so, however, required more than
individual restraint. Given the context of the Cold War and the possibility of
cheating, the two sides decided that formal international agreements provided
greater security. In contrast to the prior period, the years from late 1962 through

---

[3] Former CIA analyst and Corona manager Ray Huffstutler, quoted in Dwayne A. Day, John M.
Logsdon, and Brian Latell, eds., *Eye in the Sky: The Story of the Corona Spy Satellites* (Washington,
D.C.: Smithsonian Institution, 1998), p. 228.

[4] Ibid.

1975 witnessed the establishment of legal rules governing competition in space and limiting harmful activities. Increasingly, as declassified documents now show, Washington and Moscow began to view and treat space as an environment requiring its own special policy, moving toward a midpoint between the Antarctic and the oceans: general policies of self-restraint in weaponry but with continued competitive elements in the political, commercial, and military reconnaissance fields.

New Concerns about Nuclear Explosions in Space

By the late summer of 1962, the two sides had begun to recognize the problems of their existing course in space, but they did not seem to be able to help themselves. Politically, the Cold War continued to push the Kennedy and Khrushchev governments into a seemingly inexorable competition, with the two militaries taking the lead.

Even within the Pentagon and the Soviet General Staff, however, it was difficult not to notice the harmful effects of the ongoing nuclear tests in space and the tremendous damage that EMP radiation was causing to the space environment. U.S. leaders had already begun to worry about the impact of these tests on humans in space, as they attempted to move beyond John Glenn's mission into longer-term flights scheduled for the Mercury program. Early in the year, NASA had selected the Saturn C-V booster for the Moon mission and was now moving forward with work on precursor Gemini and Apollo flights.[5] The Soviet government approved an equivalent N-1 booster in September 1962 to mount a parallel Moon shot.[6] Both such efforts would require numerous manned spaceflights in preparation for the complexities of a lunar mission. However, no formal agreement on nuclear restraint had ever been reached. Both sides felt strong pressure from their nuclear laboratories and militaries to continue developing nuclear weapons for possible use in space. This was the way things had always been done, and there was no road map for getting off of this route. Yet, one side or the other would have to give. On balance, it seemed unlikely that a costly exploratory mission aimed mostly at prestige would win out over a military program. But other factors, such as emerging military appreciation for the value of reconnaissance satellites supported innovative thinking. Similarly, Kennedy's Science Advisor Jerome Wiesner had come to the conclusion, in a

[5] Asif A. Siddiqi, *Sputnik and the Soviet Space Challenge* (Gainesville: University of Florida Press, 2003), p. 386.

[6] Ibid., p. 387.

secret August 1962 memo to the president, that "a declaratory ban of weapons of mass destruction [in space] would not involve any real military danger to the U.S."[7] Diplomats had already been working on a possible comprehensive nuclear test ban that would include space. Could they perhaps provide the initiative?

But test-ban negotiations among U.S., Soviet, and British representatives remained deadlocked on issues of on-site inspections and how desirable it was from a military perspective to ban all nuclear tests. Hard-liners on both sides urged their governments to break off the negotiations, believing either that the other camp could not be trusted not to cheat or that war was inevitable in any case, making continued testing a military necessity. In the spring of 1962, a host of countries representing NATO, the Warsaw Treaty Organization, and the nonaligned movement—organized in the newly constituted Geneva Eighteen Nation Disarmament Conference (later the Conference on Disarmament)—had urged the three nuclear states to resume their talks.[8] Buttressed by greater confidence in seismic monitoring, the United States had reduced (but not eliminated) its requirements for on-site inspection. The Soviet side, however, had stiffened its position by calling for no international inspection system. By late summer, two distinct U.S.-U.K. draft agreements had emerged and had been submitted to the Soviet side: a comprehensive test ban and a partial test ban. The first plan included on-site seismic stations with international monitors. The second plan, because of its lower data requirements (since underground testing would be allowed), involved no joint monitoring and a reliance exclusively on remote sensors. Notably, both plans included a ban on space testing. However, as U.S. participant George Bunn notes regarding Moscow's reaction to the presentation of these two options, "The Soviet negotiator . . . promptly rejected them [both] without qualification."[9]

Back in Moscow, First Secretary Khrushchev studied the proposals more carefully. In his mind, the U.S.-U.K. proposals opened the way for a possible two-step process. More importantly, Khrushchev now had information that his diplomats had lacked: the Soviet Union had tested a successful photo-reconnaissance satellite in July. After years of research and development, these experiments had begun in late 1961. In April 1962, the four-camera Zenit 2 system

[7] Science Advisor Jerome B. Wiesner, "Memorandum for the President" (Secret), August 9, 1962, National Archives (POF 67).

[8] On this phase of the talks, see George Bunn, *Arms Control by Committee: Managing Negotiations with the Russians* (Stanford, Calif.: Stanford University Press, 1992), pp. 30–32; also Thanos P. Dokos, *Negotiations for a CTBT, 1958–1994: Analysis and Evaluation of American Policy* (Lanham, Md.: University Press of America, 1995), pp. 38–41.

[9] Bunn, *Arms Control by Committee*, p. 32.

aboard the so-called *Kosmos 4* flight[10] had returned its first images,[11] and by late July, a returning spacecraft delivered the first usable photographs of U.S. sites. While the Soviet satellite system would not have full functionality until 1964, even its initial capability represented a revolution for Khrushchev. Suddenly, he glimpsed what the United States had enjoyed for two years and recognized the potential of photo-reconnaissance to assist in treaty verification over U.S. soil without having to allow prying U.S. inspectors onto Soviet territory.

In a secret communication to President Kennedy on September 4, Khrushchev proposed, "Let us immediately sign an agreement on the cessation of nuclear tests in the atmosphere, outer space and under water."[12] The Soviet leader also pressed Kennedy for a moratorium on all underground testing while the two sides would engage in follow-on negotiations toward a comprehensive test-ban treaty. In subsequent correspondence, Kennedy noted that the United States found a moratorium untenable because of concerns related to the verification of Soviet compliance—the same problem the administration had with the complete test ban.[13] Without ground stations and on-site monitors, according to the U.S. position, the moratorium would be meaningless. On the other hand, it was not clear that the Soviets would agree to a partial test ban without the eventual promise of a comprehensive test ban, whether for substantive or perhaps propagandist reasons. Thus, progress had been made, but the high politics of working out the details of the nuclear test ban dragged on.[14]

Positions in the U.S. government ran the gamut, although an increasing number of advisors saw current, unrestricted nuclear practices as unsustainable. Harmful fallout from aboveground testing continued to spur large public demonstrations, which damaged the Kennedy administration's reputation both at home and abroad. Also, the lack of restrictions made it easier for Moscow to catch up in the nuclear arms race, something that troubled both nuclear scientists and leaders in the Pentagon. But it was hard for U.S. defense officials

[10] The Soviet Union operated a range of military missions under the deliberately ambiguous Kosmos (Cosmos) nameplate, much as the United States operated its secret Corona flights under the Discoverer moniker.

[11] On the Zenit program, see Brian Harvey, *Russia in Space: The Failed Frontier?* (Chichester, England: Praxis, 2001), pp. 110–11. This system would operate until 1967, when it was replaced by a series of increasingly sophisticated systems: Zenit types 2, 4, 2M, 4M, 4MK, MT, 6, and 8.

[12] "Informal Communication from Chairman Khrushchev to President Kennedy," dated September 4, 1962, reprinted in Thomas Fensch, ed., *The Kennedy-Khrushchev Letters: Top Secret* (The Woodlands, Tex.: New Century, 2001), p. 270.

[13] "Message from President Kennedy to Chairman Khrushchev," dated September 15, 1962, in Fensch, *Kennedy-Krushchev Letters*, p. 272.

[14] For a detailed account of these negotiations and the strategies employed by the two sides, see Nancy W. Gallagher, *The Politics of Verification* (Baltimore: Johns Hopkins University Press, 1999).

to close the door on nuclear options that some felt might be needed. No history of such agreements with communist leaders existed, and many distrusted Soviet motives.[15] Thus, despite the seemingly positive direction of U.S.-Soviet back-channel communications, progress in the official negotiations remained elusive.

Competing Visions of Space

Contradictions in U.S. space policy too had become clear by the fall. President Kennedy and a number of his top advisors had obvious doubts about the sagacity of further, unrestricted nuclear testing in space. Yet there was significant institutional momentum behind continued tests. The relevant organizations—the Pentagon and the Atomic Energy Commission (AEC)—pressed ahead. But, in their first compromise, they now accepted the notion of "staging" the tests relative to other space launches to prevent EMP radiation from harming U.S. (and Soviet) manned missions and from damaging valuable military and civilian satellites. To sort out these issues, President Kennedy held a classified discussion with a group of senior administration officials that included Secretary of Defense Robert McNamara, Secretary of State Dean Rusk, AEC Chairman Glenn Seaborg, National Security Advisor McGeorge Bundy, Science Advisor Jerome Wiesner, and NASA Chief Administrator James Webb on September 5, 1962, focusing on space nuclear testing and the planned Dominic series.[16] A NASA flight with astronaut Walter Schirra scheduled for the early fall dominated and complicated their deliberations. At the risk of calling attention to continued U.S. nuclear testing in space—which could be politically damaging, given the international politics of the test-ban negotiations—should they postpone Schirra's flight to conduct what many believed were critical nuclear tests? Or should they instead shorten the test program itself, at the risk of losing valuable military information?

After a discussion of broader issues related to the test-ban treaty, Kennedy made it clear to his top military and scientific advisors that progress in the manned space program—a key *political* commitment for Kennedy—should have priority over when or even whether specific nuclear tests in space should

[15] On opposition to a test ban and fear of Soviet cheating, see Peter Goodchild, *The Real Dr. Strangelove: Edward Teller* (Cambridge, Mass.: Harvard University Press, 2004), pp. 301, 305.

[16] This section is based on declassified transcripts of this critical meeting. See the University of Virginia's Miller Center of Public Policy, Presidential Recordings Project, "Meeting on the DOMINIC Nuclear Test Series," September 5, 1962, available in PDF as published in Timothy Naftali and Philip Zelikow, *John F. Kennedy: The Great Crises*, Volume 2, September–October 21, 1962, pp. 81–100, on the center's Web site at <http://www.whitehousetapes.org/transcripts/jfk_2_pub/09_sept05.pdf> (accessed October 2, 2006).

take place. Apparently, the failure of several space nuclear tests in the summer, the damage the July 1962 Starfish Prime shot had caused to a number of key U.S. and foreign satellites, and Kennedy's ambitious Moon effort had caused the president to change his mind about the sagacity of continued, unrestricted nuclear testing in space. This direction of thinking began to gain strength in internal U.S. debates on a test ban. Early in the fateful meeting on September 5, Kennedy took the initiative to lay out new, more restrictive parameters for nuclear testing in orbit. As he proposed to his top space and national security advisors, "Let's just set it as our policy that we will not [conduct] any tests that raise any reasonable prospect of interfering until Schirra goes. And then let's try to decide which of these [space] tests we can throw out. We don't want to do them all, if we can help it."[17]

Moreover, Kennedy worried about the possible implications on Soviet behavior and the question of U.S. credibility. Some members of the group, along with some influential scientists and members of Congress, continued to support the original space nuclear test plan designated by the AEC and the military. But Kennedy pointed out the dilemma of U.S. hypocrisy on EMP effects, saying, "There's not much use [in] our going to the Russians and telling them about the problem of electrons [in low-Earth orbit] and then going ahead and doing it ourselves and adding more electrons."[18] Secretary of State Dean Rusk agreed. Reflecting on the Starfish Prime test, he added to the skeptics in the room, "We know that some of these shots are creating a problem for us in space."[19]

Kennedy next turned the discussion to Urraca—a planned AEC space test to be exploded at a very high altitude (800 miles). Scientists had already predicted that there could be significant disturbance of the Van Allen belts given its altitude. Despite assurances from top government scientists to the contrary and AEC Chairman Seaborg's argument that the bomb's force had been reduced from 165 to 10 kilotons in yield, Kennedy seemed distrustful given the damage to satellites already caused by the lower-altitude Starfish Prime. Kennedy was clearly concerned about the radiation issue. After a pause, he confronted Seaborg directly about the Urraca test:

> Kennedy: What are you going to try to find out from this test?
> Seaborg: How to . . . ourselves . . . test if it becomes desirable in space. And to make the diagnosis from those tests that would be necessary for weapons development and how to ascertain whether the other fellow is testing.[20]

17 Ibid., p. 39.
18 Ibid., p. 101.
19 Ibid., p. 102.
20 Ibid., p. 98.

These declassified tapes show that Kennedy now realized that Urraca was an "insurance policy" meant to pave the way for possible U.S. cheating on the test-ban treaty. Despite his hawkish politics, his recent experience had changed his willingness to go along with what he considered to be vague arguments about military or scientific "necessity." With this news, he challenged the rest of the group, saying, "Well, I think what we ought to do is go around the room and everybody throw in what test they would give up, if they had to."[21] When Science Advisor Jerome Wiesner mentioned both the Thumbelina (an air-dropped bomb) and Urraca tests, Kennedy seized on Urraca and pushed the others in the group. NSC advisor Carl Kaysen also said he would drop Urraca. By the end of the discussion, Kennedy had built a consensus that Urraca was unnecessary.

Notwithstanding, the group continued to voice support for the planned Bluegill and Kingfish tests, which were seen as critical to verifying U.S. ABM technology and to acquiring further information about the effects of EMP radiation on related tracking radars and communications systems. Kennedy supported these tests too, as long as they did not interfere with astronaut Schirra's planned orbital flight.[22]

The importance of the September 5 space meeting for subsequent policy became clear two days later, when the National Security Council officially canceled the Urraca test. On October 3, 1962, Schirra successfully completed a six-orbit flight around the Earth and returned safely home. Kennedy again began to press Khrushchev for progress on a nuclear test ban, hoping to cement the partial test ban without committing himself to a broader moratorium. In a secret communication with Khrushchev on October 8, 1962, President Kennedy wrote, "It is nuclear explosions in the atmosphere, in outer space and under water which increase the percentage of fall-out and are the cause of grave concern to all peoples of the world. Such a treaty, in short, would be in itself a great step forward, both for humanitarian and political reasons."[23] But the message brought no change in the official Soviet position. With the Schirra mission out of the way and with pressure from a British-proposed moratorium on space and atmospheric nuclear tests scheduled to start on January 1, 1963, Kennedy decided to move ahead with plans to conduct a series of high-altitude and space nuclear tests in late October. Various defense department and AEC officials wanted to get in as many tests as possible before facing possibly in-

[21] Ibid., p. 99.

[22] According to Schirra's own memoirs, the radiation from the July tests had taken over a month to dissipate to a safe level. Thus, a September test would likely have delayed his flight. See Wally Schirra (with Richard N. Billings), *Schirra's Space* (Annapolis, Md.: Naval Institute Press, 1988), p. 82.

[23] "Message from President Kennedy to Chairman Khrushchev," dated October 8, 1962, reprinted in Fensch, *Kennedy-Khrushchev Letters*, p. 297.

creased international condemnation after January 1. Nevertheless, these discussions show that the seed for a norm of nuclear restraint in space had begun to germinate at the highest levels of the U.S. government.

The Cuban Missile Crisis

In the meantime, the Cuban Missile Crisis intervened. The Soviet decision to station medium-range nuclear-tipped missiles on newly communist Cuba posed a clear threat to the United States. For Soviet First Secretary Khrushchev, however, the deployment was meant to offset U.S. missiles in Turkey and U.S. bomber bases in Western Europe, as well as to bolster his credentials within the world communist movement for standing up in the defense of Cuban independence against repeated U.S. attempts to overthrow Fidel Castro.

On October 16, after President Kennedy learned of the Soviet missile deployments, tensions escalated markedly between the two sides, leading many in the administration to believe that a U.S. attack on the Cuban sites was inevitable.[24] The only question was whether a Soviet response would lead to all-out war or not. The United States had several advantages, however. As George and Meredith Friedman argue, "As a result of the Discoverer [Corona] photos, Kennedy could take actions that would have been impossible otherwise."[25] The reason was the "quality of his intelligence."[26] Thanks to prior overflights of Soviet territory, for example, Kennedy's advisors immediately knew what they were looking at in Cuba: an intermediate-range ballistic missile installation. The layout of the launchers and support buildings on the island mirrored exactly already discovered sites in the Soviet Union.[27] The administration quickly instituted a naval blockade with the aim of halting missiles en route and then negotiating for the removal of equipment already on the island. Kennedy also knew that Moscow was not in a position to launch direct strikes on U.S. territory from Soviet soil, giving the United States the ability to press its local military advantage.

---

[24] This account of the Cuban Missile Crisis and the timing of U.S. actions is based on declassified White House transcripts from the Kennedy administration's internal policy debates and discussions. See the University of Virginia's Miller Center of Public Policy, Presidential Recordings Project, White House meetings from October 22–28, 1962. These materials are now available in PDF as published in Philip Zelikow and Ernest May, eds., *John F. Kennedy: The Great Crises*, Volume III, October 22–28, 1962, pp. 1–101, on the center's Web site at <http://www.whitehousetapes.org/pages/trans_jfk3.htm> (accessed October 2, 2006).

[25] George Friedman and Meredith Friedman, *The Future of War: Power, Technology and American World Dominance in the Twenty-First Century* (New York: St. Martin's Griffin, 1996), p. 310.

[26] Ibid.

[27] Day, Logsdon, and Latell, *Eye in the Sky*, p. 226.

But with the administration's intense focus on the Cuban crisis, Kennedy and other leaders failed to rein in other parts of the government, creating a series of dangerous incidents that might have led to an inadvertent U.S.-Soviet nuclear war. Khrushchev proved no more adept at controlling the many tentacles of Soviet military and nuclear activity.[28] The United States tested an unarmed Atlas intercontinental ballistic missile (ICBM) from Vandenberg Air Force Base toward the Kwajalein test site in the Pacific (and the Soviet Union) on October 26, despite the fact that other Atlas ICBMs at the same site had been loaded with nuclear weapons in case they were needed as auxiliary delivery systems if war erupted.[29] Even more seriously, institutional inertia caused both the United States and the Soviet Union to finish planned tests of nuclear weapons at high altitude and in space. Any of these tests might have been mistaken for an attack by the other side.

While the test programs had been long planned, they had not been coordinated with the other side. Given the proven threat of radiation for manned spaceflight, the two nuclear infrastructures were trying to exploit a window in each other's manned flight schedule (which had been announced) to conduct nuclear tests in space. Although not originally planned to take place during the worst strategic crisis of the Cold War, neither did that crisis stop the tests. Responsible officials on both sides likely felt rushed to test and demonstrate the capabilities of their ABM systems in case war did break out. Space would then become the conflict's first battlefield.[30]

Following a U.S. nuclear test in space on October 20, Moscow exploded a 300-kiloton nuclear weapon above Kazakhstan at an altitude of 93 miles on October 22, 1962. By the end of the day in Washington, the Joint Chiefs of Staff had raised the U.S. nuclear alert status to Defense Condition (DEFCON) 3.

[28] For example, routine U.S. reconnaissance flights continued near Soviet airspace along the Bering Strait, despite the risk that they might be taken as the first wave of a nuclear attack, as some of these aircraft had taken on nuclear missions during the crisis. One such plane—in a test mission on August 23—had strayed inadvertently over Soviet territory and had been nervously guided back to U.S. airspace by Soviet pilots. On the so-called lost bomber incident, see Scott D. Sagan, *The Limits of Safety: Organizations, Accidents, and Nuclear Weapons* (Princeton, N.J.: Princeton University Press, 1993), pp. 73–77.

[29] Ibid., pp. 78–80.

[30] On October 20, 1962, with final planning in place for the naval blockade of Cuba (whose consequences were still far from uncertain), the United States began this dangerous game by conducting a low-yield nuclear test (Checkmate) from Johnston Island in the Pacific at an altitude of 91 miles. Early on October 26, the United States went forward with a second high-altitude nuclear test, exploding so-called Bluegill Triple Prime—with a force of 400 kilotons—at an altitude of 30 miles. On the U.S. nuclear tests from October to November 1962, see the Web site of the Nuclear Weapon Archive at <http://nuclearweaponarchive.org/Usa/Tests/Dominic.html> (accessed October 2, 2006).

Following a U.S. high-altitude shot on October 26, the Soviets conducted another 300-kiloton nuclear test in space at an altitude of 186 miles on October 28 to gauge ABM readiness. However, by evening Moscow time, Khrushchev had broadcast on Moscow Radio a clear statement accepting President Kennedy's public terms for a settlement to the crisis: withdrawal of Soviet forces for a U.S. commitment not to invade Cuba. A secret agreement stipulated the eventual U.S. withdrawal of missiles from Turkey as well. As one account summarizes the impact of space on the Cuban crisis, "Corona proved its worth as a strategic weapon barely two years after being launched by providing data that stabilized a near-war situation and prevented the United States from stumbling into war through miscalculation."[31]

As the crisis wound down, the two sides fired their last nuclear shots toward space, like exclamation points at the end of a fireworks display. On November 1, the United States conducted its third, high-altitude test of the Kingfish series with a 400-kiloton blast at 60 miles in altitude. The Soviet military responded the same day with a test of a 300-kiloton bomb at around 30 miles in altitude. Finally, on November 4, the United States closed its high-altitude test series with a 6-kiloton nuclear explosion at 13 miles in altitude.[32]

New Efforts to Achieve Space Stability

The shock of the Cuban Missile Crisis had a sobering impact on the conduct of U.S.-Soviet relations. Although it initially increased hostility, it later increased the search by the two adversaries for common ground, including in space. Kennedy became supremely interested in constructing a new relationship with the Soviet Union, even a détente. Civilian space cooperation, however, was not named by the president as an area for special progress, and the new quest for space stability did not equate with a desire to abandon competition altogether. The goal was a "standoffish" détente in which each country would tend its own garden, but in which new ground rules would be established for activities beyond one's own fence. This required first of all a new level of control over weapons and weapons testing. Satellite verification would be critical in this regard. AEC Chairman Seaborg recalls that correspondence by both leaders to one another after the events of October turned to the issue of arms control.[33] In both Washington and Moscow, there was a rethinking of past policies. As

---

[31] Friedman and Friedman, *Future of War*, p. 311.

[32] On the Dominic test series, see the Web site of the Radiochemistry Society at <http://www.radiochemistry.org/history/nuke_tests/dominic/index.html> (accessed October 2, 2006).

[33] See Glenn T. Seaborg, *Kennedy, Khrushchev, and the Test Ban* (Berkeley: University of California Press, 1981), p. 176.

William Schauer summarizes, "Both countries retreated a few paces from their previously competitive stances, paving the way for more fundamental agreements which took place during 1963."[34]

As a first step, Kennedy undertook a more concerted effort to pursue U.S.-Soviet cooperation in space science, particularly in unmanned missions. This area, administration officials believed, would be the least risky for U.S. security interests and could help test the waters. As it turned out, however, the Soviet Union—whose security fears far exceeded those of the Americans—showed little interest in the Kennedy overtures and left the invitation for scientific cooperation unanswered. Politically, Kennedy's new policy made more of a difference in assuring Americans (and much of the world) that the United States was not the aggressor in space.

In December 1962, after months of stalemate, NASA and the Soviet Academy of Sciences signed a relatively weak, interagency agreement that established cooperation in coordinating national programs in meteorology, geomagnetism, and space communications. The document also called for the creation of a twenty-four-hour "cold line" for the exchange of weather data between the two capitals. In addition, NASA and the Soviet academy representatives agreed to publish a joint study on the field of space medicine and biology.[35]

Yet the record of Soviet fulfillment of the terms of even these limited accords proved to be less than satisfactory. As the U.S. Office of Technology Assessment reported, "While some useful data were exchanged . . . the results of the agreements were disappointing."[36] Apparently, the goodwill and common interests of scientists on both sides were not enough to bring the two space programs to cooperate. In space science, there was no overriding need to cooperate and—on the Soviet side—there were still perceived risks to national security.[37]

[34] William H. Schauer, *The Politics of Space: A Comparison of the Soviet and American Programs* (New York: Holmes and Meier, 1976), p. 253.

[35] Office of Technology Assessment, *U.S.-Soviet Cooperation in Space* (Washington, D.C.: U.S. Congress, Office of Technology Assessment, OTA-TM-STI–27, July 1985), p. 19. However, because of editorial disputes, the study was not in fact published until 1975.

[36] The report continues: "Part of this was undoubtedly due to inadequate Soviet technical capabilities for processing data as well as to Soviet intransigence. The meteorological data received by U.S. scientists were late and of poorer quality than had been anticipated; no satellite data were exchanged concerning the magnetosphere; the Soviets received experimental satellite communications but declined to transmit; and the space biology and medicine study was not published until 1975, largely because of delays of up to 2 years in Soviet responses." (See Ibid.)

[37] As Harvey and Ciccoritti note, by mid-1963 there was a new stalemate in cooperative programs in space science: "After a year and a half of churning around, nothing had really happened: the USSR was still going strictly its own way in space for its own purposes and the US as a practical matter was doing the same." See Dodd L. Harvey and Linda C. Ciccoritti, *U.S.-Soviet Cooperation in Space* (Miami, Fla.: University of Miami Press, 1974), p. 105.

Over the course of the next year, however, an increasing congruence of views became apparent in the area of arms control. As the superpowers examined their most dangerous areas of competition, certain common ground became noticeable in the area of space. The reason for this newfound commonality of views related to three factors. First, U.S. and Soviet high-altitude and exo-atmospheric nuclear testing had proven to have dangerous health effects on humans in space. Second, it had proven damaging to satellite communications and photo-reconnaissance systems, particularly those operated by the two militaries. Thus, the use of these weapons in space threatened to strip the two sides of their newfound spy satellites, bringing even military leaders to support the formation of a nuclear "keep out" zone in space. Third, by early 1963, nuclear remote-sensing technology had now been developed to such a stage that government scientists recognized that there would be no need in the future for politically sensitive on-site inspections in order to reach an agreement limiting nuclear testing above ground or in space. Seismic stations could be supplemented by sensors placed on satellites, as planned in the U.S. Vela Hotel program. But space had to be kept safe for such surveillance to be possible. At the same time, technological incentives couldn't immediately overcome political obstacles to a treaty.

Distrust within the two sides' military establishments continued to be seen in decisions to move ahead with risky military test programs in space, including those intended to be used in case of a nuclear war. Seemingly oblivious to the high-level negotiations going on, the United States made its second attempt to conduct its Project West Ford experiment on May 9, 1963. Unlike in 1961, this effort succeeded in launching and dispersing some 480 million copper filaments into a 1,970-mile orbit above the Earth.[38] Unfortunately, once fully distributed, in August, the filaments proved unsuitable to their purpose—bouncing radio waves around the globe in case a highly irradiated nuclear environment disabled U.S. military communications satellites. Worse, as critics had predicted, the test then left hundreds of millions of pieces of space debris in orbit. (Many of these filaments remained in space over forty years later.[39]) But this worst-ever release of debris did not lead to collective action. The pieces were too small

[38] See online chapter from Donald Martin's book *Communications Satellites*, 4th edition, posted on the Web site of the Aerospace Corporation at <http://www.aero.org/publications/martin/martin-4.html> (accessed May 8, 2006).

[39] On the status of the Project West Ford filaments, see "Space Debris" study listed on the Web site of the Carolo-Wilhemina Technical University's Institut fur Luft- und Raumfahrtsysteme (Institute of Aerospace Systems) at <http://www.ilr.ing.tu-bs.de/forschung/raumfahrt/spacedebris/en/index.html> (accessed May 8, 2006). As they de-orbit in the coming years, these speeding needles could pose serious threats to satellites and other spacecraft, due to the difficulty of locating them via existing tracking radars and engaging in timely collision avoidance.

to track individually, and studies of the implications of orbital debris still remained very limited at this time.

Separately, the diplomatic track of U.S. nuclear policy continued as Washington sought to avoid a possible nuclear catastrophe in space and other environments. But politics continued to matter, and progress on a space test ban now became hostage to the broader—and more difficult—effort to settle on a framework for halting tests in the atmosphere, seas, and underground. U.S. and British negotiators spent the spring and early summer of 1963 pressing Moscow on the number of seismic stations needed to verify a comprehensive test ban and, particularly, the frequency of on-site inspections needed to clarify any questions (with the U.S.-U.K. team making gradual compromises to lower numbers). After its previous intransigence, Moscow seemed to be making a serious effort by the spring of 1963 to negotiate on specifics, although its team in Geneva was working to reduce the on-site inspections to only two or three per year, compared to the eight to ten demanded by the two allies.[40]

By June, the two sides had not resolved the matter and real progress toward a treaty hung in the balance. Overall, the Soviets remained convinced—being the weaker power and trying to bluff the West into believing in their greater strength—that they had more to lose from on-site inspections than to gain from them. In a private communication to President Kennedy, Soviet leader Khrushchev charged him with failing to conduct the test-ban negotiations from a "principle of equality" and insisted that "the Soviet Union will not consent to the conducting [sic] of espionage inspections."[41] Finally, Washington and London accepted the inevitable and abandoned plans for a comprehensive nuclear test ban.[42]

Civilian space accomplishments continued. In mid-June 1963, after a number of successful manned missions, the Soviets humiliated the Americans (in the social atmosphere of the time) by launching the first woman into orbit. This competitive "one upping" only strengthened American resolve to beat the Russians in the Moon race. But unlike in previous periods, the rivalry in the civilian realm did not contaminate the environment for military restraint. Washington remained convinced that halting nuclear testing in space continued to make good sense.

[40] See Dokos, *Negotiations for a CTBT, 1958–1994*, p. 42.

[41] "Letter from Chairman Khrushchev to President Kennedy," dated June 8, 1963, reprinted in Fensch, *Kennedy-Khrushchev Letters*, p. 530.

[42] Critics of the comprehensive test ban in the United States believed that seismic ground stations and satellites might not be adequate to verify small Soviet tests. They therefore required politically unacceptable on-site inspections in the USSR. Others believed that on-site measures were unnecessary to a verifiable ban. But the critics won this debate.

In July 1963, therefore, the three powers announced their readiness to go forward with a partial test ban, which would eliminate all nuclear testing in space, the atmosphere, and under water. The final draft of the treaty expressed the desire of the three powers to "put an end to the contamination of man's environment by radioactive substances."[43]

At the signing of the Partial Test Ban Treaty (PTBT) in Moscow, Secretary of State Dean Rusk—acting on Kennedy's instructions—reissued the U.S. offer for a joint Soviet-American Moon venture. Again, Khrushchev did not express any interest in the idea. Finally, in September, Kennedy himself raised the issue at the United Nations, where he emphasized the ease of cooperating in an area that—according to the 1961 U.N. resolution—should not be claimed territorially by any nation. Moscow demurred on the Moon proposal. Some new flexibility in the Soviet position on space, however, emerged in Foreign Minister Andrei Gromyko's speech, which suggested interest in a ban on weapons of mass destruction (WMD) in space. The Soviets here showed a new willingness to delink space arms control from settlements covering other fields.

The U.S. Senate hearings on the ratification of the PTBT took place in the steamy weather of August in Washington, D.C. Indicating his serious commitment to the treaty, Kennedy did his homework and recruited a crack team of experts to testify, including leading scientists, the secretaries of the defense and state departments, and the chairman of the Joint Chiefs of Staff.[44] He also contacted citizens' groups and leading business and religious figures to support the treaty. Secretary of Defense McNamara emphasized that the PTBT—by restricting Soviet nuclear testing—would help the United States maintain its technical lead over Moscow. General Maxwell Taylor, representing the joint chiefs, emphasized the need to stabilize the U.S.-Soviet arms race and the ability of the United States to withdraw from the agreement if circumstances later required testing. While conservative senators hoped that hawkish Air Force General Curtis LeMay might torpedo the agreement, his testimony concluded that the PTBT, overall, served U.S. interests. He resisted goading by a conservative senator who implied that the United States should reject the treaty out of U.S. military self-interest. LeMay: "No, sir; I think I must disagree with you there. We have a broader duty, I think, to the country than just considering military questions. . . . We must consider political factors in the solution of our

---

[43] For the full text of the Partial (or Limited) Test Ban Treaty, see the state department Web site at <http://www.state.gov/t/ac/trt/4797.htm> (accessed May 7, 2006).

[44] For an excellent description of the Senate hearings, see Seaborg, *Kennedy, Khrushchev, and the Test Ban*, pp. 265–74.

military problems, because they are important, and they do have a bearing on our solutions."[45]

But a minority of influential nuclear scientists did oppose the treaty, including Dr. John Foster, the director of Lawrence Livermore National Laboratory. These and other opponents testified in separate hearings held by the Preparedness Subcommittee of the Senate Foreign Relations Committee. Dr. Edward Teller—a leading participant in both the Manhattan Project and the U.S. hydrogen bomb program—supported Foster in coming out strongly against the pact for reasons of national security. As he stated in his memoirs about his opposition to the PTBT, "I testified to the Senate that even a limited test ban would impede our development of active defenses against nuclear weapons. Atmospheric tests were the way that we could best find out how to destroy an approaching nuclear missile. The atmospheric test ban made testing this possibility much more difficult."[46] Besides supporting more U.S. nuclear testing for ABM weapons, Teller also remained convinced that the PTBT contained a verification "loophole" in regard to space, which he believed the Soviet Union would try to exploit.[47] In contrast to the Foreign Relations Committee's report, the Preparedness Subcommittee passed the treaty on to the full Senate with a six to one *negative* recommendation.[48] But Kennedy stood firm and rallied other forces. He pledged budget subsidies to the national weapons laboratories for implementation of the treaty.[49] Also, the opening of the treaty to membership by other countries at the United Nations put pressure on the senators to make good on Kennedy's signature or risk a U.S. loss of prestige internationally.[50] Finally, on September 24, 1963, the Senate voted 80 to 19 to approve the PTBT.

The inclusion of space in the 1963 PTBT served a number of U.S. and Soviet interests and marked the first step since the informal exchanges in August 1962 toward settling problems of space competition through negotiation rather than bluster. This agreement and the reduction of space weapons research that followed marked the beginning of limited learning in U.S. and Soviet space policy. Some leading scientists today point out that the PTBT's significance from an arms control perspective has been overrated, because underground testing proved fully adequate to the development of both sides' nuclear arsenals.[51]

[45] Gen. LeMay quoted in Seaborg, *Kennedy, Khrushchev, and the Test Ban,* p. 271.

[46] Edward Teller, *Memoirs: A Twentieth-Century Journey in Science and Politics* (Cambridge, Mass.: Perseus, 2001), p. 468.

[47] Author's interview with Wolfgang Panofsky, Stanford, California, March 9, 2006.

[48] Bunn, *Arms Control by Committee,* p. 40.

[49] Ibid., p. 41.

[50] Seaborg, *Kennedy, Khrushchev, and the Test Ban,* p. 297.

[51] Author's interview with Wolfgang Panofsky, Stanford, California, March 9, 2006.

However, the PTBT's importance to *space* was critical, as the absence of such a treaty would have prevented the further development of space and a sanctuary from being created for satellite reconnaissance and manned missions. Ironically, civilian space cooperation proved harder to achieve, given the symbolism tied up especially in the two manned programs.

Given these dynamics, Kennedy's September 1963 U.N. proposal regarding a joint U.S.-Soviet Moon mission met with a highly mixed response in Congress. Soviet noninterest and lack of enthusiasm for past bilateral projects combined with a popular desire among Americans to establish U.S. space supremacy made the proposal, by and large, unpopular. Some members argued that Kennedy was now undercutting his own worthy goal of 1961 to beat the Soviets to the Moon. Although leading liberal opinions in the press supported Kennedy's cooperative gestures, those who distrusted the Soviets and opposed cooperation were more politically powerful, enough so to attach a rider to the 1964 NASA appropriations bill that specifically stipulated that no U.S. funds could be spent on a joint U.S.-Soviet manned lunar mission without specific congressional approval.

But support for limitations on the respective military space activities began to increase in the fall. At the United Nations, the two countries that still constituted the world's only space powers reached agreement on a resolution pledging not to orbit any weapons of mass destruction.[52] The formal passage of G.A. Resolution 1884 on October 17, 1963, thus expanded the breadth of banned military space activities beyond testing to forbid the action of placing these weapons in space in the first place.[53] For the United States, its willingness to move forward with restraint was bolstered by the October 1963 launch of the first two Vela Hotel nuclear detection satellites.[54] The constellation expanded over time and provided additional certainty to U.S. military planners against nuclear tests in space. While Soviet progress toward a space-based early-warning capability would have to wait until 1972,[55] Moscow abandoned earlier efforts to protest U.S. satellite overflights, as it recognized the growing advantages of satellite photo-reconnaissance for its own military purposes.[56]

This change in attitude helped lead to additional progress within COPUOS

---

[52] See Arthur M. Schlesinger Jr., *A Thousand Days: John F. Kennedy in the White House* (New York: Houghton Mifflin, 1965), pp. 920–21.

[53] On this resolution, see Gennady Zhukov and Yuri Kolosov, *International Space Law* (New York: Praeger, in cooperation with Novosti Press, 1984), p. 34.

[54] See Jeffrey T. Richelson, *America's Space Sentinels: DSP Satellites and National Security* (Lawrence: University of Kansas Press, 1999), p. 79.

[55] See Nicholas Johnson, *Soviet Military Strategy in Space* (London: Jane's, 1987), p. 70.

[56] On this point, see also John Lewis Gaddis, *The Long Peace: Inquiries into the History of the Cold War* (New York: Oxford University Press, 1987), pp. 203–4.

as the U.N. General Assembly moved toward passage of the Declaration of Legal Principles Governing Activities of States in the Exploration and Use of Outer Space (G.A. Resolution 1962). Although the resolution did not achieve final passage until December 13, 1963, the groundwork had been laid by the more tentative 1961 resolution and, over the past year, by the growing Kennedy-Khrushchev rapprochement. The new measure declared that outer space and the celestial bodies should be "free for exploration and use by all States" and that all space activity must be in "accordance with international law."[57] The document also stipulated that the celestial bodies were not subject to national appropriation. Other clauses held nations liable for any damage caused by their spacecraft, but declared that astronauts themselves were to be treated as "envoys of mankind"—deserving aid in case of emergency and safe return to their home country in case of an accident or off-target landing. Although noting that "States shall be guided by the principle of cooperation and mutual assistance," the resolution stopped short of explicit restrictions on military activities. The agreement signified progress toward a new understanding of space security from that of the previous period.

Unexpected Leadership Changes in Washington and Moscow

But domestic political events soon slowed this process of building a space rapprochement internationally. President Kennedy's assassination on November 23 in Dallas removed a key player from the cooperative formula, leaving in his place Vice President Johnson, who had staked out a reputation as a proponent of at least civilian space competition. Johnson's space policy would be characterized by three trends: (1) a strong commitment to the U.S. manned program; (2) a drive to achieve U.S. leadership in space internationally; but also (3) a reduction of extraneous military space programs, under the sharp eye of Defense Secretary McNamara. More so than the Kennedy administration, Johnson's team decided that nonessential military competition with the Soviets in space should be halted, to keep the country's focus on beating the Soviet Union to the Moon.

The understandings of 1962 and the formal agreements of the following year had put a "floor" of limited cooperation under the U.S.-Soviet space relationship, but they remained limited primarily to the security field. Expanding the

[57] For this and the subsequent quotes, see text of U.N. General Assembly Resolution 1962, "Declaration of Legal Principles Governing Activities of States in the Exploration and Use of Outer Space," December 13, 1963, reprinted in Harvey and Ciccoritti, *U.S.-Soviet Cooperation in Space*, pp. 166–67.

range of bilateral space cooperation into other fields would take increased communication and political interest on both sides. Such trends would not emerge until after 1968. In fact, during his five years as president (November 1963–January 1969), Johnson did not make a single direct, personal overture to the Soviet Union in the area of space cooperation.[58] Johnson already had what he needed: a stable military situation that would allow him to concentrate on beating the Soviets in the civilian sector and reestablishing international recognition and respect of the United States as technological leader.

President Johnson remained true to the pro–civilian space reputation he had established for himself during his late-1950s chairmanship of the Senate space committee. By 1965, NASA's budget would soar to $5.1 billion, now a 74 percent share of total U.S. space spending (civilian and military).[59] Yet Johnson's military space policy emphasized negotiations rather than new weapons. Although thirteen tests of the Program 505 ASAT system had occurred by December 1963, these tests now ended. Nonnuclear proximity tests of the Thor-based Program 437 took place periodically.[60] But in the end, Defense Secretary McNamara decided to deploy only two of these ASAT interceptors and only at Johnston Island, providing a very limited operational capability. Within the top circles of the Pentagon, it was now well understood that "a nuclear explosion in space was just as likely to damage or disable US satellites in its vicinity as was the Soviet target."[61] As one observer noted in a 1964 book, "What the military space advocates want most right now is a manned orbiting space station."[62] Sharing these sentiments, the same author criticized the administration for continuing "to give our civilian space program priority over any military efforts."[63]

Concern for the safety of U.S. space assets had grown considerably because of the achievements of U.S. commercial satellites by this time. After the success of *Telstars 1* and *2*, NASA had launched *Syncom 2* into geostationary orbit (22,300 miles up) in July 1963 to provide the first synchronous video and voice communications not limited by orbital period (as with prior low–Earth orbit satellites).[64] The satellite used solar panels to power its operations for a planned lifespan of a year, which it exceeded.[65] It would be followed by other U.S. sys-

[58] Harvey and Ciccoritti, *U.S.-Soviet Cooperation in Space*, p. 141.

[59] Paul Stares, *The Militarization of Space: U.S. Policy 1945–1984* (Ithaca, N.Y.: Cornell University Press, 1985), Appendix I, Table 1, p. 255.

[60] Stares, *Militarization of Space*, p. 124.

[61] Ibid., p. 96.

[62] Bergaust, *Next Fifty Years in Space*, p. 106.

[63] Ibid., p. 105.

[64] William E. Burrows, *This New Ocean: The Story of the First Space Age* (New York: Random House, 1998), p. 269.

[65] For more on the Syncom satellites, see "Syncom: The World's First Geosynchronous Com-

tems until the first *INTELSAT I* reached orbit in 1965. These spacecraft would rapidly accelerate the speed and frequency of global communications.

Very early in the Johnson administration, senior defense officials also began to question the need for extensive ABM defenses, believing that they would spur the nuclear arms race and lead to crisis instability.[66] Fear of possible Soviet deployments, however, led the uniformed military and defense supporters in Congress to push for continued research on the so-called Nike X program, an ABM concept that combined nuclear-tipped Nike-Zeus missiles with smaller Sprint interceptors in a two-tiered system. The program aimed at eventual coverage of a number of U.S. cities against Soviet attack. China's first nuclear test in 1964 spurred support for Nike X, as its backers believed the radical communist leadership in Beijing could not be deterred by the threat of nuclear retaliation. Secretary of Defense McNamara only reluctantly supported the ABM effort, and would continue to try to convince others of Nike X's limitations and possibly harmful strategic side effects.[67] His guidance during the Johnson administration fostered continued U.S. military restraint toward space.

Instead, McNamara pushed forward on a variety of fronts to improve the U.S. ability to quietly gather information from space. Not all of the programs, however, had achieved the success of Corona. The military's SAMOS program had met with repeated problems in trying to deliver useful radio-transmitted images to ground stations, while the MIDAS early-warning satellites remained plagued with technical difficulties related to false alarms. Without a sophisticated onboard computer—such as would be developed later in the decade—it proved difficult to distinguish explosions, cloud refractions of sunlight, and other infrared heat sources from missiles launches.[68] Still, work to perfect these systems continued behind the scenes of international diplomacy.

---

munications Satellite," on the Web site of the Boeing Corporation at <http://www.boeing.com/defense-space/space/bss/factsheets/376/syncom/syncom.html> (accessed October 31, 2006).

[66] Herbert York credits the idea of limiting defenses to ARPA's Jack Ruina in 1964. Outside the Pentagon, Soviet officials also failed to grasp the logic behind ABM restraint. On these early discussions, which included Track 2 meetings in the context of the Pugwash conferences of U.S. and Soviet scientists, see Herbert F. York, *Making Weapons, Talking Peace: A Physicist's Odyssey from Hiroshima to Geneva* (New York: Basic, 1987), p. 223.

[67] In 1966, McNamara barely succeeded in beating back a concerted attempt by the Joint Chiefs of Staff to deploy the Nike X system. He argued that a large-scale ABM deployment would only encourage further deployment of Soviet offensive forces at a time when the United States was trying to stabilize its nuclear arsenal. See Donald R. Baucom, *The Origins of SDI: 1944–1983* (Lawrence: University of Kansas Press, 1992), p. 28.

[68] See Richelson, *America's Space Sentinels*, p. 18; also, Norman Friedman, *Seapower and Space: From the Dawn of the Missile Age to Net-Centric Warfare* (Annapolis, Md.: Naval Institute Press, 2000), pp. 92–93.

Leadership stability and frequent, direct communications between Kennedy and Khrushchev before, during, and after the Cuban Missile Crisis had contributed to U.S.-Soviet cooperative agreements in 1963. Changes within the Soviet leadership in October 1964, however, added an undesired fluidity to the international environment affecting space relations. The Politburo's deposal of Khrushchev—for damaging the Soviet economy, disrupting the Communist Party structure, and engaging in "harebrained scheming" in foreign policy—led to a period of domestic retrenchment and a shift toward Soviet conservatism. The coalition that formed around new Communist Party First (later General) Secretary Leonid Brezhnev and Chairman of the Council of Ministers Alexei Kosygin moved gradually to isolate the Soviet Union from foreign influences and to redress lagging Soviet military capabilities after Moscow's humiliation at the hands of the United States in the Cuban crisis. In their view, national priorities did not require a withdrawal from existing space cooperation, but neither did they demand a more forthcoming policy.

Despite these political changes, on January 25, 1964, the two sides achieved some limited success in joint tracking and data exchanges from the U.S. *Echo II* meteorological satellite. However, Soviet fulfillment of the terms of this earlier agreement did not lead to anything more in the arena of international cooperation.

New developments in communications satellites led to U.S. efforts to transfer the benefits of this technology to other states. After lengthy negotiations among the nations in the International Telecommunications Union (ITU), the United States and a large number of interested countries agreed in July 1964 to regulations for the formation of the International Telecommunications Consortium (INTELSAT). Over criticisms by some Europeans, the new system gave a majority share in INTELSAT to the U.S.-controlled COMSAT company in exchange for access to the use of U.S. technology. For American businesses, the agreement meant a lion's share of INTELSAT contracts for at least the next several years, since no other country could yet produce proved satellite communications systems and launch vehicles.[69] Internal Soviet Foreign Ministry reports seethed that the United States planned to use COMSAT "for its own political, military, and economic goals."[70]

---

[69] In the 1970s and 1980s, the INTELSAT system would change substantially as the nations of Western Europe fought for a more equal share in the system and more favorable access to INTELSAT contracts. Eventually, the difficulties of managing the system would result in the agreement losing market share to competing systems in other areas of satellite telecommunications.

[70] Yu. Zonov, "Spravka," April 16, 1964, report to the deputy head of the Department of International Organizations, Soviet Ministry of Foreign Affairs, Russian Foreign Ministry Archive, Fond 47, Opis' 10, Poz. 45, Papka 116, Lis. 73–78.

Meanwhile, a report from NASA Chief Administrator Webb to President Johnson presented the case that the United States should adopt a more limited approach toward space cooperation with the Soviets than even that attempted by Kennedy, setting aside any joint lunar projects and working instead toward cooperation only on unmanned missions and data exchanges.[71] Webb stressed the lack of Soviet fulfillment of existing agreements and urged a wait-and-see policy on the president. NASA would continue to pursue such cooperation via the NASA–Soviet academy channel only. For Johnson, this back-burner approach was tantamount to putting the issue aside. Judging from his conduct in the 1964 presidential campaign, the president seemed more comfortable emphasizing the competitive aspects of U.S. space activity, at least to his domestic audience.

The Soviet Space Program (1964–1966)

Within the Soviet program, a collective sigh of relief could be heard with the removal of Khrushchev. Although he had driven the Soviet program to initial success, this mercurial Ukrainian had irritated many top scientists in his desire for spectacular space firsts, regardless of scientific priorities. Korolev, his chief director, struggled considerably under Khrushchev's high-handedness and unwillingness to listen to advice. In the interests of maintaining a civilian, scientific facade surrounding the Soviet program, Khrushchev had refused to give Korolev any public mention (much less credit) for all of the work he and his team had done to promote an image of Soviet power and scientific know-how, passing the popular acclaim instead to Academician Sedov.[72]

Still, while now receiving a greater say in space decision making, the chief designer and his staff continued to face severe pressures to maintain their popularly perceived "lead" over the United States. Some evidence of progress could be seen in a more rationalized organization of the space program and a strengthening of its scientific aspects. In 1965 a new, centralized Institute for Space Research (IKI) emerged within the Academy of Sciences to provide a nexus for civilian space science activities, including cooperative programs with

---

[71] "Report on Possible Projects for Substantive Cooperation with the Soviet Union in the Field of Outer Space," January 31, 1964; cited and partially reprinted in Harvey and Ciccoritti, *U.S.-Soviet Cooperation in Space*, pp. 137–39. Similar to McNamara's line on defense, this report can be viewed as a sort of "flexible response" policy for space cooperation.

[72] On Korolev's understandable bristling over this issue, see account in Leonid Vladimirov, *Sovetskiy Kosmicheskiy Blef* [The Soviet space bluff] (Frankfurt, Germany: Possev-Verlag, 1973), p. 15.

the West (and later with Eastern Europe).[73] The Soviet leadership appointed Academician Georgiy I. Petrov to run the new institute, a position he would hold until 1973.[74] IKI became the scientific center of Soviet civilian space activities, especially in the area of unmanned flight and scientific exploration. However, despite these reforms, the first meeting between Korolev and the new Brezhnev-Kosygin leadership proved disappointing to the chief scientist. His new political handlers informed him that the space program would continue to proceed (as under Khrushchev) not on the basis of *scientific* priorities, but "on the principle of 'any cost to beat the Americans.'"[75] The focus also shifted from civilian to military space developments.

The new Soviet leadership transformed the special state committee Khrushchev had organized in 1961 to manage space and other emerging technology issues into the State Committee on Science and Technology (GKNT) in 1965. As with other Soviet space-related organizations, former military-industrial personnel soon populated its top leadership.[76] The Soviet government began construction of a secret military Soviet launch facility 600 miles north of Moscow in Plesetsk. This higher latitude provided better orbital paths for satellite overflight of North America and quickly became the most active launch facility in the USSR. Nevertheless, for propaganda reasons, the new Soviet leadership refused to admit Plesetsk's existence.

As preparations for the twentieth anniversary of the United Nations took shape in early 1965, Johnson asked NASA Deputy Chief Administrator Dryden to prepare a report outlining possibilities for space cooperation with the Soviet Union. None of the proposals, however, appeared in Johnson's speech in San Francisco (the site of the United Nations' founding). The new official view on cooperation seemed to be best summed up by Dryden in a letter to White House Science Advisor Donald Hornig: "Experience has taught us that the Soviets prefer general discussion and agreements which relieve them of the need to be specific and allow them to gain credit for international cooperation without partaking of its substance."[77]

---

[73] In the 1960s, its staff numbered approximately one thousand.

[74] Siddiqi, *Sputnik and the Soviet Space Challenge*, p. 436. He would be succeeded in this position by Academician Roald Z. Sagdeev, who headed IKI from 1973 to 1988. Albert A. Galeev succeeded Sagdeev as IKI's director in 1988 and still holds the position (as of fall 2006).

[75] Ibid., p. 17.

[76] David Holloway, "Innovation in the Defense Sector," in Ronald Amann and Julian Cooper, eds., *Industrial Innovation in the Soviet Union* (New Haven, Conn.: Yale University Press, 1982).

[77] Dryden letter of January 1965, quoted in Harvey and Ciccoritti, *U.S.-Soviet Cooperation in Space*, p. 146.

A revealing incident exposing the political problems faced by Soviet scientists seeking greater cooperation with their U.S. counterparts took place in May 1965 in Argentina, where a meeting of the Committee on Space Research—a body of the International Council of Scientific Unions (ICSU)—was being held.[78] There, Blagonravov admitted in a private conversation with Dryden that Soviet weather satellite photographs had not been exchanged previously simply because of the embarrassingly poor quality of the images. Furthermore, Blagonravov admitted that the meteorological service in the USSR was not under the control of the Academy of Sciences.[79] Like the rest of the space program at the time, it too fell under the control of the Soviet military.

Apart from these access problems, however, larger geostrategic interests also affected the Soviet position on space cooperation with the United States. With the escalating U.S. military presence in the war in Vietnam, the Soviets felt pushed by world communist opinion not to appear too close to the United States. Thus, a seemingly harmless invitation by Webb for a Soviet academy representative to attend the launch of *Gemini 6* scheduled for the fall of 1965 was turned down by Soviet officials to make a political statement.[80] Moscow also resisted U.S. efforts to move forward with a formal rescue and return agreement for spacefarers who had fallen outside their landing zones and had to be rescued by foreign countries. Internal Soviet documents make clear the strategy of building a self-righteous coalition of states at the United Nations opposed to U.S. insistence for the return of all spacecraft.[81] Soviet interlocutors continued to take the supposed political high road of blocking an agreement that would force the return of military spacecraft and personnel "launched with clearly hostile aims," despite the Soviet Union's own extensive military space programs and use of air force fighter pilots.

More information on the limits under which Soviet academy officials participating in international exchanges worked appeared in early 1966. Responding to Webb's request for Soviet data from their recent Venus probes, Blagonravov wrote in reply that he did not have the necessary authority to release the information and therefore could not comply with the American request.[82] After

[78] This committee had been set up originally during the International Geophysical Year.

[79] Harvey and Ciccoritti, *U.S.-Soviet Cooperation in Space*, pp. 153–54.

[80] Ibid., p. 155.

[81] Yu. Ribakov, "O rabote 4-oi sessii Pravovogo podkomiteta Komiteta OON po kosmosu" [On the work of the fourth session of the Legal Subcommittee of the U.N. Committee on Space], December 7, 1965, report to the head of the Department of International Organizations, Soviet Ministry of Foreign Affairs, Russian Foreign Ministry Archive, Fond 47, Opis' 11, Poz. 31, Papka 124, Lis. 34–54.

[82] Ibid., p. 156.

other overtures in letters that spring, Webb stopped trying, and U.S.-Soviet co-operative programs in space science remained in limbo throughout the next year.[83] The only bright spot in the rest of 1966 was the Soviet transmission of some very poor-quality satellite pictures over the "cold line" in September.

The Emergence of the Outer Space Treaty (1967)

Since 1963, there had been no progress of any sort in the legal subcommittee of COPUOS. Within the U.S. government, however, the Department of State had begun sounding out other branches of the government on the potential for a space treaty.[84] Although both the NSC and the Department of Defense insisted on provisions that would protect American rights to use space for military purposes and to abrogate the treaty if national security demanded it, the state department succeeded in persuading other agencies of the usefulness of the Antarctic Treaty (1959) model, which would rule out the emplacement of weapons on the Moon and planets but not affect passive military use of non-terrestrial regions of space. Beginning in mid-1966, the United States began an international effort to codify in a formal treaty the legal issues previously contained in the 1963 U.N. resolution by having U.S. Representative Arthur Goldberg submit a new U.S. draft treaty to COPUOS and by opening negotiations in Vienna with the Soviet Union. These moves appeared to be inspired partially by the growing intensity of the Moon race and a U.S. desire to keep the Soviets from exploiting any territorial or other advantages should they beat the Americans. The Soviets, motivated by reciprocal concerns and the rapid progress of the U.S. manned program, soon submitted their own remarkably similar version. Since both documents differed little from the 1963 U.N. resolution, negotiations at first proceeded apace.

Chief Soviet negotiator Platon D. Morozov, however, soon stalled the talks by inserting a Soviet claim to equal access to U.S. space tracking facilities in countries such as Kenya and Australia, where Washington had negotiated successfully for the construction of stations.[85] Soviet strategy documents note Moscow's goal of using space to raise the Soviet profile among developing countries in order to promote their global struggle against international capitalism.[86] Again, the prospects for an agreement appeared bleak. Yet, in October

---

[83] After the death of Deputy Chief Administrator Dryden in December 1965, Chief Administrator Webb himself had taken charge of cooperative overtures to the USSR.

[84] Walter A. McDougall, . . . the Heavens and the Earth: A Political History of the Space Age (New York: Basic, 1985), p. 415.

[85] See McDougall, . . . the Heavens and the Earth, pp. 416–17.

[86] V. A. Demen'tev, "Rabochaya gruppa komiteta OON po kosmosu (18–25 Yanvarya 1966 goda):

1966, as other nations rallied to the American side and the futility of Moscow's position became clear, the Soviets finally agreed to drop this demand and enter into serious negotiations.[87]

By the end of the year, all of the main, substantive issues had been settled, save for the questions of specific liability limitations for damage caused by spacecraft and funding for the rescue and return of fallen spacecraft. The various parties agreed, however, that a treaty could be signed covering the issues where a consensus had been achieved. In a unanimous vote in the General Assembly (G.A.) in December, the United Nations adopted G.A. Resolution 2222, the Treaty on Principles Governing the Activities of States in the Exploration and Use of Space, Including the Moon and Other Celestial Bodies. The three principal space-faring nations—the United States, the Soviet Union, and the United Kingdom (also an original party to the PTBT)—signed the treaty in simultaneous ceremonies in Washington, Moscow, and London on January 27, 1967.[88] Among its provisions, the treaty formalized most of the features of the 1963 U.N. resolution, including the ban on territorial claims by any nation on the Moon or other planets and the celestial bodies. But it went further in the military area, drawing on the Antarctic Treaty in stipulating that "the Moon and other celestial bodies shall be used by all States Parties to the Treaty exclusively for peaceful purposes. The establishment of military bases, installations and fortifications, the testing of any type of weapons and the conduct of military manoeuvres on celestial bodies shall be forbidden."[89]

As an enforcement mechanism, the treaty formalized an earlier pledge that all stations, installations, and vehicles on the Moon and celestial bodies "shall

kratkaya spravka" [The working group of the U.N. Committee on Space (18–25 January 1966): a short summary], February 15, 1966, report to the head of the Department of International Organizations, Soviet Ministry of Foreign Affairs, Russian Foreign Ministry Archive, Fond 47, Opis' 12, Poz. 36, Papka 131, Lis. 23–30.

[87] Soviet Embassy report from Washington in late 1966 noted President Johnson's description of the treaty in a Texas press conference as "the most important event in the field of arms control since the 1963 Limited Test Ban Treaty." See A. Kokorev, "Ob otklikakh v SShA na dostizheniye dogovorennosti v OON o zaklyuchenii Dogovora o printsipakh deyatel'nosti gosudarstv po issledovaniyu i ispol'sovaniyu kosmicheskogo prostranstvo, luny i drugikh nebesnykh tel" [On comments in the USA regarding U.N. progress toward an understanding on completion of a Treaty on the principles of activity by states in research on and the use of space, the Moon, and the other celestial bodies], December 19, 1966, report to the Soviet Foreign Ministry [recipient unnamed], Russian Foreign Ministry Archive, Fond 47, Opis' 12, Poz. 31, Papka 136, Lis. 200–203.

[88] After some delay in the ratification process in the U.S. Senate and the Supreme Soviet, the treaty came into effect on October 10, 1967.

[89] See text of the Outer Space Treaty, reprinted in Johnson, *Soviet Military Strategy in Space*, Appendix 2, pp. 236–40.

be open to representatives of other States Parties to the Treaty."[90] Thus, the agreement showed that a new consensus had been reached between Moscow and Washington on the desirability of preventing military confrontation over potential minerals, territory, and strategic assets throughout terrestrial space, not just the Moon. But the agreement fell short of banning all weapons in space by failing to include the area between the celestial bodies, including near-Earth space.[91] Nevertheless, by creating an infinitely large WMD "keep out" zone, it went a long way toward limiting the scale of any potential space conflict, marking a significant expansion of U.S.-Soviet negotiated agreements regarding space.

Despite its clear benefits for space stability, the treaty still required the ratification of the U.S. Senate. This was not a foregone conclusion, compared to the Supreme Soviet's rubber-stamping of the agreement. Conservative senators feared that certain clauses might be interpreted to ban existing U.S. military activities or, because of ambiguous wording, to force the U.S. government to provide Third World countries with access to space.[92] Active lobbying by Ambassador Goldberg and other administration officials (including Secretary of Defense McNamara) succeeded in quelling this potential opposition.[93] On April 25, 1967, the Senate approved the treaty with a resounding vote of 88 to 0. The fact that the treaty had been in the planning stages for several years helped it to gain approval. But in the end, effective lobbying proved important in securing this major step to provide legally binding rules preventing militarization of the Moon, claiming of its territory and resources, and the orbiting of WMD in space.

[90] Again, this feature came from the Antarctic Treaty's inspection mechanism. Of course, in space, the practical aspects of inspection would be much more complicated, as compatible docking mechanisms would have to be constructed, thus requiring some cooperation from the "inspected" country. But these details were pushed off to the future. (See Ibid.)

[91] Two former U.S. government lawyers from the time argue that the Outer Space Treaty implicitly did ban weapons in space through the "peaceful uses" clause and the requirement for consultations before conducting experiments that might cause "harmful interference" with other spacecraft. See George Bunn and John B. Rhinelander, "Outer Space Treaty May Ban Strike Weapons" (Letter to the Editor), *Arms Control Today*, Vol. 32, No. 5 (June 2002), available via the Arms Control Association Web site at <http://www.armscontrol.org/act/2002_06/letterjune02.asp?print> (accessed October 8, 2006).

[92] In Article I, the agreement—drawing on the 1963 U.N. resolution—stated, "The exploration and use of outer space, including the Moon and celestial bodies, shall be carried out for the benefit and in the interests of all countries, irrespective of their degree of economic or scientific development, and shall be the province of all mankind." (See Ibid.)

[93] As McDougall writes of Goldberg's efforts, "He assured the [Senate] committee that the major clauses were not new, but drawn from past UN resolutions." (See McDougall, . . . *the Heavens and the Earth*, p. 418.)

Despite this progress, the treaty's conclusion had failed to settle the ongoing U.S.-Soviet dispute over ABM systems, whose deployment and use could also destabilize space. In his January 1967 State of the Union address, President Johnson had announced his intention to pursue a treaty to ban ABM systems.[94] In addition, Secretary McNamara remained convinced that—regardless of Soviet intentions—this route meant a new stage in the arms race. But Johnson's sense of his political opposition in Congress led him to include $375 million in his budget proposal to deploy an ABM network in case the Soviets rejected these negotiations and continued to move forward with their Galosh antimissile system around Moscow (and possibly other ABM deployments).

Johnson, the Pentagon, and the Apollo Program

Throughout the mid-1960s, the U.S. space program was preoccupied with one overriding goal: reaching the Moon. NASA continued to receive the truly massive budgets it required to meet Kennedy's difficult goal, thanks to enduring congressional support and willingness to face down competing demands in the federal budget. By 1965, for example, NASA's outlays surpassed the defense department's space budget by over three times.[95]

With the growth of NASA under Johnson, the U.S. military felt under increasing pressure to cement a role for itself in the U.S. manned space program. Under Kennedy, Air Force General Curtis LeMay had lobbied hard for the Dyna-Soar manned space bomber (X-20). However, there were few objective military requirements for the Dyna-Soar and considerable technical problems facing its development. In the first stages of a battle against his own uniformed personnel that lasted throughout McNamara's tenure, Johnson weighed in on the side of his defense secretary and, in December 1963, cancelled the program. Concerned with Pentagon waste in a variety of other military space programs, McNamara succeeded over the course of 1964 in halting a number of other redundant and destabilizing programs, including SAINT. Politically, Johnson withstood tirades from congressional conservatives (including Republican presidential candidate Senator Barry Goldwater) for these defense cuts. As John Lewis Gaddis observes regarding evolving trends in U.S. ASAT development, "The capabilities of these systems were severely limited . . . : there was never any more than two launch sites, and only a handful of launchers were actually produced. The pattern of the early 1960's was one, then, of refraining from de-

---

[94] Baucom, *Origins of SDI*, p. 32.
[95] Stares, *Militarization of Space*, Appendix 1, Table 1, p. 255.

veloping the full-scale anti-satellite systems that could have been put in place by that time."[96]

After Johnson's reelection, the uniformed military continued to fight for a role in space. To quell a possible mutiny, Secretary McNamara extended funding for the air force's Blue Gemini project, despite his doubts about its importance. According to plans approved in August 1965, the Blue Gemini missions now called for a Manned Orbiting Laboratory (MOL) to be launched and occupied in 1968 by air force personnel, who would be rotated in and out on two-week Earth observation tours by Gemini spacecraft. For a while, there was peace among the air force, McNamara, and NASA.

As human spaceflight became the main focus of U.S. space spending and attention under Johnson, scientists for the first time began to focus greater attention on the possibility of impacts between spacecraft and man-made orbital debris. According to space physics, such hazards—although still few in number—could represent deadly obstacles to reaching the Moon. As it prepared for the Apollo missions in 1966, NASA began to model and catalogue space objects, although calculations at this point still indicated a relatively low probability of collisions with manned spacecraft.[97] But the issue had now been added to the figurative and literal radar screens of U.S. government scientists.

Meanwhile, despite competing NASA and defense department demands, Johnson continued to fund the increasingly expensive NERVA nuclear-powered rocket program, showing his commitment to expand the space race beyond the Moon. Space enthusiasts, including Walter Dornberger, believed that such engines would make interplanetary travel possible and even routine within fifty years.[98] Eventually, the AEC would conduct ground-based tests of twelve reactors at Jackass Flats in Nevada in anticipation of a manned mission to Mars planned for sometime in the 1970s.[99]

But these boom years for space could not go on forever. In domestic politics, Johnson's Great Society programs in particular had caused social spending to soar. Even in peacetime, the simultaneous drains on the federal budget of space

[96] Gaddis, *Long Peace*, p. 202.

[97] Nicholas L. Johnson and Darren S. McKnight, *Artificial Space Debris* (Malabar, Fla.: Orbit, 1987), p. 69.

[98] Walter R. Dornberger, "Foreword," in Bergaust, *Next Fifty Years in Space*, p. xiv. Bergaust's own predictions foresaw even more widespread use of nuclear power in space. He also saw such capabilities emerging in the Soviet Union, observing regarding apparent Russian interest in similar technology, "The race for atomic rockets seems to be in high gear" (p. 199).

[99] John W. Simpson, *Nuclear Power from Underseas to Outer Space* (La Grange Park, Ill.: American Nuclear Society, 1995), p. 147.

programs and expansive new social programs were bound to cause eventual fiscal strain. With the growing cost of the Vietnam War, which had by 1968 seen the U.S. presence on the ground jump to a force of over half a million men, cuts had to be made somewhere.[100]

In the decision making for planned launches in the early 1970s and beyond, the dual pressures from liberals favoring more social spending and conservatives favoring more military spending began to catch NASA in a scissors. As scientists and engineers worked to achieve NASA's greatest ever accomplishments, the administration's leaders were already under pressure from Congress to cut back on unnecessary programs. Fiscal 1967 brought the first major cut in NASA's budget since the organization's inception, as Congress and the president together trimmed Webb's original $5.58 billion request by a half a billion dollars.

The expansion of manned missions to prepare for the coming Moon shot did not come without a human cost, underlining the dangers in space exploration that had been largely hidden from public view. On January 27, 1967, the United States suffered its worst disaster up to that time when a fire broke out in the *Apollo I* capsule as it sat on its launch pad during a routine inspection and simulation.[101] In the all-oxygen atmosphere of the sealed capsule, a spark ignited and caused astronauts Gus Grissom, Ed White, and Roger Chaffee to suffocate as they struggled unsuccessfully to escape the flames. Although an investigation failed to identify the specific cause of the blaze, most signs pointed to faulty wiring and earlier NASA decisions not to go with the more complex requirements of operating a mixed nitrogen-oxygen environment. The incident also highlighted other problems, such as a manually operated escape hatch that had proved too slow to save the trapped astronauts. The fire set back the lunar program by over a year and a half as NASA worked to redesign the Apollo capsule. After a series of unmanned Apollo tests, *Apollo 7* finally took three astronauts (Walter Schirra, Don F. Eisele, and R. Walter Cunningham) into space on October 11, 1968. An accelerated flight schedule followed as NASA worked to make up for lost time, hoping that the Soviets would not surpass their efforts and reach the Moon first.

Soviet Struggles in Space (1966–1968)

Despite Soviet attempts to adapt organizationally to the more challenging space environment after Kennedy's initiation of the Moon race in 1961, Moscow

[100] As Manno notes, "America's involvement in Vietnam ultimately affected all aspects of US space activity" (Manno, *Arming the Heavens*, p. 120).

[101] On the accident, see Burrows, *This New Ocean*, pp. 405–12.

found itself by the mid-1960s slipping behind the massive U.S. effort in manned spaceflight.[102] Domestically, the sudden death of Korolev in 1966[103] and serious problems with the Soviet heavy-lift N-1 booster in 1967 and 1968[104] led to a slowdown in the Moon effort and a recalibration of technical considerations.[105] Korolev's absence, in particular, created a leadership vacuum at the very time when the Soviet space program needed critical resources, top leadership support, and skilled management. His successor (Vasiliy Mishin) proved unable to meet these challenges in the face of a full-court U.S. press. In April 1967, the Soviets received a brutal reminder of the dangers of haste in space when their new *Soyuz I* failed to deploy its three-canopied parachute fully after re-entry, killing veteran cosmonaut Vladimir Komarov as his spacecraft hit the ground at high speed. This tragic event had a sobering effect on both Soviet space officials and the Kremlin. No Soviet cosmonauts would fly again until 1969.

The 1967 U.S. and Soviet space accidents served as a trigger event for a major new cooperative initiative, just as the combination of the Starfish Prime nuclear test and the subsequent Cuban crisis had helped spur earlier efforts. In late

[102] Although there are no hard budget figures available for the Soviet space program in the 1960s, the consensus of government and academic analysts who have studied this issue is that the United States far outspent the Soviet Union in space throughout the decade. William H. Schauer (*Politics of Space*, p. 166) suggests that one reason the Soviets lost the Moon race was budgetary. He presents the case that the beginning of the Moon race in 1961 caught them in the midst of Khrushchev's ambitious seven-year plan (1959–65) in which the government had committed itself to major expenditures toward improving other areas of the domestic economy. Soviet accounts support this view but also point to internal problems, such as the loss of Korolev and some mistaken technical decisions.

[103] According to James Harford (who interviewed Korolev's daughter in 1991), Korolev went in after the January 1 holidays for what seemed to be a relatively simple lower-bowel operation for polyp removal. But the surgeon (a high-ranking Soviet medical official, long out of practice) detected cancer of the colon. During his attempts to cut out the cancer, Korolev died on the table from blood loss. See James Harford, *Korolev: How One Man Masterminded the Soviet Drive to Beat America to the Moon* (New York: Wiley, 1997), pp. 277–81; also Burrows, *This New Ocean*, pp. 402–5.

[104] On these problems, see Asif A. Siddiqi, *The Soviet Space Race with Apollo* (Gainesville: University of Florida Press, 2003), pp. 646–50; also, M. Rebrov, "A delo bylo tak: sovershenno sekretno; Trudnaya sud'ba proekta N-1" [This is the way it was: top secret; the problematic fate of the N-1 project], *Krasnaya Zvezda*, January 13, 1990, p. 4.

[105] At the time (and even as recently as the mid-1980s) there was considerable debate in the West as to whether the Soviets were actually trying to beat the United States to the Moon, in part because of later Soviet denials of any intentions. But later revelations in the Gorbachev-era press, personal interviews by the author with former Soviet space officials, and recent accounts—such as Harford's *Korolev* and Siddiqi's *Soviet Space Race with Apollo*—confirm that the Soviets *were* involved in a Moon race. Among other evidence, former IKI Director Roald Sagdeev said in a speech regarding Chief Designer Korolev's intentions, "He was extremely eager to send a man to the Moon" ("*Perestroika* and the Academic Community," Regents Lecture at the University of California, Berkeley, July 20, 1989).

1967 and early 1968, the two sides began to move toward an agreement on providing aid for spacefarers in distress and on guaranteeing the return of fallen spacecraft. The Soviet side now dropped its prior insistence that Moscow be able to deny the return of "hostile" spacecraft and astronauts (i.e., those serving military or intelligence functions) and began serious negotiation toward a final document. The eventual Agreement on the Rescue and Return of Astronauts and Space Objects was opened for signature at the United Nations in April 1968 and went into effect that December.

But the Brezhnev-Kosygin leadership had not abandoned space competition. Instead, stung by growing problems in their now less-successful "civilian" program, it began to shift funds from manned and space scientific activities to new military programs.[106] The Soviets feared that U.S. efforts to deploy missile defenses (such as the Nike-X) could be combined with the continued U.S. lead in strategic nuclear forces to bully Moscow in a future nuclear crisis. Some speculated that Washington might believe it could credibly carry out a disarming first strike and clean up remaining Soviet forces with ABM interceptors. The United States had no program, however, to defend against strikes coming from the south. This led Soviet military planners to begin tests of a fractional orbital bombardment system (FOBS) in 1967.[107] The plan called for missiles armed with nuclear weapons to be launched from Baikonur in a southeasterly direction (toward Antarctica) and then approach the continental United States from the southeastern Pacific, finally crossing into U.S. airspace from Mexico. Accuracy would suffer considerably and the Soviets had no capacity for multiple launches from Baikonur, but the test program had the strategic advantage of keeping Americans from excessive nuclear confidence. Nine nonnuclear FOBS tests occurred in 1967, although none of them using a wartime trajectory or flying toward the United States. Fearful American defense officials speculated that the Soviets might decide to station such nuclear weapons in orbit in case of war, but this never occurred. Apart from direct launch costs and maintenance, such actions would have violated the Outer Space Treaty, thereby opening the door both to negative international publicity and to possible reciprocal U.S. violations.

A second Soviet space-related military program emerged in 1968 in the area of ASAT technology, using a plan similar to the U.S. SAINT program. The So-

[106] On these budget cuts in space science, see Victor Yevsikov, *Re-Entry Technology and the Soviet Space Program (Some Personal Observations)* (Falls Church, Va.: Monograph Series on the Soviet Union, Delphic Associates, December 1982).

[107] For a detailed account of the Soviet FOBS program, see Johnson, *Soviet Military Strategy in Space*, esp. pp. 131–48.

viet tests—which had mixed success—involved sending a satellite packed with conventional explosives into an orbit close to a target satellite and then maneuvering it into a position (using a homing device) where it could explode a fragmentation warhead. According to a recent Russian history of the program, "The antisatellite system included a ground command center in the Moscow region, a space interceptor control station, and six launchers in Baykonur."[108] By 1971, the Soviets developed the capability to reach targets in orbits ranging from 250 to 1,000 kilometers,[109] although in only one case during the 1968–71 test program did the encounter damage the target satellite, emitting four detectable fragments. The rest of the 490 pieces of trackable debris from these seven tests came from the breakup of the interceptors following their release of their single fragmentation warheads,[110] which apparently were not detonated but tested only for proximity. But the nontrackable debris was undoubtedly much greater. Like the U.S. Program 437 system, the Soviet co-orbital ASAT had limited capacity for handling multiple targets and therefore provided little clear military advantage. Why did the program move forward? As York says, based on interviews with Soviet scientists, it was self-interested support by those who designed the system: "The Soviet ASAT was less the result of an action-reaction mechanism, or a calculation of a military need, than a case of an interesting technology sold to a gullible and poorly informed leadership by aggressive salesmen."[111]

In response to President Johnson's and Secretary McNamara's comments on the desirability of ABM limitations, the Soviets began to consider further limits on space weapons. ABM defenses became a central theme during the Glassboro, New Jersey, summit held with Soviet Premier Kosygin in June 1967. But Kosygin's likely uncertainty about the exact status of the Soviet program and reflexive opposition to specific limitations prompted his impassioned exclamation to the Americans: "Defense is moral! Offense is immoral!"[112] Thus, while the topic was broached and considered, little progress came out of the summit. Given his own pledge to Congress, President Johnson now had no choice but to put in motion a U.S. ABM deployment. In November, Secretary McNamara gave a long speech in which he initially downplayed the utility of ABM systems

---

[108] Pavel Podvig, ed., *Russian Strategic Nuclear Forces* (Cambridge, Mass.: MIT Press, 2001), p. 433. See also Johnson, *Soviet Military Strategy in Space*, pp. 148–57.

[109] Podvig, *Russian Strategic Nuclear Forces*, p. 434; Johnson and McNight, *Artificial Space Debris*, p. 15.

[110] Johnson and McKnight, *Artificial Space Debris*, p. 15.

[111] York, *Making Weapons, Talking Peace*, p. 275.

[112] Kosygin, quoted in Baucom, *Origins of SDI*, p. 34.

for use against the Soviet Union but finally called for a limited defensive system for U.S. cities against a possible Chinese attack, given China's recent test of its first hydrogen bomb.[113] But the system, now named Sentinel, would not be scheduled for deployment until the early 1970s, thus providing time for further U.S.-Soviet negotiations.

By the late fall, the Soviet Union's own efforts to build ABM defenses had hit a turning point. Although extensive research had been conducted at the Sary-Shagan test site in Soviet Central Asia, the results showed that the nuclear-tipped system had fundamental flaws and would be unable to intercept multiple warhead missiles (which Moscow rightly believed the U.S. military planned to deploy).[114] In late 1967, the Soviets decided against a full-scale deployment of the Galosh (A-35) system. In 1968, a government commission amended the criteria for the system to defend against limited or single nuclear strikes, thus allowing the program to proceed with a scaled-down deployment of sixty-four interceptors (instead of several hundred).[115] The shift presaged a growing Soviet realization that foreseeable ABM systems would have limited value relative to emerging offensive weapons, making movement toward an agreement with the United States on ABM technologies possible.[116]

Finally, Moscow had achieved some noteworthy successes in the military support arena, after lagging well behind the United States since the first few years of the space age. In late 1967, the Soviets had launched a radar satellite (*Kosmos 198*) able to provide information on the movements of the U.S. Navy,[117] a critical gap in Moscow's prior ability to track U.S. forces that might threaten its territory. The Soviet military had also developed and launched its first electronic intelligence (Tselina) satellites into low–Earth orbit.[118] It gradually built a constellation of military spacecraft capable of identifying U.S. radar sources to improve missile penetrability and intercepting communications (especially from U.S. ships).[119] In the area of reconnaissance, in contrast to the U.S. preference for increasingly sophisticated and longer-lived satellites, the Soviets continued to use cheaper, short-mission spacecraft that deorbited and parachuted their cameras and film together to Earth, thus allowing the cameras to be reused.

[113] Ibid., p. 37.

[114] Podvig, *Russian Strategic Nuclear Forces*, p. 414.

[115] Ibid., p. 415; see also Baucom, *Origins of SDI*, p. 30.

[116] Podvig, *Russian Strategic Nuclear Forces*, p. 415.

[117] Stares, *Militarization of Space*, p. 141.

[118] Johnson, *Soviet Military Strategy in Space*, p. 60.

[119] On these capabilities of the various Tselina versions, see Friedman, *Seapower and Space*, p. 60.

Nixon, Détente, and Expanded Space Cooperation

With Johnson's withdrawal from the 1968 presidential race over his frustrations with the Vietnam War[120] and with Americans eager for new ideas to address this seemingly endless and increasingly costly conflict, former Republican Vice President Richard Nixon emerged as the nation's choice to lead the country. Nixon brought with him a new agenda for U.S.-Soviet relations. At the advice of his National Security Advisor Dr. Henry Kissinger, Nixon proposed an easing of tensions with Moscow. Kissinger argued that approximate nuclear parity with the Soviet Union was now inevitable and that Washington should seek Moscow's assistance in managing other global problems, such as pressuring North Vietnam to end its support for the war in the South. But the Soviet invasion of Czechoslovakia in the summer of 1968 and Moscow's brutal crackdown on dissent raised significant new political difficulties for near-term negotiations. In the meantime, the Nixon administration would begin to establish new priorities for its planned defense and space policies. Although a realist by nature, Kissinger's focus on stability and risk reduction meant that elements of global institutionalism now found their place in U.S. policies in both the space and arms control fields. Symbolic firsts over the Soviets, by comparison, seemed less important to the incoming administration.

Before 1969, four factors had enabled NASA to weather its budgetary battles intact. First, the legacy of Kennedy's pledge to reach the Moon and popular and congressional support for this goal had largely suppressed debates on the cost and relevance of the space program. Second, the personal interest and solid lobbying of President Johnson had been an important asset in keeping NASA requests intact and raising political costs to possible opponents. Third, NASA Chief Administrator Webb had a good relationship with the White House and a proven ability to run a huge and complex scientific bureaucracy. And fourth, the enduring hostility of overall U.S.-Soviet relations had proved invaluable for rallying domestic support behind the continued series of costly civilian missions, lest the United States fall behind in the worldwide political struggle against communism for hearts and minds. Under President Nixon, all of these factors would change.

For the Nixon/Kissinger team, space was a sideshow to a larger game. Despite his strong anticommunist reputation from the 1950s, Nixon (under Kissinger's influence) had come to the conclusion that all-out competition with a nuclear-

---

[120] The shock of the North Vietnamese-led Tet offensive on January 31, 1968, discredited the administration's war effort and helped convince Johnson that the war was unwinnable. On the failure of the United States to anticipate this tide-changing attack, see James J. Wirtz, *The Tet Offensive: Intelligence Failure in War* (Ithaca, N.Y.: Cornell University Press, 1991).

armed Soviet Union no longer served U.S. interests. Within this framework, the role of space competition in fanning U.S.-Soviet competitive flames now became counterproductive. The Nixon administration wanted stability in space, not an arms race, and it now benefited from the availability of ample Corona data on the Soviet nuclear arsenal. In a way, Nixon's first term allowed him to have the best of both worlds. At no cost to him, he could bask briefly in the glory of the fruition of Kennedy's competitive pledge—and Johnson's costly expenditures—to put an American astronaut on the Moon by the end of the decade. Meanwhile, he could work behind the scenes to cut expenditures and work out new cooperative arrangements for managing geopolitical superpower conflicts, including in space.

Early on, the Nixon administration risked upsetting the air force by canceling its planned Manned Orbital Laboratory (MOL). The competitive dynamics of the 1960s had changed because of the evolving détente relationship with the Soviet Union, making concepts of direct space warfare seem outmoded. By this time, American photo-reconnaissance technology had also progressed to such a level that putting military personnel into space stations was simply a waste of money, since reconnaissance could be performed far more efficiently by unmanned spacecraft. Armed with these assets, the United States seemed on the brink of a major tactical change in its space priorities: from military superiority to stability.

But the Nixon administration could not be accused of being dovish on the Soviet Union. It also knew that the upcoming arms control negotiations with the Soviets would not be easy. For this reason, Kissinger supported at least a limited U.S. ABM deployment so that he would have something to make Moscow more serious about agreeing to offensive arms reductions, which was his real goal.[121] Thus, stability in both strategic arms and space would at first be pursued by threatening to disrupt these conditions. In March, President Nixon called for an expanded, nationwide ABM system now named Safeguard. Unlike Johnson's Sentinel, the emphasis of the new network would be on protection of U.S. ICBM sites, not cities, thus putting additional pressure on the Soviets to back both ABM and offensive nuclear limitations.

The Moon Landing: Mission Accomplished

On July 20, 1969, NASA's lunar module *Eagle* separated from the *Apollo 11* capsule and brought Neil Armstrong and Buzz Aldrin to their historic first steps on the surface of the Moon. As the U.S. flag was planted, many Americans felt their greatest national pride since the end of World War II and, for younger

[121] Henry Kissinger, *White House Years* (New York: Little, Brown, 1979), pp. 208–9.

viewers, perhaps ever. Suddenly, it didn't seem to matter anymore what the Russians might achieve in space, as the United States had already won its most coveted prize. This new perspective refocused national thinking regarding space competition and put into question plans within NASA to continue the racing mentality of prior U.S. policy. These conditions opened a window for Nixon and Kissinger to put a greater priority on space cooperation with the Soviet Union. Indicators of an emerging policy shift came in the sudden expansion of U.S. cooperative overtures, the reduction of hostile rhetoric, and the scaling back of military space expenditures, especially for research. But the other victim of this new thinking was NASA itself. Ironically, NASA's greatest accomplishment resulted directly in its gradual decline over the next fifteen years.[122] The Moon landing would mark the end of an era for Moscow too, changing its view on the sagacity of further full-scale civilian competition.

Early into his administration, Nixon instructed a Space Task Group under Vice President Spiro Agnew to write a report detailing the options facing the U.S. space program in the post-Apollo era. The committee members concluded that the "experience over the past ten years makes clear that the central problem in developing space cooperation is political rather than technical or economic."[123] Soon, in light of its own political, economic, and military priorities, the new administration would begin to move forthrightly to bridge some of these barriers.

The Agnew report also offered three options for NASA itself, based on different levels of funding availability. The first and least ambitious plan called for a small U.S. space station, a shuttle program to serve it, and eventually, a manned mission to Mars. At the next level, the report proposed a slightly larger space station, a shuttle, and a Mars mission in the late 1980s. Finally, the most ambitious option called for a fifty-person space station, a primary shuttle, a secondary shuttle program for Earth-Moon operations, a Moon base, and a trip to Mars set for the early 1980s.[124]

When the study appeared, both the White House and Congress balked at the multibillion dollar price tags proposed by the largely pro-NASA panel. As Nixon considered the options, it became quickly apparent that the administration's primary concerns lay elsewhere, especially as domestic protests against the Vietnam War escalated. By November, 250,000 demonstrators were marching on Washington, and the news of the My Lai massacre had set even erstwhile supporters of the war on edge. In this new environment, Nixon had neither the

[122] As McDougall sums up American thinking, "NASA had whipped the Soviets," and now "The hope that rode on Apollo [as Americans turned back to Earth's problems] was the hope for human adequacy in the face of awful challenges" (. . . the Heavens and the Earth, p. 413).

[123] Agnew report, quoted in Harvey and Ciccoritti, U.S.-Soviet Cooperation in Space, p. 214.

[124] On these options, see Manno, Arming the Heavens, pp. 112–13.

interest nor the money required for a 1960s-style space program. His eventual
choice was a watered-down version of even the lowest-level option from the
study: a space plane without the Mars trip. As NASA moved into the 1970s, its
prospects for repeating the achievements of the 1960s had been erased. The only
major mission that raised any public interest was the planned *Skylab* space station, which would go on to success in flights with three long-duration manned
missions from 1973 to 1974.[125]

Given the administration's priorities, the new Office of Management and
Budget (OMB), headed by George Shultz,[126] made immediate cuts in the space
budget. It sensed correctly that, after the Moon landing, the Apollo program
would begin to drift from the public consciousness. Indeed, few Americans
followed the post–*Apollo 11* series of Moon missions and their accompanying
research. Politically speaking, there seemed to be no point served by spending
more money simply to fulfill the narrow, scientific goals of NASA scientists and
engineers. Beyond this rationale, Nixon also seemed to harbor a built-in bias
against NASA, which he saw as the creation of the man who had defeated him
in the 1960 election.[127] In early 1970, OMB clipped four planned lunar missions
from the space budget along with funds for their massive Saturn V boosters.
But civilian missions were not the only targets. Military space projects, too,
faced Shultz's ax, suffering even greater cuts.[128]

As a result of the new consensus on reaching an accommodation with Moscow, the U.S. military space budget fell from over $2 billion in 1969 to $1.4
billion by 1972.[129] (Not until 1977 would the defense department's space budget
surpass its 1969 level, even in dollars not adjusted for inflation.) Major military programs enacted in the Nixon years emerged only in the relatively less
costly and less provocative unmanned area, particularly in the field of satellite
communications and reconnaissance, where the administration took the first
steps toward developing a real-time battle management network to be deployed
in the 1980s. While the NASA budget suffered severe cuts from the heights of
the 1960s, it maintained a spending level roughly double that of the defense
department's space effort through the mid-1970s.[130]

---

[125] For more details on the *Skylab* program, see Burrows, *This New Ocean*, pp. 443–45.

[126] Shultz later served as secretary of state under Ronald Reagan.

[127] As Trento notes, "NASA was Kennedy's agency and Nixon would give it as little support as he
could get away with politically." See Joseph J. Trento, *Prescription for Disaster: From the Glory of Apollo
to the Betrayal of the Shuttle* (New York: Crown, 1987), p. 99.

[128] Stares, *Militarization of Space*, p. 157.

[129] Ibid., Appendix 1, Table 1, p. 255.

[130] These figures cast doubt on Manno's alarmist claim that "the militarization of . . . American
space activities was nearly complete by the mid-1970s" (Manno, *Arming the Heavens*, p. 139).

But other changes were taking place in NASA that did not bode well for the organization. As its budget declined and missions dried up, the quality of personnel that NASA could attract suffered as a result. The retirement of Johnson's last chief administrator, Thomas Paine, in March 1971—due to frustrations with the low level of funding given the shuttle program and its dependency on air force design specifications—led to Nixon's appointment of Dr. James Fletcher. Although skilled in technical issues, Fletcher's management practices would lead him into trouble. Relying on ties from his days as the president of the University of Utah and his Mormon loyalties, Fletcher helped Morton Thiokol—a small Utah chemical company with no major space booster experience—to win the major solid rocket booster contract for the shuttle.[131] Fletcher also disrupted the routine of strict accountability by NASA's regional centers to the central administration in Washington. As one analyst describes these changes, "Against the advice of old NASA hands, Fletcher decided to divide the management of the program among the various NASA centers."[132]

Over the course of the 1970s, quality control and morale began a steady decline, as many top personnel from the 1960s retired in disgust over management changes and budget cuts. Among other reductions, NASA canceled the $1.5 billion NERVA nuclear propulsion program in 1973, after determining that it would not be undertaking manned missions to Mars after all.[133] In contrast to both NASA programs and space weapons, the administration rated passive military systems as critical to national assets. Given their role in the coming arms control process, the administration invested heavily in them. As William Burrows describes, "The space reconnaissance program was the equivalent of the Apollo program in magnitude and effect. Unlike Apollo, though, it was officially cloaked in absolute secrecy."[134] Funding now focused on much more sophisticated satellites than those of the 1960s. By 1971, the intelligence community's KH-9 ("Big Bird") spy satellites (40 feet long and 10 feet in diameter) had begun traversing low–Earth orbit to gather valuable data for the U.S. military.[135] These new photo-reconnaissance satellites could develop film on board and quickly retransmit the images back to Earth, as well as deliver more detailed images by deorbiting exposed film as in the past. Their cameras had a resolution of several inches from more than 100 miles up, offering detailed informa-

---

[131] The company's only experience in booster design consisted of having built small retro-rockets for the Mercury and Gemini flights.

[132] Trento, *Prescription for Disaster*, p. 129.

[133] Simpson, *Nuclear Power from Underseas to Outer Space*, p. 149.

[134] Burrows, *This New Ocean*, p. 531.

[135] Philip Taubman, *Secret Empire: Eisenhower, the CIA, and the Hidden Story of America's Space Espionage* (New York: Simon & Schuster, 2003), p. 347.

tion on Soviet missile systems and making U.S. verification of possible limits in offensive arms much more effective.[136] Meanwhile, the United States had by 1972 orbited sophisticated geostationary signals intelligence satellites under the code name Rhyolite.[137] These spacecraft, operated by the defense department's National Security Agency, allowed the military to eavesdrop on Soviet radio communications and, perhaps most importantly, intercept telemetry information from their missile tests. To guard against any possible attacks, the air force launched and successfully deployed the first geostationary early-warning satellites as part of the Defense Support Program (DSP) in 1971.[138] These long-lived, infrared detection satellites could process information on board to distinguish false alarms from actual launches.

The New View from Moscow and
Moves Toward Cooperation

The international repercussions from the successful U.S. Moon landing in 1969 dealt a major propaganda defeat to the Soviet side, nearly tantamount to *Sputnik*'s effect on America. The fear of additional U.S. victories in space caused a sobering of opinions on the viability of a continued policy of racing against the technologically more advanced United States.[139] Clearly, a new strategy was needed, one that did not abandon competition but that added a new element of cooperation to soften the blows of possible future defeats. After numerous costly failures in the last stages of the Moon race, Moscow sought to cut its space-related expenditures. Internationally, the willingness of the Nixon administration to treat the Soviets as coequals in arms control negotiations meant that the world would not view their willingness to cooperate as "capitulation" to the capitalist West. They were now a recognized superpower in their own

[136] For a detailed analysis of the U.S. photo-reconnaissance program at this time, see William E. Burrows, *Deep Black: Space Espionage and National Security* (New York: Random House, 1986); on the KH-9, see esp. pp. 236–43. See also Curtis Peebles, *Battle for Space* (New York: Beaufort, 1983), p. 24.

[137] On the Rhyolite satellites, see Burrows, *Deep Black*, pp. 191–92; and Friedman and Friedman, *Future of War*, pp. 322–23.

[138] Richelson, *America's Space Sentinels*, pp. 64–66.

[139] Several weeks before the *Apollo 11* launch, the Soviets had tried to get in a sample return mission to the Moon using their massive N-1 rocket, hoping to steal publicity from the planned U.S. manned mission. But the rocket exploded violently a short distance from the pad, damaging the facility. (On this incident, see Siddiqi, *Soviet Space Race with Apollo*, pp. 690–91.) On the eve of the Apollo launch, the Soviets made one more attempt, this time to launch an unmanned lunar station. But the mission had been hastily planned, and although the spacecraft flew successfully to the Moon, it crashed onto the lunar surface at high speed because of miscalculations about the altitude. (See Siddiqi, *Soviet Space Race with Apollo*, p. 696.) These were costly failures in both financial and psychological terms.

right. In addition, the Soviets could use the goodwill fostered by space cooperation to move the United States toward a more broad-ranging détente in areas
of greater Soviet concern, including trade and broader security issues. Finally,
within the communist world, the Cultural Revolution in China had put the
attempted maintenance of good relations with the other communist giant beyond the pale of Soviet concerns. By contrast, within Eastern Europe—following the unpopular Soviet crackdown in Czechoslovakia—there were unstated
popular and governmental pressures for a rapprochement with the West.

Buoyed by these trends, U.S.-Soviet arms control negotiators met in November 1969 in Helsinki to begin work on a first-of-its-kind treaty limiting offensive nuclear arms. As part of the discussions between U.S. Arms Control and
Disarmament Agency (ACDA) Director Gerard C. Smith and Soviet Ambassador Vladimir Semenov, various alternative scenarios were vetted to sound out
what the other side might be thinking. The Soviets seemed open to a number
of options on the ABM issue: large-scale deployments, limited sites only (which
would likely grandfather the existing Galosh system), or even a total ban. The
flexibility of Soviet Ambassador Semenov surprised the U.S. representative. As
ACDA's General Counsel George Bunn describes, "It quickly became evident
that the Soviets have changed their minds about ABM systems."[140]

In the fall of 1969, however, Congress had moved to approve funding for
the first two ABM sites. Trying to keep up the pressure on Moscow, President
Nixon called for an expansion of these deployments to eight sites in early 1970.
Indeed, it may now have appeared to the Soviets that the United States had
changed course on the ABM issue, but in the other direction. Although ABM
opponents in the U.S. Senate attempted to block funding for the expanded deployment, the bill passed by the end of the summer, thanks to administration
lobbying about the need for a "bargaining chip" in its negotiations with the
Soviet Union.[141] Despite this U.S. effort to prove its military resolve, the rapprochement with Moscow continued in other areas.

In October 1970, the first serious, high-level U.S.-Soviet meeting in five years
on space cooperation brought together Dr. Robert Gilruth of NASA and Soviet
Academician Petrov. At the meeting, NASA officials sensed a changed Soviet attitude. The meeting resulted in surprising progress, with an agreement to set up
three working groups to hold regular talks over the course of 1971.[142] By March
of the next year, negotiators had expanded the area under discussion and established working groups in two other fields. These sessions culminated in a series

[140] Bunn, *Arms Control by Committee*, p. 115.

[141] Baucom, *Origins of SDI*, p. 61.

[142] Early discussions included plans for a joint *Skylab/Salyut* rendezvous in 1973, but the *Soyuz
11* disaster in June 1971 set back the Soviet program too far to include the upcoming *Skylab* for any
joint mission.

of specific agreements in early 1972. Gradually, the standoffish position of the Soviet Union was beginning to ease. One sign of the new Soviet willingness to cooperate came in the cessation of Soviet ASAT tests, an issue of considerable prior concern to American military space planners. Paul Stares comments that "Soviet ASAT tests abruptly ceased in 1971, apparently in response to the new climate of detente between the superpowers."[143]

A second area of new cooperation emerged in the commercial realm, where the two sides joined other nations in 1971 in forming the International Frequency Registration Board of the ITU. This new body assumed responsibility for ensuring that countries would not interfere with one another's commercial space communications, by allotting specific frequency bands to each state for its transmissions. But the Soviets also moved to shore up their weak position in international commercial programs by establishing a communist-area satellite communications network—Intersputnik—in 1971 as a counterweight to the U.S.-dominated INTELSAT.[144]

A third step toward greater cooperation appeared in the area of manned spaceflight, where the United States and the Soviet Union succeeded in negotiating an unprecedentedly ambitious agreement, the Apollo-Soyuz Test Project (ASTP). Rather than a limited exchange of information, the ASTP called for a highly complicated, joint manned rendezvous and docking mission, requiring close contact between U.S. and Soviet technicians and on-site visits to one another's previously off-limits space facilities. In the deal signed on April 6, 1972, the two governments agreed to a series of planning meetings and technical exchanges in expectation of a mission scheduled for 1975.

For the Soviets, however, these cooperative discussions could only partially mask a series of setbacks at home. Indeed, the early 1970s proved to be a time of bitter failure in Soviet domestic efforts to reconstitute their ailing manned space program. Their attempt to build and occupy the first manned space station (*Salyut I*) seemed destined for success after three cosmonauts had occupied the station for a month in June 1971. But while parades and media fanfare awaited the cosmonauts in Moscow, the returning capsule suffered a malfunction in a crucial pressure valve as it re-entered the Earth's atmosphere.[145] Although the capsule landed safely, the recovery crew found its occupants dead. Abruptly, the optimism of the previous few weeks dissolved, and this second

[143] Stares, *Militarization of Space*, p. 135.

[144] Agreements for Intersputnik were signed in November 1971, and the new organization went into effect in July 1972.

[145] As an investigation revealed, the manual override for the valve took longer to turn by hand than the time it took for oxygen to leak from the capsule.

major space tragedy threw the Soviet people into a period of national mourning. For the Brezhnev government, the mission was an unprecedented public relations disaster.[146] Again, the space program had to delay its planned series of manned launches as it worked to rectify its mistakes.

The following month, Soviet and U.S. arms control negotiators met in Helsinki to discuss more specific plans for a future treaty linking offensive missile limits to restrictions on ABM deployments in the so-called Strategic Arms Limitation Talks (SALT). However, the two sides remained far apart after the Soviet delegation rejected a U.S. proposal to allow ABM protection of either a national capital or four ballistic missile sites.[147] For the Soviets, this meant giving the United States a clear opportunity to trump Moscow's already deployed Galosh system. For the Americans, the Soviets were attempting to protect what was in effect a dual-use system, since some three hundred ICBMs ringed Moscow and could conceivably be protected by the ABM network.[148] The physical implausibility of a scenario in which the Galosh's multiple overhead nuclear explosions would "protect" these ICBMs, the city of Moscow, or the communications required by the Soviet command structure seemed to matter little. Despite their shared goal of an eventual agreement, both sets of negotiators had high stakes in trying to win any "advantages" they could under the watchful eyes of their respective domestic political audiences, especially ones that might help them appear tough on the other side.

Within the U.S. government, a heated debate raged during the Helsinki talks about what types of ABM systems to allow under a proposed treaty. Given the ongoing Safeguard program, U.S. negotiators assumed that fixed, land-based systems would be allowed, at least for some interim period. But an internal state department memorandum outlined *three* disparate positions on other ABM technologies among representatives to the U.S. Verification Panel Working Group:[149] ACDA opposed any "non-fixed land-based [i.e., mobile] ABMs . . . and all future ABM systems"; the Office of the Secretary of Defense (OSD) would allow future, fixed ABM systems based on new technologies; but the Joint Chiefs of Staff wanted no limitations of any sort.[150] Notably, the memo-

---

[146] In terms of its traumatizing effects on the Soviet population and its raising of serious questions regarding the space program, this accident parallels the later U.S. *Challenger* disaster.

[147] Bunn, *Arms Control by Committee*, p. 122.

[148] Baucom, *Origins of SDI*, p. 67.

[149] Ronald I. Spiers, "SALT: Future ABM Systems—Verification," Briefing Memorandum to the Under Secretary of State, August 5, 1971 (declassified January 6, 1999); available through the Web site of the National Security Archive, George Washington University, at <http://www.gwu.edu/~nsarchiv/> (accessed October 7, 2006).

[150] Ibid.

randum stated that ACDA and OSD *had* agreed on proposed language for banning "sea-based, air-based, space-based, or mobile land-based ABM launchers
or ABM radars."[151] Officials believed that deployment of nationwide defenses
using these technologies would create tremendous pressure for an expansion
of the nuclear arms race.[152] A week later, President Nixon instructed the U.S.
delegation in Helsinki to insert the ACDA/OSD language into the draft treaty,
including the ban on all space-based ABM systems.[153] This language would
eventually gain favor with the Soviet side as well.

The unresolved question of the number of land-based systems to allow continued through the next stage of negotiations in Vienna that November. The
Soviets slowed progress by attempting to keep their Moscow system and allowing each side only one new site to protect an ICBM deployment.[154] However, Soviet flexibility on their heavy SS-9 missiles and on deployments of sea-
launched ballistic missiles allowed the talks to make progress toward offensive
limitations.

By early 1972, the Soviets had clearly signaled their willingness to accept restrictions on their largest missiles, but only in return for prior U.S. restraint
on ABM defenses,[155] which Moscow feared might be more effective than its
own.[156] Notably, Moscow had also finally adopted the U.S. position—based on
the August OSD recommendation—that the only new ABM technologies that
would be allowed under the treaty in the future would be for fixed, land-based
systems.[157] (This debate would later reemerge, however, under the Reagan administration.) In March, President Nixon offered the Soviets a deal now per-
mitting each side two ABM sites: either two ICBM bases (the U.S. preference)
or an ICBM base and the national command authority (the apparent Soviet

[151] Ibid.

[152] The state department memorandum concluded, "It is easier to control weapons before they
come into being than after." (See Ibid.)

[153] State Department Telegram (Top Secret) to the U.S. SALT Delegation, Subject: "SALT Guidance," Text from the White House, August 12, 1971 (declassified July 26, 2001); available through the
Web site of the National Security Archive, George Washington University, at <http://www.gwu.
edu/~nsarchiv/> (accessed October 7, 2006).

[154] Baucom, *Origins of SDI*, p. 68.

[155] Kissinger, *White House Years*, p. 815.

[156] The U.S. side also had an advantage in new multiple warhead missiles, but the Soviets
planned eventually to deploy their own, as well, on missiles with heavier carrying capacity.

[157] The historical record on this important issue can be traced back to NSDM 127 from the
summer of 1971. See Stares, *Militarization of Space*, p. 169. Also, the August 12, 1971, telegram (cited
earlier) to the U.S. delegation makes it clear that the United States wanted to keep this (and only
this) mode of adding new technologies.

choice). In April, the Soviets countered with the formula that would appear in the final treaty: two sites, but only one of which could be an ICBM base. In effect, this cemented a Soviet advantage, since the U.S. government—for political reasons—had no intentions of building a site around Washington. However, it left open the option to do so, and Moscow had not yet built a second, ICBM-oriented site either. Both sides would agree to drop their second-site options in 1974.

With the Moscow summit nearing in May 1972, progress achieved in space science cooperation also began to loom large on the political horizon. As Harvey and Ciccoritti note, "Space, as it turned out, had been the first field in which successful negotiations between the US and USSR had proved possible after the policy shifts inaugurated by President Nixon."[158] Although the main focus of the summit related to arms control and the Basic Principles Agreement signed by Nixon and Brezhnev,[159] space cooperation played an unprecedented role in this direct meeting of U.S. and Soviet leaders. On May 24, Nixon and Chairman of the Soviet Council of Ministers Kosygin signed the Agreement Between the United States of America and the Union of Soviet Socialist Republics Concerning Cooperation in the Exploration and Use of Outer Space for Peaceful Purposes. This pact called for cooperation in numerous fields of space science, including programs relating to near-Earth space, the Moon, and space medicine and biology. In the optimism of the time, the agreement could be renewed by mutual consent every five years. Most importantly, it also finalized plans for the Apollo-Soyuz Test Project. Article 4 of the document pledged both sides, more generally, to encourage "international efforts to resolve problems of international law in the exploration and use of outer space for peaceful purposes with the aim of strengthening the legal order in space."[160]

Meanwhile, the signing of the ABM Treaty on May 26 had an impact on the space environment that exceeded the terms for Earth, where two fixed, ground-based systems would be allowed. By contrast, no ABM systems would be permitted in space. Specifically, the final ABM Treaty language included the nearly verbatim U.S. ACDA/OSD text from August 1971, requiring that "Each

---

[158] Harvey and Ciccoritti, *U.S.-Soviet Cooperation in Space*, p. 228.

[159] This document, entitled "Basic Principles of Relations Between the United States of America and the Union of Soviet Socialist Republics" set out a list of guidelines intended to govern the new détente relationship. It included common goals of "peaceful coexistence" and the avoidance of military confrontation. The document also rejected efforts by either side to "obtain unilateral advantage at the expense of the other." See the full text, as printed in *American Journal of International Law*, Vol. 66, No. 4 (October 1972), pp. 920–22.

[160] Ibid.

Party undertakes not to develop, test, or deploy ABM systems or components which are sea-based, air-based, space-based, or mobile land-based."[161] The treaty further stipulated conditions for at least a limited ban on ASATs, stating, "Each Party undertakes not to interfere with the national technical means of verification of the other Party."[162] The SALT I Interim Agreement,[163] which laid out ceilings for both sides in the deployment of strategic nuclear launchers, also noted its reliance on satellite verification and contained identical language on noninterference with those means. Thus, the agreements codified in legally binding terms constraints in space that had previously existed only as norms. From 1972 on, with the exception of a few laser test activities,[164] space satellites from both sides would be treated de facto as out-of-bounds of hostile activity, establishing a powerful consensus for international behavior. These treaties established a firm link between safe access to space for treaty verification and the implementation of U.S.-Soviet nuclear security on the ground. (These agreements and their protections would later be extended to the post-Soviet states after 1991.)

Part of what had made this possible was Moscow's success during this period in improving its own passive military capabilities. Indeed, the late 1960s and early 1970s witnessed a surge of investment in remote sensing satellites to build the Soviet side's capability to keep track of U.S. long-range missiles and other military forces. As Stares notes, in the early 1970s "the number of [Soviet] electronic reconnaissance, communication, and meteorological satellites more than doubled."[165] Moscow also now employed a fleet of nuclear-powered Radar Ocean Reconnaissance Satellites (RORSATs)[166] to follow U.S. ships around

[161] See text of the ABM Treaty, reprinted in Johnson, *Soviet Military Strategy in Space*, Appendix 5, pp. 253–57.

[162] Ibid. and see discussion of this portion of the treaty in J.E.S. Fawcett, *Outer Space: New Challenges to Law and Policy* (Oxford, England: Clarendon Press, 1984), p. 111.

[163] The full name of this treaty was the Interim Agreement Between the United States of America and the Union of Soviet Socialist Republics on Certain Measures with Respect to the Limitation of Strategic Offensive Arms.

[164] These exceptions included alleged Soviet "dazzling" of U.S. early-warning satellites in a few incidents beginning in 1975. The nondestructive tests did not target lower and faster-moving reconnaissance satellites.

[165] Stares, *Militarization of Space*, p. 143.

[166] The Soviets initially had trouble constructing efficient solar cells and had therefore opted for nuclear reactors. While perfectly legal according to the Outer Space Treaty, these spacecraft would later cause environmental damage to space due to leaks of their sodium coolant and occasional forced re-entries, such as a crash in 1978 that spread plutonium across a field in Canada (described in Chapter Five).

the world's oceans. Finally, in 1972 it had begun to launch its first dedicated early-warning satellites into elliptical (Molniya) orbits,[167] beginning with *Kosmos 520*, to provide timely information about both tests and possible attacks.[168] Elements of the system would later be deployed into geostationary orbit. The Soviets had clearly become a more mature space power since the Khrushchev days and had adopted a strategy more closely parallel to that of the United States in emphasizing passive military operations.

The Apex of Space Détente and Shifting U.S.-Soviet Politics

The period from 1973 to 1975 marked the high-water point of space cooperation until at least the late 1980s, particularly in the political dynamics evident in the superpower space relationship. Cooperation for the first time became a leading and regular element of space policy in both countries. During these years, groups of U.S. and Soviet space officials, scientists, and technicians held an unprecedented number of substantive meetings, including visits to once secret space facilities, especially in the Soviet Union. Exchanges of program information and technical data also reached new heights. Not coincidentally, this period followed the most hopeful years of the détente era, when it was believed that the Basic Principles Agreement had established ground rules for general superpower behavior and for cooperation in conflict management worldwide.

More space science cooperation took place in 1973 than in all the years of space activity leading up to it. In March, the two countries conducted the first joint U.S.-Soviet scientific experiment involving equipment from both sides in a meteorological test over the Bering Sea. In addition, there was considerable progress toward settling problems in the publication of a now-expanded, joint three-volume study on space medicine and biology. In exchanges, the two sides traded data gathered by the Soviet *Mars 2* and *3* probes and the U.S. *Mariner 9* spacecraft. A visit by a U.S. team in October to the closed Baikonur launch facility marked the first time that Western officials had seen the facility since a brief trip by French President Charles de Gaulle in the early 1960s. The timing of the visits was not coincidental. Consistent with long-term Soviet policies,

---

[167] Johnson, *Soviet Military Strategy in Space*, p. 71. Molniya orbits, used frequently by Soviet satellites, had a high apogee and a low perigee, thus providing long "hang time" over targets in the United States (such as missile sites).

[168] Podvig, *Russian Strategic Nuclear Forces*, pp. 428–29.

the opening of the space program to Western eyes had come only in the context of progress on arms control, specifically, the recent SALT I and ABM treaties. These agreements had only come about (in part) because of the new ability of the Soviet Union to keep track of the U.S. military from space. As Khrushchev had feared years earlier, site visits exposed Soviet weaknesses in spaceflight that surprised American officials.[169] However, the Brezhnev leadership was willing to accept these disclosures because it now could monitor the other side reliably from space, allowing it to enjoy the other benefits of détente without as much worry about its security.[170] Both sides now enjoyed safer access to orbit than ever before by ruling out a series of harmful practices and potentially destabilizing technologies that had seemed predestined for deployment only a decade before.

At the international level, Soviet and U.S. negotiators had worked within the context of the United Nations to iron out complex issues of liability for damage by spacecraft, requiring compensation to cover damage occurring on the Earth, to aircraft in flight, and to spacecraft of other nations.[171] On March 29, 1972, the Convention on International Liability for Damage Caused by Space Objects was opened for signature by other countries in ceremonies in Moscow, London, and Washington.[172]

Efforts continued in the United Nations on other problems related to commercial programs. Over the course of 1974, negotiators formalized provisions for the registration of all objects launched into space. Up to this time, although many countries voluntarily supplied such data as the orbital parameters, date of launch, and the spacecraft's function, others did not or only reported selected features. The new convention required signatory nations to supply to the U.N. secretary general an array of specific information: "the name of the launching state or states, the appropriate designation of the space object or its number, the date, the territory or location of the launching, the basic parameters of the orbit, including the nodal period, inclination, apogee, and perigee, and the

[169] For example, NASA technicians learned of the lack of computerized rocket controls—something long considered "essential" equipment on U.S. flights—on the Soyuz spacecraft during their visit to Baikonur. Then-current Soviet practice for exo-atmospheric rocket burns turned out to be shockingly primitive: "They timed it with a stopwatch, then waited for the ground station to tell them if the manoeuvre had been successful." From James Oberg, *Red Star in Orbit* (New York: Random House, 1981).

[170] Soviet officials visiting NASA facilities tried to purchase ten Apollo spacesuits for the Soviet space program. The U.S. officials involved refused both their request and the Soviet wish to keep the offer secret. (See account in Schauer, *Politics of Space*, p. 216.)

[171] For more on the liability convention, see Zhukov and Kolosov, *International Space Law*, pp. 101–8.

[172] The agreement entered into force on August 30, 1972.

general purpose of the space object."[173] On January 14, 1975, the Convention on Registration of Objects Launched into Outer Space was opened to signatures at the United Nations.[174]

By the mid-1970s, the Soviet manned space program finally began to get back on track. Following a major investigation after the *Salyut I* disaster, the Soviets instituted a number of reforms, including Mishin's replacement as chief designer by Valentin Glushko (a highly respected missile engineer).[175] Although *Salyut 2* and an unnamed successor both failed in orbit and burned up in untimely (and embarrassing) re-entries into the atmosphere, *Salyut 3* began a series of successful Soviet space station flights after 1974.[176] These missions gradually overtook the U.S. *Skylab* records in long-duration spaceflight and proceeded with over a decade of progress in materials processing and biomedical research. The Soviets also deployed the *Progress* tanker and service ship to provide reliable, unmanned service flights to their space stations.[177] This and the more advanced *Soyuz T* manned vessel were legacy spacecraft of Chief Designer Mishin and would see service for decades ahead.

Perhaps the greatest moment in U.S.-Soviet manned space cooperation occurred in midsummer of 1975. After the launch of the *Apollo 18* and *Soyuz 19* spacecraft in their respective countries, the two vessels docked and exchanged crews on July 17. In this internationally televised event, the astronauts and cosmonauts shook hands, shared a meal, and conducted a number of joint experiments in the course of the next two days. The meeting marked the first U.S.-Soviet docking in history and symbolized the fruition of five years of planning, meetings, and exchanges.

Conclusion: The Denouement of Détente

By the time the Apollo-Soyuz mission actually took place in July 1975, however, the political relationship that had formed the background for this improvement in space cooperation had largely disintegrated. The Watergate

---

[173] Zhukov and Kolosov, *International Space Law*, p. 88. One matter that remained a subject of debate, despite the agreement, was the issue of marking space vehicles. Thus, the convention left it up to states whether or not they wished to mark spacecraft, but it required that those bearing markings be so registered with the United Nations.

[174] The document went into force on September 15, 1976.

[175] Harvey, *Russia in Space*, p. 13.

[176] One of these reforms was to cut the number of cosmonauts in the Soyuz vessels from three to two and to restore the spacesuits used in previous spacecraft during re-entry. However, the Soviets also designated some of the Salyut missions for military reconnaissance, given their comparative lag in unmanned technology. On these flights, see Burrows, *This New Ocean*, pp. 509–11.

[177] Siddiqi, *Soviet Space Race with Apollo*, p. 839.

scandal, the fall of Vietnam, the revolutions in Mozambique and Angola, and the congressional linking of U.S.-Soviet trade levels to Moscow's policies on Jewish emigration (in the Jackson-Vanik amendment) had destroyed the hopes of 1972. The "great handshake in space" would not be followed by any similar joint missions, but the arms race had largely been stabilized and the two sides had formalized means for keeping nuclear war at bay thanks to space assets that few individuals in the public at large even knew about.

Ideas for space weapons had risen and fallen at various times during this period, suggesting the powerful continuing forces within both governments to react with traditional military responses to perceived (if at times inflated) space threats. But the two leaderships cut back and eventually canceled plans for nearly all of these weapons programs, especially those whose testing and deployment might destabilize or threaten access to space. They also acted increasingly according to specific rules of noninterference that helped separate space activity from other environments of U.S.-Soviet conflict.

In terms of treaties, the PTBT, the Outer Space Treaty, the ABM Treaty, and SALT I created a legacy of cooperation and restraint that would help keep space free of orbiting weapons and direct military confrontation. These agreements were buttressed by the shared understanding by both leaderships that deploying and using space weapons—unlike in various parts of the world where superpower arms were proliferating—would cause them more harm than good. Indeed, in October 1975, the bargaining chip of the U.S. Safeguard system was shut down by Congress, just a day after being declared fully operational. The multibillion dollar system had proved to be unnecessary and even harmful, with its multiple nuclear weapons pointed at space.[178] Given its cost, members of Congress expressed great regret at shutting it down, but did so anyway.

The next decade of space relations would witness a decline in a number of U.S.-Soviet cooperative programs, especially in the area of space science.[179] Although disappointing for many of the scientists involved, the slide did not destroy all U.S.-Soviet agreements. Scientists remained in communication and programs that lost funding waited until more favorable times might emerge. The positive memories of cooperation also helped stabilize the rocky period ahead. Scientific study of space began to affect policies on orbital debris, although formal efforts at restricting it still remained some years off. Part of the reason had to do with the halt in Soviet ASAT tests in 1971, and part because of

[178] According to Pentagon studies, Safeguard's radars would be blinded on the first nuclear explosion from its own interceptors (which carried huge, 5-megaton weapons).

[179] However, established personal contacts remained in place and helped alleviate tensions between the two sides in space, while facilitating the resumption of contacts in the mid- to late 1980s.

political sensitivities about sharing this information within the context of the ongoing Cold War. But the delay also stemmed from the limited catalogue of existing debris. At higher altitudes, debris fragments were hard to track using existing radars and computer models. Even at lower altitudes, the data had to be identified and then attributed to a likely cause. In this regard, what we now call "situational awareness" in space still remained quite limited even by the mid-1970s. As NASA's chief scientist for orbital debris studies Donald Kessler noted later, regarding U.S. civilian satellite fragmentations at the time, "Few people were aware that any breakups had occurred" and "only a fraction of the fragments had been catalogued."[180] Thus, orbital debris had not yet evolved as a science to a level where it had saliency in political and military policymaking. This would begin to change in the 1980s.

What is most important for our story here is that a number of areas of military space restraint became institutionalized during the 1962–75 period. These measures had evolved from tacit understandings in the late summer of 1962 (from emerging knowledge about the collective harm caused by nuclear testing) to a series of formal treaties and agreements by 1963 and beyond. In the weapons area, loopholes remained, some intentional. But those programs that survived constituted the exception rather than the rule. The consensus around this restraint-based framework in space, however, would soon face one of its most serious challenges. Hopes for global institutionalism in space would soon yield, again, to renewed space nationalism.

[180] Donald Kessler, "Foreword," in Johnson and McKnight, *Artificial Space Debris*, p. vii. The causes of these civilian fragmentations had to do with technical problems and fuel accidents.

CHAPTER FIVE

Challenges to Space Security
and Their Resolution

1976–1991

The years from the end of Gerald Ford's administration to the Soviet breakup in December 1991 witnessed various attempts to challenge the existing framework of military space restraint. Renewed Soviet anti-satellite (ASAT) testing, the rise of U.S. space security concerns, and the emergence of the Strategic Defense Initiative (SDI) could easily have led to serious problems for safe access to space and the right of free passage. Support for cooperative forms of space security and continued restraint in terms of deployed weapons, however, remained engrained in enough critical nodes of power within both the U.S. and Soviet governments to maintain existing agreements and norms. The administration of Ronald Reagan represented a unique challenge in that the president pledged both to deploy a vast network of space defenses and to save the world from nuclear weapons,[1] even as his advisors split on whether the SDI program should be a bargaining chip or a means of achieving some form of military-led space control. Because of the president's radical agenda for weaponizing space and his unfamiliarity with existing norms, the period from 1983 to 1986 saw the unlearning of past lessons. Plans for SDI envisaged the deployment of thousands of orbital weapons. But this bold project made no mention of the likely harmful implications of this massive military test and deployment program on the space environment. It seemed that a new form of U.S. space nationalism—abandoning the lessons of the past—had taken hold because of overriding national security fears.

Events turned out differently from all apparent trends of the early 1980s. One small bellwether event came from outside the SDI program in 1985, when the air force decided to test a kinetic-kill ASAT weapon. Over the objections of

---

[1] On this point, see Paul Lettow, *Ronald Reagan and His Quest to Abolish Nuclear Weapons* (New York: Random House, 2005).

NASA scientists and other experts who warned about the negative implications of debris, the test went ahead. But the scientists' concerns proved justified, and instead of moving forward with a series of similar tests, the United States began a second major stage of space learning as a result of this event. Within a year, the U.S. defense department would adopt its first debris mitigation guidelines. NASA would later begin a process of educating and drawing in foreign space agencies and commercial actors to begin the process of internationalizing debris management principles, a process that continues today. This social interactionist approach driven by a small "epistemic community" of NASA experts and associated U.S. debris scientists had begun to make a difference, although military tests did not cease completely.

As for SDI, restrictions placed on the administration's actions in space by the Democratic Congress, the weight of the Nixon-era Anti-Ballistic Missile (ABM) Treaty, and Soviet General Secretary Mikhail Gorbachev's unwillingness to participate in what he saw as a technological folly that would destabilize the arms race held the program (and a small, little-known, pre-Gorbachev Soviet counterpart effort) in check. Eventually political changes in the Soviet Union, technical problems, and cost concerns related to SDI pushed this renewed effort at military-led space security to the back burner, while Gorbachev stopped the Soviet program in its tracks. The new Soviet general secretary's dramatic domestic and foreign policy reforms moved Soviet policy in a fundamentally different direction from that of his predecessor, shocking the relatively unprepared United States in his bold arms control initiatives and his willingness to relinquish the Soviet empire in Eastern Europe. Meanwhile, Soviet space policy lurched from a heavily military-driven program to one now characterized for the first time by global institutionalism. The disconnect between U.S. and Soviet space trends had perhaps never been greater. But the logic of Soviet political and economic reforms eventually led to something Gorbachev didn't expect: the breakup of the Soviet Union, as domestic events moved beyond his control by late 1991.

At the same time, the increasing capabilities of military support programs showed that broader Cold War security continued to hinge on safe access to space for peaceful uses. U.S. remote sensing satellites would help prevent inadvertent nuclear war in an incident in 1979. Further development of space-based surveillance technologies would assist the negotiation of new arms control treaties by the late 1980s, when political relations began to be repaired following Gorbachev's reforms.

Meanwhile, new commercial space services began to develop as never before. The emergence of the U.S. Global Positioning System (GPS), the expansion of international launch services, and growth of satellite systems to service many new regions brought the benefits of space to an increasingly large portion of the world's population. It also led to the development of devastatingly accurate, precision-guided conventional munitions for the U.S. military.

Finally, in manned space exploration, the 1986 U.S. loss of the shuttle *Challenger* with its entire crew and NASA's focus on restoring the agency's credibility and safety record reduced incentives for racing in space and led to an increasing understanding of the risks involved in human spaceflight. For the Soviet space program, ironically, the rapid Soviet reform process opened unprecedented opportunities for cooperation with the United States by the late 1980s. Such links had all but been extinguished in the period from 1976 to 1986. By the end of the period, détente in space would return. This new rapprochement in space was not oriented toward additional treaties but instead toward greater transparency, stability, and practical cooperation between the two superpowers.

The Unraveling of Space Cooperation
Under Gerald Ford

Considering the dynamic space cooperation envisaged in the early 1970s, the results of U.S.-Soviet space relations under the Ford administration proved highly disappointing. Ford had entered the Nixon administration as vice president only in 1973, following the ignominious resignation of Spiro Agnew over tax evasion charges. He emerged as president in August 1974 following the even more troubling resignation of Richard Nixon over the Watergate scandal. Unfortunately, the congenial, trusted, and experienced Michigan representative lacked the leadership credentials and the acquired loyalty among government subordinates and the American people that an elected president would normally possess, with the result that his administration had the distinct air of a caretaker government. Moreover, President Ford had the disadvantage of presiding over the decline of Nixon's perhaps oversold policies of détente, whose flaws and underlying problems had now begun to emerge.

From the U.S. perspective, a number of Soviet policies by the late 1970s displayed a failure by Moscow to live up to its détente commitments: the tightening of Soviet policy on Jewish emigration, the stationing of mobile SS-20 nuclear-tipped missiles in Europe, and the resumption of ASAT weapons testing in 1976.

From Moscow's vantage point, it was U.S. actions that exerted a negative impact on bilateral relations and, as a subset, prospects for continued mutual self-restraint in space. Soviet leaders saw their continued efforts to promote communism in Africa and Southeast Asia—like U.S. interventions in Latin America—as fully compatible with détente. Similarly, the Russians believed that despite their acquiescence to the Helsinki Accords in 1975, human rights should remain primarily a domestic issue and not something Washington should be pressing with its "equal" partner. Finally, in space, the Soviet reading of reports in the U.S. press about renewed ASAT work in 1975 may have stimulated those in Moscow who were pushing for a resumption of their own test program.[2] Thus, the fundamental incompatibilities of the two countries' perspectives began to become clear as Kissinger's *realpolitik* faded into the background and traditional American concerns for democratic and religious freedoms were reasserted by Congress.

In regard to space, new concerns emerged late in the Ford administration about Soviet activities, including at least two incidents of suspected laser illumination of U.S. satellites in late 1975 from a facility in Central Asia.[3] The Ford administration—in the person of new Secretary of Defense Donald Rumsfeld—stated in a December 1975 press conference that a subsequent investigation had ruled out this suspected interference.[4] But Moscow's resumption of its co-orbital, explosive ASAT program in February 1976 and execution of four tests led to new worries about Soviet intentions in space.[5] The administration therefore commissioned a report by a panel chaired by the former head of the Defense Science Board, Solomon Buchsbaum, to study the ASAT problem.[6] The panel's findings noted the possible vulnerability of U.S. military satellites to attack. Consistent with established views on space security, it also stated that a U.S. ASAT system "could not function as a deterrent to Soviet [ASAT] use."[7] Nevertheless, it did not rule out that such a program might serve as a "bargaining chip,"[8] along the lines of Nixon's ABM policy.

Just before leaving office in January 1977, President Ford issued a mixed di-

---

[2] On this point, see Paul Stares, *The Militarization of Space: U.S. Policy 1945–1984* (Ithaca, N.Y.: Cornell University Press, 1985), p. 172. As Stares notes, "It should not be overlooked that [U.S.] research was already under way—including identifiable elements of the current US [F-15 ASAT] system—*before* the renewed Soviet effort" (p. 173, italics in original).

[3] On these charges, see Nicholas Johnson, *Soviet Military Strategy in Space* (London: Jane's, 1987), p. 174.

[4] Jeffrey T. Richelson, *America's Space Sentinels: DSP Satellites and National Security* (Lawrence: University of Kansas Press, 1999), p. 77.

[5] Johnson, *Soviet Military Strategy in Space*, Table 6, p. 154.

[6] Stares, *Militarization of Space*, p. 169.

[7] See the Air War College's summary of the report in "Nixon and Ford Years: 1969–1976," posted on the Air War College Web site at <http://www.au.af.mil/au/awc/awcgate/au-18/au18003e.htm> (accessed November 3, 2006).

[8] Ibid.

rective (National Security Decision Memorandum 345) ordering the Pentagon to develop an operational ASAT capability, while also calling on the state department to study possible arms control options with the Soviet Union.[9] Ford had no desire to break out of existing space agreements or violate the norm of orbital weapons restraint, but he did not want to see the United States left behind either. Still, official U.S. policy did not overemphasize the Soviet threat or indicate a shift in U.S. responses. As Secretary of Defense Rumsfeld testified before Congress in January 1977, "Current US space defense policy is to abide by our space treaties, exercise our rights to the full and free access to space, and limit our use of space to nonaggressive purposes."[10] But Rumsfeld noted the administration's decision to proceed with additional passive military assets and nonoffensive defenses to improve satellite survivability.[11]

The relative confidence of the Ford administration was supported by the 1976 U.S. deployment of the KH-11 reconnaissance satellite, a major technological innovation that had improved already strong U.S. capabilities to monitor Soviet military developments.[12] This satellite provided the U.S. intelligence community with highly accurate, real-time digital images of Soviet military locations.[13] No longer would the United States have to rely either on grainy video images transmitted by radio waves or film dropped from the satellite and delivered (at best) hours later to officials. Instead, it could now receive timely electronic data with a resolution good enough to read Soviet license plates.[14] These capabilities would help ensure Washington's continued ability to verify arms control agreements. Later versions of the satellite launched in the 1980s would add infrared imaging to facilitate surveillance at night or in poor weather, and maneuvering capabilities to avoid possible ASAT attacks,[15] thereby creating a highly valuable platform.

In civilian space activity, arguments supporting new efforts to expand space

[9] Lt. Col. (USAF) Peter L. Hays, *United States Military Space: Into the Twenty-First Century*, Occasional Paper No. 42 (Colorado Springs: U.S. Air Force Academy, Institute for National Security Studies, September 2002), pp. 102–3. See also Stares, *Militarization of Space*, p. 171.

[10] Rumsfeld, quoted in Stares, *Militarization of Space*, p. 174.

[11] Ibid., p. 175.

[12] On the U.S. deployment of the KH-11, see Jeffrey T. Richelson, *Spying on the Bomb: American Nuclear Intelligence from Nazi Germany to Iran and North Korea* (New York: Norton, 2006), p. 279; and Brian Harvey, *Russia in Space: The Failed Frontier?* (Chichester, England: Praxis, 2001), p. 115.

[13] On the KH-11's capabilities, see George Friedman and Meredith Friedman, *The Future of War: Power, Technology and American World Dominance in the Twenty-First Century* (New York: St. Martin's Griffin, 1996), p. 316.

[14] William E. Burrows, *Deep Black: Space Espionage and National Security* (New York: Random House, 1986), p. 248.

[15] Ibid., p. 317. Burrows quotes an expert indicating a likely two-inch resolution regarding the KH-11's still-classified capabilities.

cooperation with the Soviet Union became suspect. But while high-profile space cooperation with Moscow declined after the 1975 Apollo-Soyuz docking, exchanges of data in the field of planetary research continued. Topics included such fields as solar wind research and magnetospheric plasma physics research. U.S. and Soviet scientists also traded considerable data from their respective Mars and Venus programs. The U.S. side expressed particular satisfaction with the rapid Soviet provision of data from the 1975 Soviet *Venera* spacecraft, which landed on Venus, took readings, and transmitted the first images of the planet's surface. The subsequent U.S. *Pioneer* mission—which conducted a fly-by mapping of Venus—benefited considerably from the Soviet data.[16]

Jimmy Carter and Unfulfilled Hopes for Strengthened Space Security

Compared to President Ford, Jimmy Carter had greater optimism about the prospects for space arms control with Moscow. His space security strategy made no moves to withdraw from existing treaties and indeed tried to strengthen them through formal negotiations with the Soviet Union. But Carter would also continue existing research programs into possible U.S. space defenses, although at relatively low levels of funding. For political reasons, his administration would treat U.S.-Soviet cooperation in civilian space activity as "sending the wrong signal" regarding other Soviet actions he opposed (such as human rights policy), with the result that space became quickly tied up in the complexities of "linkage" politics.

Carter was an outsider with little diplomatic experience; his relationship with the Brezhnev leadership in Moscow started badly and only got worse. In the administration's first meeting with the Soviets in March 1977, Carter's Secretary of State Cyrus Vance attempted to rewrite arms control understandings that reached as far back as President Ford's 1974 Vladivostok meeting on SALT II in order to redress areas of seeming Soviet advantage. This infuriated the Brezhnev leadership. As the 1970s wore on, Moscow grew increasingly frustrated with Carter's criticism of its domestic policies, his move toward establishing diplomatic relations with arch–Soviet rival China, and his decision in December 1979 to deploy nuclear-armed Pershing II and ground-launched cruise missiles into Europe. Moscow saw the United States as vulnerable after the debacle of Vietnam, which provided it with an incentive to exploit possible U.S. weaknesses, although not through direct confrontation.

[16] Office of Technology Assessment, *U.S.-Soviet Cooperation in Space* (Washington, D.C.: U.S. Congress, Office of Technology Assessment, OTA-TM-STI-27, July 1985), p. 74.

In civilian space cooperation with the Soviet Union, the Carter administration parted company with the Nixon and Ford administrations. Carter and his Polish-born National Security Advisor Zbigniew Brzezinski believed that cooperating with Moscow would tacitly indicate U.S. support for the Brezhnev leadership before the rest of the world concerning Soviet activities both in the security area and in the human rights field, where the Carter team sought to make its mark. As a 1985 report by the U.S. Office of Technology Assessment noted about this period, "Along with other measures intended to show displeasure with Soviet actions, the United States severely curtailed cooperation in space with the USSR."[17] In contrast to the 1960s, it was the Americans, not the Soviets, who decided to limit space cooperation as a means of showing their displeasure with the other's policies.

One of the first fallouts after 1976 was the canceling of discussions on a proposed major mission, a joint Shuttle/Salyut project involving the exchange of personnel. Other missions soon also felt the political pinch. Although the Brezhnev and Carter administrations did renew the 1972 U.S.-Soviet space cooperation accords when they came up for their five-year review in 1977, there was no real progress toward expanding space exchanges.[18] Most of the ambitious plans cited in the text of the accords—including joint experiments using future U.S. and Soviet manned spacecraft as well as a proposed "international space platform"—never reached fruition.[19] In the second phase of the agreement, the main projects involved completion of earlier missions or "low-level scientist-to-scientist cooperation" outside official governmental channels.[20]

One significant program for multinational cooperation, however, did come into being during this period in the commercial area: the 1979 agreement on the formation of the COSPAS/SARSAT rescue system.[21] Notably, given the limited

---

[17]See Office of Technology Assessment, *U.S.-Soviet Cooperation in Space*, p. 32.

[18] Meeting in Geneva on May 18, 1977, U.S. Secretary of State Cyrus R. Vance and Soviet Foreign Minister Andrei Gromyko signed the agreement, based on a test drawn up in NASA meetings with the Soviet Academy of Sciences on May 11. See the Congressional Research Service report prepared for the Senate Commerce Committee, *Soviet Space Programs: 1976–80*, Part I (Washington, D.C.: U.S. Government Printing Office, 1982), pp. 216–18.

[19] See "Agreement Between the United States of America and the Union of Soviet Socialist Republics Concerning Cooperation in the Exploration and Use of Outer Space for Peaceful Purposes," as printed in Congressional Research Service, *Soviet Space Programs*, pp. 391–94.

[20] See Office of Technology Assessment, *U.S.-Soviet Cooperation in Space*, p. 33. One of these "leftover" missions that proved successful was the simultaneous launching in 1978 of U.S. and Soviet sounding rockets to compare data on the ionosphere. This close operational cooperation involved a U.S. launch from its Wallops Island facility and a Soviet launch from a nearby ship.

[21] COSPAS was the Russian acronym for "Space System for the Search of Vessels in Distress." SARSAT stood for "Search and Rescue Satellite-Aided Tracking," which was the joint U.S.-Canadian-French element of the system.

reach of each nation's satellite coverage, this pact served both sides' mutual interest in using space to improve navigational safety. The planned system envisioned a worldwide cooperative rescue operation involving the use of transmitters aboard ships and aircraft linked to a joint satellite network. Distress signals would be received and located (using the Doppler effect) and the rescue information patched through to the nearest network country to provide aid.[22]

Another area of close cooperation during this period was in space biomedical research. As one U.S. participant noted, it was one of the most poorly funded areas and therefore below the government's radar screen.[23] For U.S. scientists, however, it provided the only means of forwarding biomedical research in a period when no U.S. spacecraft were available.[24] The first mission (under the 1972 agreement) took place in 1975 aboard the Soviet *Kosmos 782* satellite, which carried eleven U.S. space biology experiments. The series continued throughout the Carter administration and into the early Reagan years with a 1977 flight on *Kosmos 936* (seven U.S. experiments), a 1979 flight on *Kosmos 1129* (fourteen U.S. experiments),[25] and a 1983 flight on *Kosmos 1514* (four U.S. experiments).[26] Although U.S. export control officials in the state department at times raised technology transfer issues with scientists and required substitutions in some instances,[27] the biosatellite program proved to be one of the last remnants of the détente relationship.

NASA's budget as a whole expanded only slowly under Carter, barely keeping up with inflation. It rose from $3.2 billion in 1977 to only $4.9 billion in 1981, but much of this funding was taken up with shuttle preparation work.[28] In the face of an oil shortage then gripping the United States, the Iranian hostage crisis, and worsening relations with the Soviet Union, space also suffered from a lack of political attention. In fact, no U.S. astronauts would fly into space

[22] The agreement went into effect in 1980.

[23] Author's interview with bioscientist Richard Mains in Berkeley, California, January 1988. This and other interviews from the 1987–89 period cited in this chapter are drawn from James Clay Moltz, "Managing International Rivalry on High Technology Frontiers: U.S.-Soviet Competition and Cooperation in Space" (Ph.D. dissertation, University of California at Berkeley, 1989).

[24] Ibid. As Mains argued, the advantages of cooperation from the U.S. side were the low cost of the projects, their use of Soviet launchers, and the relatively low level of the technologies involved.

[25] Although OTA 1985 lists the date of this launch in two places as 1978, it in fact did not take place until September 25, 1979. (See Congressional Research Service, *Soviet Space Programs*, p. 441.)

[26] The typical practice involved U.S. packaging of experiments with instructions for care of animals and Soviet return of data and specimens.

[27] According to export control regulations in the U.S. International Traffic in Arms Regulations (ITAR), space technology fell under the classification "implements of war." Thus, space equipment was "assumed to be militarily sensitive until it is shown not to be." (See Office of Technology Assessment, *U.S.-Soviet Cooperation in Space*, p. 86.)

[28] Stares, *Militarization of Space*, Appendix 1, Table 1, p. 255.

under Carter. Part of the reason had to do with the difficulty of bringing the shuttle program to fruition at such low levels of funding. But part too was the built-in bias of the Carter administration against the manned program itself. As an engineer, Carter respected the space program's accomplishments, but he saw no reason to spend vast sums for manned missions when unmanned research could serve national purposes better and at much lower cost.[29] In addition, Carter's vice president, former Minnesota Senator Walter Mondale, had been a longtime foe of the space program in the Senate, where he had led an unsuccessful fight to cancel the shuttle. But Carter's respected and influential Secretary of Defense Harold Brown *favored* the shuttle's deployment and, with some help from the air force (which still planned to use it for selected satellite releases and repairs),[30] convinced Carter to back it as well. Still, to fit within Carter's budgets, the shuttle's construction would have to be accomplished on the cheap.

In comparison with NASA's, the Pentagon's space budget grew at a slightly faster rate (from $2.4 billion to $4.8 billion) during the Carter administration.[31] Some of this spending was directed at the new U.S. ASAT program, a direct-ascent, kinetic-kill weapon to be launched from an F-15 aircraft. But the bulk of it went into improving the navy's NAVSTAR navigational network and planning for such future projects as the new MILSTAR military communications network, the Lacrosse radar imaging satellite, and the new KH-12 reconnaissance platform to be deployed in the 1980s.

The Soviets were willing to engage in limited civilian space cooperation and in general military restraint as a quid pro quo for Cold War peace and space stability, but in the face of a U.S. dormancy they used this period to recoup their past civilian losses. The successful orbiting of *Salyut 7* in September 1977 allowed the space program to rack up considerable international publicity and technical expertise while the U.S. manned program remained on the ground. Unprecedented capabilities for automatic resupply flights, the docking of two spacecraft at the station, and sophisticated space construction built up a new Soviet lead in manned spaceflight.[32] Notably, all of these missions came off without fatalities, restoring the Soviet reputation for safety.

[29] As Trento describes Carter's welcoming of his new NASA chief Dr. Robert A. Frosch, "In 1977, during one of Frosch's first meetings with the president, Jimmy Carter asked his new NASA administrator to examine the shuttle program to determine if it should be shut down." (See Joseph J. Trento, *Prescription for Disaster: From the Glory of Apollo to the Betrayal of the Shuttle* [New York: Crown, 1987], p. 149.)

[30] Also supporting Brown's side in the argument was Air Force Secretary Dr. Hans Mark.

[31] Stares, *Militarization of Space*, Appendix 1, Table 1, p. 255.

[32] Asif A. Siddiqi, *The Soviet Space Race with Apollo* (Gainesville: University of Florida Press, 2003), p. 839; and Congressional Research Service, *Soviet Space Programs*, p. 323.

Moscow also began to address the large technological gap with the United States in the passive military sector. By 1977, it had deployed a partially operational space-based early-warning network of three satellites in Molniya orbits for regular coverage of missile sites in the continental United States.[33] The system would finally be completed in the early 1980s with nine satellites, although it continued to lag behind U.S. capabilities and needed to be supplemented with ground-based radars.

During the Carter years, the Soviets conducted six co-orbital ASAT tests with a more advanced infrared sensor, although at least four of them failed.[34] The tests generated considerably less debris than the first series, owing to a change in the "Soviet intercept profile" that involved relatively rapid de-orbiting of the interceptor.[35] But despite Moscow's poor success rate with these tests, the United States worried about possibly emerging Soviet capabilities. The Pentagon had dismantled its clumsy (and self-defeating) nuclear-tipped ASAT system (Program 437) on Johnston Island in the Pacific in 1972[36] without a clear replacement system, short of redirecting nuclear-tipped missiles slated for strategic roles.

Despite this challenge, much of the air force remained relatively unconcerned and even uninterested in moving back down the route toward a deployed ASAT capability. The failure of past programs had combined with previous administrations' decisions to remove manned military programs from space to create an environment in which there was—in the words of one analyst—"often resistance to new space proposals"[37] within the air force.

The Carter Effort to Negotiate an ASAT Ban

As Soviet ASAT tests continued, Carter administration officials began to consider the desirability of a complete ASAT ban. Indeed, negotiations on codifying a formal and specific prohibition on ASATs became the subject of unprec-

---

[33] Johnson, *Soviet Military Strategy in Space*, p. 71.

[34] On the sensor, see Howard Ris, "ASAT Test Ban Would Favor U.S." (Letter to the Editor), *Wall Street Journal*, September 25, 1985. Ris was then executive director of the Union of Concerned Scientists. On the test results, see Johnson, *Soviet Military Strategy in Space*, p. 154, and Ris's comment that all six failed.

[35] Nicholas L. Johnson and Darren S. McKnight, *Artificial Space Debris* (Malabar, Fla.: Orbit, 1987), p. 15.

[36] Although the official deactivation date was June 1973, the nuclear warheads had been separated from the missiles long before then, weather had degraded the missiles and infrastructure, and Hurricane Celeste in August 1972 had destroyed much of the launch facilities and computers, rendering them inoperable. See Lt. Col. (USAF) Clayton K. S. Chun, *Shooting Down a 'Star': Program 437, the US Nuclear ASAT System and Present-Day Copycat Killers*, Cadre Paper No. 6 (Maxwell Air Force Base, Ala: Air University Press, April 2000), pp. 30–31.

[37] Stares, *Militarization of Space*, p. 176.

edented U.S.-Soviet negotiations in 1978, with the aim of expanding the reach of space security cooperation. The Partial Test Ban Treaty already prevented the *use* of nuclear ASATs in space. Moreover, the Outer Space Treaty banned the orbiting of nuclear ASATs, while the SALT I Treaty prohibited the use of any ASAT against a military arms-control monitoring satellite, and the 1973 International Telecommunications Convention prohibited disturbance of communications satellites. Yet, both sides felt that there were large enough loopholes to make a separate ban on ASATs desirable. A problem existed, however, in the fact that both the United States and the Soviet Union had present or residual ASAT weapons, meaning that ASAT technologies would have to be "put back into the bag." In a series of talks held in Helsinki (1978) and Bern (1979), the difficulties of arriving at a verifiable ban without politically sensitive on-site inspections became apparent. As a result, the discussions narrowed to the simpler issue of a ban on the *use* of ASATs (which, as shown above, already existed at least for verification satellites). By the final round of talks in Vienna in the late spring of 1979, the two sides seemed to have reached a working agreement on the possibility of proceeding with such a first-step framework.[38] Traditional Soviet complaints about dual-use systems (the U.S. space shuttle) and exceptions for hostile acts by satellites stood in the way of a deal, as did concerns of the Joint Chiefs of Staff about the likely U.S. loss of eavesdropping bases in Iran with the shah's ouster in January 1979, which raised the stakes of possible Soviet ASAT cheating. Subsequently, outside political developments intervened to derail the negotiations altogether.

With concerns over events in Iran mounting, the U.S. "discovery" of a Soviet combat brigade of twenty-five hundred troops in Cuba in the summer of 1979 became a near crisis as President Carter (likely spurred by newly influential hardliners in his administration) initially declared its presence "unacceptable." Only when U.S. intelligence reports declared that the intelligence community had known about the brigade's presence for years and that it posed little threat to U.S. security did President Carter retreat from his comments. But the event further damaged already shaky U.S.-Soviet relations. In late October, the Carter administration's decision to admit the cancer-stricken shah of Iran into the United States for medical treatment spurred radical Iranian students to seize the U.S embassy in Iran in early November and take its staff as hostages. The event fostered a "siege mentality" in the Carter White House as the administration struggled to obtain their safe release.

In the midst of this turmoil, on November 9, 1979, a potentially catastrophic event occurred within the U.S. nuclear complex. The mistaken loading of a

[38] Ibid., pp. 198–99.

training tape into the U.S. Strategic Air Command's computer system led to missile crews receiving warning signals that the United States was under a massive Soviet nuclear attack. Only when decision makers quickly reviewed data from U.S. early-warning satellites to verify the launch of the Soviet missiles did they finally recognize the mistake.[39] Once again, U.S. passive military assets in space had prevented an accidental nuclear war. Given Moscow's lack of knowledge of this frightening event, however, it had no positive spillover effects on broader U.S.-Soviet relations.

In December 1979, the Soviet Union invaded Afghanistan to prop up a failing Communist-led government that had seized power in 1978. The Soviets clearly underestimated the expected reaction from the Carter administration, which, guided by National Security Advisor Brzezinski, mistook the move as the first stage in an offensive Soviet drive to the Persian Gulf. U.S.-Soviet relations suffered accordingly. The administration pulled SALT II from the Senate in January 1980, moved to reinstitute draft registration in the United States, and in regard to space, postponed ASAT negotiations indefinitely, despite the preliminary agreement reached in Vienna.

President Carter's failure to reverse these negative trends in foreign policy, his seeming inability to redress economic problems (inflation and rising oil prices), and the desire of many Americans for a stronger hand in the White House led to the election of conservative Republican Ronald Reagan in the 1980 presidential campaign. Prospects for U.S.-Soviet space cooperation dimmed even further.

Ronald Reagan's Attempted Return
to Military-Led Space Security

President Reagan came into office committed to major policy change.[40] He had run on a platform of redressing a perceived decline in U.S. military capabilities relative to the Soviet Union, which, while debatable in substance, seemed hard to argue with in terms of international perceptions and momentum. Accordingly, Reagan initiated a large-scale U.S. military buildup. Quickly, U.S.-Soviet relations declined to perhaps their lowest point since 1962, with President Reagan now referring to Washington's erstwhile partner in détente as the "evil empire." Reagan's main fear was the increasing number of warheads

[39] On this incident, see Geoffrey Forden, Pavel Podvig, and Theodore A. Postol, "False Alarm, Nuclear Danger," *IEEE Spectrum*, Vol. 37, No. 3 (March 2000).

[40] On Reagan's agenda, see John Arquilla, *The Reagan Imprint: Ideas in American Foreign Policy from the Collapse of Communism to the War on Terror* (Chicago: Ivan R. Dee, 2006).

aboard the Soviets' large arsenal of heavy ICBMs (particularly the ten-warhead SS-18). This outcome had resulted from mutual mistrust and the failure of the two superpowers to negotiate a ban on multiple, independently targetable re-entry vehicles (MIRVs) in the late 1960s, largely because of the need for on-site verification. Now, conservative analysts in Washington feared a possible Soviet "disarming" first strike against U.S. nuclear forces. While critics noted the likely survivability of many U.S. bombers and the entire U.S. submarine fleet, the momentum to restore America's military might from its post-Vietnam malaise carried the administration forward. In regard to space, the administration argued that Moscow was gearing up for possible space war with its ASAT test program and its new radar system at Krasnoyarsk, whose location, Washington argued, represented an effort to construct a nationwide missile defense in violation of the ABM Treaty. These developments, in the view of top Reagan administration officials, put a new priority on restoring U.S. space defense capabilities, regardless of the cost or treaty implications. These policies marked a major shift in U.S. space security strategy and would put U.S. efforts on a collision course with Moscow.

Soon after Reagan's election, the conservative Heritage Foundation think tank in Washington sponsored a study by Lieutenant General (Air Force, ret.) Daniel O. Graham on future space priorities. Entitled *High Frontier: A New National Strategy*, the published report urged the creation of a three-level space-based defensive barrier employing a network of 432 orbital missile platforms.[41] Over the advice of the White House science staff, Graham and his team had succeeded in convincing Reagan that a revolutionary defensive shield could once and for all do away with U.S. vulnerability to Soviet land-based nuclear weapons. However, after consultations with the Joint Chiefs of Staff, who opposed the project, *High Frontier* was temporarily shelved.

But the general thrust in policy had been set, and Reagan's shift in space priorities from civilian to military programs began to be seen in space expenditures. From 1981 to 1983, the defense department's space budget jumped from $4.8 billion to $8.5 billion,[42] surpassing NASA's budget for the first time since 1960. Much of this funding, however, would go into military support programs, whose advanced technology made them increasingly expensive.

In an effort to stimulate U.S. civilian space activity, the Reagan administration made a similarly dynamic mark on U.S. commercial space policy. With its economic ethos derived from the free-market school of the University of

[41] Donald R. Baucom, *The Origins of SDI: 1944–1983* (Lawrence: University of Kansas Press, 1992), p. 164.

[42] Stares, *Militarization of Space*, Appendix 1, Table 2, p. 257.

Chicago, the administration took a strong ideological line against NASA's traditional dominance of American commercial space policy. In the Reagan years, a number of key policy changes cut out NASA's exclusive privileges in the use of U.S. space facilities and its preferential pricing policy for satellite launches. Against the wishes of NASA officials, the new policy directives encouraged the formation of a private space industry in a variety of fields, but particularly in the all-important launch area. To supervise and support these developments, the Department of Transportation opened a new Office of Commercial Space Transportation. In a parallel initiative, the Department of Commerce formed the Office of Commercial Space Programs to help new companies overcome bureaucratic obstacles in their efforts to market other space-related services (for example, in the area of packaging space experiments for launch) to American and foreign businesses.

Meanwhile, NASA itself responded to these organizational reforms by setting up its own Office of Commercial Programs in 1984 to market NASA space products (including research space on the shuttle, satellite launch services, and satellite retrieval services) to private businesses and foreign governments. Another significant Reagan initiative in this area included the Landsat Commercialization Act of 1984. This transferred the Landsat remote-sensing system to the private EOSAT (Earth Observation Satellite) Corporation of Lanham, Maryland. The White House also moved to privatize the weather satellite METSAT (Meteorological Satellite), but Congress rejected the proposal.[43]

Although NASA registered a major accomplishment in April 1981 with the launch of the shuttle *Columbia*, the agency was clearly suffering from budget problems. As one writer who reported on the launch describes the contrasts now evident at Cape Canaveral from the boom years of the 1960s, "Along U.S. Route 1 abandoned businesses and buildings were commonplace. It was like visiting the battle site of an old war fought by another generation."[44]

On the international front, the early 1980s saw no progress in space arms control and, consistent with the administration's plans, considerable backsliding and a return to unilateral military policies. Paul Stares writes, "Throughout President Reagan's first term, initiatives for space arms control came solely from the Soviet Union."[45]

To make matters worse, the leadership crisis in Poland and the growing power of the trade union Solidarity brought new difficulties for space. The

[43] See James Everett Katz, "New Directions Needed in U.S. Space Policy," in *International Space Policy*, Daniel S. Papp and John R. McIntyre, eds. (New York: Quorum, 1987), pp. 52–53.

[44] Trento, *Prescription for Disaster*, p. 187.

[45] Paul B. Stares, *Space and National Security* (Washington, D.C.: Brookings Institution, 1987), p. 148.

declaration of martial law by General Wojciech Jaruzelski, and his perceived link to Kremlin strategists, led the Reagan administration in 1982 to reject the five-year renewal of the original 1972 agreement on U.S.-Soviet space cooperation, thereby ending even the limited ongoing program of scientific cooperation. The Reagan administration also moved forward with an elevation of space within the air force by creating a separate Space Command in September 1982 to coordinate the service's growing realm of responsibilities for military support operations from space, as well as possible future offensive missions.[46] The air force believed that the Soviet Union was preparing to seize the offensive in space, although critics doubted the evidence.[47] As Under Secretary Edward Aldridge stated in 1982, "If we need weapons in space, we will go get them."[48] Treaties now seemed to matter for very little, although new military space programs would still need to go through Congress.

As his new NASA chief administrator, President Reagan appointed former NASA employee and aerospace contractor James Beggs. Although business experience helped him win the appointment, the record of the next few years would show that Beggs lacked the necessary political skills to master Washington conservatives, who rode roughshod over the agency in their efforts to promote the military space program. As Joseph Trento notes, "While Beggs' people were trying to figure out how to keep a manned program of large proportions going on a tight budget, defense planners were already looking at the civil program's budget as a possible source of revenue."[49] But NASA managed to continue many aspects of its purported main job—space science research—including technical work on the growing problem of orbital debris. Early in the administration, NASA moved forward with a ten-year study program to assess the threat and the risks it might pose to spacecraft.[50] In July 1982, NASA also held its first national conference on orbital debris at the Johnson Space Center, which included representatives from the Department of Defense (DOD), industry, and outside U.S. research institutions.[51]

---

[46] William E. Burrows, *This New Ocean: The Story of the First Space Age* (New York: Random House, 1998), p. 548.

[47] Johnson, *Soviet Military Strategy in Space*, p. 159.

[48] Aldridge, quoted in Johnson, *Soviet Military Strategy in Space*.

[49] Trento, *Prescription for Disaster*, p. 187.

[50] Johnson and McKnight, *Artificial Space Debris*, p. 85.

[51] Notably, at this time, DOD did not yet believe the problem would affect its own missions until at least the year 2000. On this point, see Johnson and McKnight, *Artificial Space Debris*, p. 86. On the findings of this technical conference, see NASA, *Orbital Debris*, Proceedings of a workshop held at NASA Lyndon B. Johnson Space Center, Houston, Texas, July 27–29, 1982 (compiled by Donald J. Kessler and Shin-Yi Su) (Washington, D.C.: NASA, Scientific and Technical Information Branch, 1985).

In early 1983, *High Frontier* had been revived by President Reagan and a tight circle of close associates (including National Security Advisor Robert McFarlane). While McFarlane later argued that the intention was to create a "bargaining chip" for arms control talks with the Soviet Union,[52] Reagan clearly had invested himself much more heavily in the program.[53] Against technical advice, Reagan viewed SDI as a literally heaven-sent answer to America's future security problems with the Soviet Union: an impenetrable space shield and a means of effecting nuclear disarmament.[54] Among the Joint Chiefs of Staff, Navy Admiral James Watkins proved the most supportive and bought into Reagan's vision as the only way out of a "strategic valley of death" that Watkins believed otherwise favored the Soviet Union.[55]

For many supporters of space-based defense within the U.S. weapons laboratories, such as Lawrence Livermore physicist Edward Teller, these technologies also represented a promising means of reversing the downturn in funding and recruitment that had plagued the laboratories since the late 1970s and the recent rise of the "nuclear freeze" movement.[56] Young physicists and engineers had sought to avoid the protesters that now crowded the gates to U.S. weapons facilities, which led them instead to opt for more lucrative careers in industry.[57] By being able to offer extensive research funding in technologies that would perhaps save the world from nuclear weapons, laboratory officials believed they could turn the tables on the protest movement in the struggle for the hearts and minds of young scientists. They could also bring massive new funding into the laboratories on the scale of the Manhattan Project.

In a special address to the nation on defense and national security on March 23, 1983, Reagan announced SDI to the nation. The speech outlined a bold plan to "counter the awesome Soviet missile threat with measures that are defensive" relying on the "very strengths in technology that spawned our great industri-

---

[52] Brock Brower, "Bud McFarlane: Semper Fi," *New York Times Magazine*, January 22, 1989, p. 32.

[53] As Robert Joseph argues, critics "failed to realize that SDI was intended to transform the Soviet-American strategic relationship." See Joseph's chapter, "The Changing Political-Military Environment," in James J. Wirtz and Jeffrey A. Larsen, eds., *Rockets' Red Glare: Missile Defenses and the Future of World Politics* (Boulder, Colo.: Westview Press, 2001), p. 60.

[54] See Lettow, *Ronald Reagan and His Quest to Abolish Nuclear Weapons*.

[55] On Watkins's support for SDI, see Lettow, *Ronald Reagan and His Quest to Abolish Nuclear Weapons*, pp. 93–96. See also Hays, *United States Military Space*, pp. 89–90.

[56] The freeze movement—started by Boston-based activist Randall Forsberg and colleagues—aimed at halting the nuclear arms race by calling for an immediate freeze on U.S. and Soviet development of new weapons followed by negotiations for their reduction.

[57] On this point, see William J. Broad, *Star Warriors: A Penetrating Look into the Lives of the Young Scientists Behind Our Space Weaponry* (New York: Simon & Schuster, 1985), pp. 24–25.

al base and that have given us the quality of life we enjoy today."[58] President Reagan's aim was clearly transformational, as he stated his own personal conviction "that the human spirit must be capable of rising above dealing with other nations and human beings by threatening their existence."[59] He therefore proposed what he described as a "long-term research and development program to begin to achieve our ultimate goal of eliminating the threat posed by strategic nuclear missiles."[60] But, in an interesting caveat, Reagan expressed his own recognition of the interdependence of space activity and arms control in that defenses paired with offenses could be "viewed as fostering an aggressive policy, and no one wants that."[61] Therefore, he outlined an idealistic course of tying SDI's deployment to eventual "arms control measures to eliminate the [offensive] weapons themselves."[62]

The SDI program caused immediate debate in the media, in Congress, and within the administration itself. Fearing that internal criticisms of the plan could torpedo it before he could reach the American public, Reagan had kept this part of his speech hidden from many top advisors, even separating it from the rest of the speech in draft notes and taking special care to alter several sections of the draft speech with his own handwritten notes. Few officials even in the defense department had even heard of the program before its announcement.[63] But SDI (dubbed "Star Wars" after the popular 1977 movie and its 1980 sequel) soon occupied national debates on space security and fiscal decision making.

Existing funding for the projects included in the program totaled some $1 billion (in then current dollars), although by summer Congress would be debating a $1.8 billion presidential request (of which $1.6 billion would be approved). This funding stream would more than double by Reagan's last year in office.[64]

The SDI program combined a series of preexisting research efforts with a range of exotic new technologies aiming to push the technological envelope

[58] Printed and hand-marked copy of President Reagan's speech catalogued as "President Reagan's Backup Copy: Address on Defense," March 23, 1983, Final Version, 8:00 PM, Ronald Reagan Library, Files of White House, Office of Speechwriting, SP 735, Folder 1, p. 19.

[59] Ibid., p. 18.

[60] Ibid., p. 22.

[61] Ibid., p. 21.

[62] Ibid., p. 22.

[63] As Edward Reiss notes: "The joint chiefs of staff, the Secretary of Defense and the Secretary of State had only two days' warning of its contents." (See Edward Reiss, *The Strategic Defense Initiative* (New York: Cambridge University Press, 1992, p. 38.) These points support Paul Lettow's notion that the concept of bringing high technology forward as the SDI program was an idea Reagan had fostered himself. See Lettow, *Ronald Reagan and His Quest to Abolish Nuclear Weapons*, pp. 82–83.

[64] "Historical Funding for MDA FY85–07," Web site of the Missile Defense Agency, U.S. Department of Defense, at <http://www.mda.mil/mdalink/pdf/histfunds.pdf> (accessed October 13, 2006).

well beyond the realm of what might normally be possible in a more conservative research and development program. With presidential support and what would soon be a soaring budget, the program began to investigate technologies that might earlier have been rejected as impractical or too costly. Some had only tepid support from the White House, and there were divisions within the administration as to what to fund. Livermore physicist Lowell Wood and his mentor Teller had been nursing a small research program for several years called Excalibur: a space-based, nuclear-pumped, X-ray laser reportedly capable of the kind of high-power, multiple shots needed to destroy large numbers of nuclear-tipped missiles in their boost phases, before they deployed their more dangerous multiple warheads.[65] They criticized *High Frontier*'s conventional interceptors as requiring too many launches and too high a price tag,[66] believing their own nuclear system to be cheaper. While tests of the program would require withdrawal from the Partial Test Ban Treaty and the Outer Space Treaty, the program now received new funding, although the main emphasis eventually turned to a related program for chemical-powered, space-based lasers.[67]

Consistent with the president's aims, however, the main emphasis of the SDI program consisted of a variety of high-technology, nonnuclear weapons: magnetic rail guns that would hurl metal projectiles at high speed into rising ballistic missiles, orbital space-based interceptors (kinetic-kill vehicles) that would collide with their targets, ground-based interceptors for midcourse interception, and orbital space-based interceptors. The most advanced of the kinetic-kill programs included the Exoatmospheric Reentry Vehicle Interceptor System (ERIS), intended to hit Soviet warheads before they plunged back into the atmosphere, and the High Endoatmospheric Defense Interceptor (HEDI), which would aim at final attacks on warheads shortly after their re-entry.[68] Adding to its costs, the system would require an array of space-based and ground-based sensors, computers, and communications links needed to support the significant identification, tracking, targeting, homing, and verification requirements of an effective missile defense system.[69]

---

[65] On the roots of the Excalibur program, see Broad, *Star Warriors*, pp. 105–27; Peter Goodchild, *Edward Teller: The Real Dr. Strangelove* (Cambridge, Mass.: Harvard University Press, 2004), pp. 328–44; and Frances Fitzgerald, *Way Out There in the Blue: Reagan, Star Wars and the End of the Cold War* (New York: Simon & Schuster, 2000), pp. 128–30.

[66] Baucom, *Origins of SDI*, p. 156.

[67] Neither Reagan nor his Science Advisor George ("Jay") Keyworth backed this program, Reagan because of its nuclear character, which conflicted with his goals for SDI, and Keyworth because of anticipated technical problems. On these points, see Lettow, *Ronald Reagan and His Quest to Abolish Nuclear Weapons*, p. 82.

[68] See Fitzgerald, *Way Out There in the Blue*, p. 374.

[69] On the weapons and sensor technologies involved, see Sanford Lakoff and Herbert F. York,

Separate from SDI, the administration also announced plans to move forward with a National Aerospace Plane (NASP), a concept that would replace the earlier Dyna-Soar. Air force officials hoped to find a role for the next generation of pilots—in space.[70] Like Dyna-Soar, however, technical and mission-related problems kept funding for the NASP at comparatively low levels before Congress canceled it altogether.

In many respects, the SDI program became a kind of "blind man's elephant." For laboratory and industry supporters, it was a cash cow and financial savior in a time of relatively weak government support for space. For hard-liners, it represented an opportunity to build strong defenses to supplement perceived weakened U.S. offensive capabilities, thus restoring a potent U.S. deterrent against the Soviet threat. For moderates in the administration, SDI was a means to an end. Like the U.S. ABM system and ASAT capabilities, it was to be a tool used mainly for future arms control bargaining. For opponents, however, it represented a new stage in the U.S.-Soviet arms race and a serious threat of its extension into space. In any case, the system represented a clear rejection of the policies of collective military restraint followed by both of the superpowers in space since 1963 and an attempt instead to use a weapons-led approach to achieve the same end. If the administration were to be believed, the deployment of SDI would come about only with the simultaneous elimination of nuclear weapons. Getting from here to there—and handling the major transaction costs of such a shift—would not be trivial and risked a major destabilization of the space environment.

Gradual Changes in Soviet Space Policy

During the late 1970s, the Soviets had begun construction of a new, larger, and more advanced *Mir* (peace) space station. This long-duration spacecraft, which eventually incorporated multiple modules, reached orbit in 1986.[71] The project marked the fruition of a decade and a half of steady progress within existing organizations and a focus on lower-profile missions after the politically successful but technologically costly emphasis on spectacular firsts under Khrushchev. Yet Brezhnev and his immediate successors continued to use space as a political tool, albeit to a lesser extent than Khrushchev. For example, the

---

A Shield in Space? Technology, Politics, and the Strategic Defense Initiative (Berkeley: University of California Press, 1989), pp. 84–131.

[70] See "X-30 National Aerospace Plane (NASP)," on the Federation of American Scientists Web site at <http://www.fas.org/irp/mystery/nasp.htm> (accessed November 3, 2006).

[71] On the architecture of the *Mir* station, see Siddiqi, *Soviet Space Race with Apollo*, p. 840.

Soviets "pre-empted" the 1983 flight of the first American woman (Dr. Sally Ride) into space by sending up their second female cosmonaut. Also, a space-walk by a third female cosmonaut was ordered soon after the United States announced plans to conduct the first female EVA (extravehicular activity).

In the passive military sector, following earlier U.S. advances with the KH-11, the Soviet military had been eager to develop its own digital imaging capabilities. In December 1983, the Yantar 3KS1 system (launched as *Kosmos 1426*) provided this new asset to strengthen the existing Soviet reconnaissance program with real-time imaging.[72] Unlike previous-generation reconnaissance vehicles, the Yantar 3KS1 technology aimed at year-long missions conducted in approximately circular obits at 140 miles up.[73] However, Soviet manned spacecraft continued to be called upon for reconnaissance functions, suggesting that some gaps in the Yantar system remained. The Soviets also retained a ready-launch capability for old-style photo-reconnaissance missions using film return. This "assembly line"[74] approach to space meant that while Moscow would rarely outclass American technology, a tried-and-true system was always ready to be thrown into space in a crisis. This capability caused some in the United States to worry that expensive, multimission U.S. spacecraft would be more vulnerable in case of a superpower war.[75]

But Moscow's military space activities continued to experience periodic accidents. In 1978, a Soviet nuclear-powered military reconnaissance satellite (*Kosmos 954*) had suffered problems while attempting to boost to a final parking orbit at the end of its service life. Instead, it crashed into the atmosphere, eventually breaking up and spewing highly enriched uranium fuel and other debris across a wide swath of territory in northern Canada.[76] The growing Soviet deployment of nuclear-powered RORSATs continued to raise serious environmental and safety concerns, particularly after a second RORSAT accident in 1983 over the Atlantic Ocean.[77]

[72] Harvey, *Russia in Space*, p. 115.

[73] Ibid.

[74] Nicholas Johnson, quoted in Burrows, *Deep Black*, p. 258.

[75] These persons included officials from Ford's Pentagon, including Secretary Rumsfeld and the Director of Defense Research and Engineering Malcolm Currie. See Stares, *Militarization of Space*, p. 173.

[76] Johnson, *Soviet Military Strategy in Space*, p. 62. According to the terms of the 1972 Liability Convention, the Soviets later (in a 1981 settlement) agreed to pay $3 million to Canada toward the costs of the cleanup, although they did not admit fault and failed to cover the estimated $14 million in total damages requested by Canada.

[77] The second Soviet RORSAT accident spread radioactive material across a wide section of the Atlantic Ocean. See Johnson, *Soviet Military Strategy in Space*, p. 63. By 1986, when the program was discontinued by the Soviet Union, twenty-eight nuclear reactors from RORSATs had been success-

In the early 1980s, however, some hints of the weakening of military dominance within the Soviet space program began to emerge, as Soviet leaders began to see the negative effects internationally of their military space activities and the inadequacies of their commercial space program. The first result was a shift in arms control policy, involving a cutoff of ASAT testing in 1982 and other new overtures under Soviet General Secretaries Yuri Andropov and Konstantin Chernenko. The second was a more open attitude toward overall space cooperation with the West, especially in space science.

In a reversal of the 1960s, the Soviets now led moves to reduce security tensions in space. In August 1981, Moscow proposed at the United Nations a complete ban on the orbiting of any weapons in space.[78] As Stares notes,[79] the new proposal included one Soviet concession from its previous position in the dropping of consideration of the U.S. space shuttle as an ASAT system.[80] However, the Reagan administration was not moved. It viewed the proposal as simply another attempt by the Soviet Union to keep its existing ASAT system (which was not "stationed in orbit") and to prevent America from developing a rival system. Soviet efforts to bring the United States to the bargaining table continued through 1982. The Soviets conducted three ASAT tests during 1981 and 1982, although two of them failed.[81] The Soviet military also continued development on two other space weapons programs begun in the 1970s as a hedge against a possible U.S. breakout. Although originally intended as counters to expected U.S. military use of the space shuttle, this research and development effort consisted of work on a possible space-based, laser-equipped battle station (Skif, or system 17F19) and a similar space-to-space missile station (Kaskad, or system 17F111).[82]

After Reagan's March 1983 speech and plans for SDI, the Soviets followed a classic dual-track strategy: their secret military programs accelerated while Soviet arms control efforts became more vocal and urgent. In August 1983, Soviet General Secretary Andropov finally announced a unilateral Soviet moratorium on launching ASATs into space. In itself, this could be seen as a purely political

fully boosted to parking orbits of 900 km. Unfortunately, they will remain in space indefinitely and are still there as of fall 2006 in only slightly decaying orbits.

[78] "Soviet Draft Treaty Submitted to the U.N. General Assembly: Prohibition of the Stationing of Weapons of Any Kind in Outer Space," August 11, 1981.

[79] See Stares, *Space and National Security*, p. 148.

[80] Previously, the Soviets had claimed that because of the shuttle's maneuverable arm—which can retrieve satellites for repair—it must be considered an active ASAT system.

[81] Johnson, *Soviet Military Strategy in Space*, Table 6, p. 154.

[82] On these programs, see Konstantin Lantratov, "The 'Star Wars' That Never Happened: The True Story of the Soviet Union's Polyus (Skif-DM) Space-Based Laser Battle Stations" (Part I), *Quest*, Vol. 14, No. 1 (2007).

move since the Soviets already possessed a relatively well-tested ASAT system.[83] But a formal U.N. proposal delivered by Foreign Minister Andrei Gromyko soon after showed that the Soviets were prepared to make serious concessions to bring about an agreement. Their new draft treaty proposed a ban on testing and deployment of all "space-strike" weapons, and a prohibition on testing and use of manned ASATs (in which they included the U.S. space shuttle, because of its manipulator arm and cargo bay) or other military systems. Most significantly, they called for the destruction of *existing* ASAT systems. U.S. negotiators, however, declined to take the Soviets up on their offer, arguing that since Washington did not know how many Soviet ASATs already existed, it could never be sure that Moscow had destroyed them all. U.S. views on Soviet cheating on the ABM Treaty at Krasnoyarsk also had an impact on U.S. negotiators.[84] Given U.S. disinterest and its pledge to develop SDI, the military "hedge" part of the Soviet strategy now moved into higher gear. The Soviet approach would not be a symmetric one aimed at countering U.S. ballistic missiles, but instead a simpler, asymmetric effort aimed at taking out the U.S. SDI interceptors based in space.[85]

While it is not yet known if the United States knew about these secret programs, it did know that the Soviets had been steadily expanding other military capabilities that could be used in a time of war, should such aggressive intentions eventually become part of Moscow's plans for space. In September 1984, Moscow orbited a new signals intelligence spacecraft (*Kosmos 1603*) to bolster its capabilities to intercept telemetry from U.S. tests.[86] It also continued research into ground-based lasers, some of which might have been capable of being used against satellites. The Soviets made the plausible claim that the laser station's main work involved space tracking, not weapons research.[87] In 1984, however, the Pentagon charged that Moscow was conducting laser research at the Sary Shagan facility in then-Soviet Tadzhikistan for use against space objects in low–Earth orbit.[88] Later visits by U.S. experts to these sites after the Cold War,

---

[83] The Soviet military had declared its co-orbital ASAT system operational in July 1979, although its capabilities still remained limited, since its attack profile involved hours of transparent tracking. See Pavel Podvig, ed., *Russian Strategic Nuclear Forces* (Cambridge, Mass.: MIT Press, 2001), p. 434.

[84] Soviet negotiators, of course, argued that the United States had cheated on its ABM commitments too by upgrading its radar systems at Thule and Flyingdales.

[85] Lantratov, "'Star Wars' That Never Happened," p. 7.

[86] Burrows, *Deep Black*, p. 268.

[87] On this dispute, see Burrows, *This New Ocean*, pp. 547–48. According to longtime journalist Burrows, "Since the United States never had any good intelligence on Soviet laser ASATs, it fell into the trap of basing its estimates on its own growing laser weapon capability, which was better than the enemy's" (p. 548).

[88] Johnson, *Soviet Military Strategy in Space*, p. 174. Earlier charges had been made in 1981 about an alleged facility 30 miles from Moscow. These too remained unproven, suggesting the possible

however, cast doubt on the veracity of at least some of these claims. Certain capabilities had been achieved, but attempts to place them in an orbital station ran into some of the same problems that occurred in the SDI program—too complicated and simply too heavy.[89]

The rapid transition of Soviet leaders—due to old age and death—caused Moscow's tentative effort at space arms control to stall. By 1984, the Soviet space program was in many ways outdated. Decision making still required direct intervention by the Politburo, channels of communication among enterprises remained blocked by both secrecy requirements and by bureaucratic controls, and there were no civilian enterprises in place within the space program to provide technological spin-offs for the domestic economy. The Soviet military used the space program as it needed for a variety of support missions and for expensive research and development on next-generation weapons, but the nation in general had little to show from its space prowess save international scientific prestige. Behind the scenes, the planned economy was proving increasingly inefficient and the costs of military programs (in part because of the war in Afghanistan) had begun to add up. Still, the system teetered on in its old dysfunctional way.

The Soviets did follow through on their contribution to and participation in the COSPAS/SARSAT program initiated under Carter. Owing to the program's success, by 1985 the four original members—the United States, the USSR, France, and Canada—had been joined by Norway, Great Britain, Sweden, Finland, and Bulgaria. With the failure of the first U.S.-launched SARSAT in 1984, the system relied heavily on Soviet satellites. Ironically, the Soviet Union benefited very little from the system because of the small number of transmitters carried initially by its ships and aircraft compared to the United States, where two hundred thousand civilian aircraft and six thousand ships were linked to the system by the late 1980s.[90]

At around the same time, under new orders from the Reagan White House, the United States began to construct a civilian version of the previously closed NAVSTAR satellite navigational network, which used a much more accurate technology employing multiple satellites and triangulation. The new GPS program aimed at eventual deployment of twenty-four satellites by the end of the

---

impact of threat inflation by the administration. William Burrows argues that claims in such U.S. government documents as the Pentagon's 1984 *Soviet Military Power* that the United States lagged behind the Soviet Union in laser research were "pure rubbish." (See Burrows, *This New Ocean*, p. 540.)

[89] Lantratov, "'Star Wars' That Never Happened," p. 7.

[90] See Office of Technology Assessment, *U.S.-Soviet Cooperation in Space*, p. 113.

decade and planned to have a more accurate (and classified) military signal.[91] This technology would eventually generate a wave of space-related commercial applications in the 1990s, when GPS receivers would begin to become available for boats, cars, and hikers.

However, just at this time, the U.S. government began to express renewed interest in cooperating with the Soviets in civilian space activity. In June 1984, President Reagan noted in a speech the desirability of new cooperative programs. Congress quickly supported this move with the passage of Public Law 98–562, which called for the renewal of the 1972/1977 space cooperation agreement with the Soviet Union. The fall 1984 presidential campaign also brought to light administration goals of a nuclear arms control agreement with the Soviets, the first serious mention of the topic since 1981.

As the 1984 U.S. elections neared, Soviet General Secretary Chernenko began to hint at a Soviet compromise on its previously firm policy of refusing to consider space arms control separately from other security issues, such as the now-moribund SALT II agreement.[92] Indeed, prior Soviet space weapons programs were now seen as counterproductive by Moscow, as the United States had announced its SDI in response to prior Soviet ASAT testing.[93] Thus, the new focus in space began to take into account other, less threatening means of competing with the United States.

Following Reagan's reelection, this policy shift became clearer, paving the way for a U.S. response early in the next administration.

The Second Reagan Term: "Trust but Verify"

Having won a second term, President Reagan began to modify the official U.S. view of the Soviet Union, easing the harsh rhetoric of his first term and reducing the level of tension in the U.S.-Soviet relationship. Indeed, a second track within U.S. SDI policy began to be developed at this time. While one side of the administration had picked up on the necessity of testing and deployment, the other focused on moving toward the overall strategic objectives for

[91] On these plans for the GPS system and its roots in the NAVSTAR program, see Norman Friedman, *Seapower and Space: From the Dawn of the Missile Age to Net-Centric Warfare* (Annapolis, Md.: Naval Institute Press, 2000), pp. 266–67.

[92] On this change in Soviet policy, see Howard M. Hensel's "Alternative Western Interpretations of Soviet Ballistic Missile Defense and Space Policy," in Daniel S. Papp and John R. McIntyre, eds., *International Space Policy: Legal, Economic, and Strategic Options for the Twentieth Century and Beyond* (New York: Quorum, 1987), p. 277.

[93] Author's interview with Academician Roald Sagdeev at the Space Research Institute in Moscow, April 4, 1988.

disarmament established by Reagan himself. Although the press still clung to the dangers of the Reagan vision, the administration maintained a consistent policy of emphasizing SDI's stabilizing effects in a planned *bilateral* deployment of the system. In early 1985, Secretary of State George Shultz reemphasized in a speech that a prerequisite for any SDI deployment would be the prior establishment of "agreed ground rules" with the Soviet Union. Shultz continued, "The alternative—an unconstrained environment—would be neither in our interest nor in theirs."[94] Thus, while critics charged that Reagan's pledge of sharing the technology had been insincere, the administration seemed to be backing up this policy. Moreover, in February, the administration announced three criteria, developed by Reagan advisor and longtime defense policy expert Paul Nitze, that would govern SDI's deployment: (1) the program must be effective; (2) it must be survivable; and (3) it had to be "cost-effective" on the margin (in other words, cheaper than the Soviet construction of offensive countermeasures).[95]

This modification of the hard-line policy followed by the administration since 1981 had to do with a number of factors. Part of the reason lay in the increasing difficulty of convincing Democrats in Congress to continue supporting SDI. While Republicans had seized the Senate in 1980, Democrats continued to control the House. Later, in the 1986 midterm elections, Democrats took back the Senate, granting them considerable influence over both fiscal and policy matters during Reagan's second term.

SDI faced Democratic congressional opposition on three grounds. First, there was the issue of expense, which even conservative estimates put at a deployed cost of hundreds of billions of dollars. As the federal budget deficit reached record figures, Congress began to crack down on the high-priced defense policies of the early 1980s. In this evolving fiscal environment, space "racing" against the Soviets in any area would be impossible. Yet, ironically, while the total cost of SDI would amount to over $17.5 billion from 1983 to 1989,[96] cost alone was not the determining factor in its demise. During the 1980s, in fact, the United States would actually spend far more—approximately $50 billion—on research, development, and deployment of passive military space systems.[97]

---

[94] Secretary of State George Shultz, speaking in Austin, Texas, March 28, 1985. (Cited in John A. Jungerman, "The Strategic Defense Initiative: A Primer and Critique," Institute on Global Conflict and Cooperation, University of California, San Diego, 1988, p. 32.)

[95] Lettow, *Ronald Reagan and His Quest to Abolish Nuclear Weapons*, p. 155.

[96] Figure calculated by author from estimated missile defense project figures for FY 1983 ($1 billion), known FY 1984 appropriations ($1.6 billion), and FY 1985–89 figures, provided in "Historical Funding for MDA FY85–07," Web site of the Missile Defense Agency, U.S. Department of Defense, at <http://www.mda.mil/mdalink/pdf/histfunds.pdf> (accessed October 13, 2006). All figures are in then-current (unadjusted) dollars.

[97] Burrows, *Deep Black*, p. 201. These unadjusted dollar figures combine intelligence procure-

Second, a growing body of technical information gathering within the U.S. government began to conclude that the SDI program was unachievable at any cost. The chemical lasers intended for the program were too heavy to launch. Space-based mirrors—to redirect beams sent from ground-based lasers—seemed to be the answer, but aside from weather concerns, technicians could not figure out how to grind them fine enough to ensure direct hits, as a few inches meant the difference between success and total failure. Another complication was the dilemma of how to fold the so-called battle mirrors for launch and then open them in orbit. Other systems faced serious obstacles as well. Given the large number of Soviet missiles, Brilliant Pebble interceptors would have to deal with the "absentee" problem—up to three thousand would need to be deployed in orbit to ensure an adequate number over key targets at any given launch time. With the shuttle grounded and a number of other launchers having been discontinued, the United States lacked the wherewithal to put that many required interceptors into orbit.[98] Other parts of the SDI system would be even harder to put into space. Moderate voices within the administration now began to emerge, urging some compromises regarding SDI in the face of popular and congressional criticism.

Finally, as prototype systems—such as chemical lasers—began to enter their testing phases, congressional criticism of SDI became more vocal on the grounds of a potential treaty violation. The administration had announced after a study conducted in 1985 that by its interpretation, Agreed Statement D of the ABM Treaty banned only existing ABM systems and that weapons based on "new physical principles" could be developed and tested.[99] The actual clause in the annex stated that "specific limitations on such systems would be subject to discussion," thus seeming to allow them only with Soviet approval. Critics, such as ABM Treaty conegotiators Albert Carnesale, John Rhinelander, and Gerard Smith (now out of the government), stated that the Reagan administration's interpretation clearly violated the treaty's intended meaning in 1972.[100] However, without congressional action, the Reagan administration forged ahead with SDI according to its own view of the treaty.

Separate from the SDI program, the administration and the air force continued to move forward to overcome a perceived U.S. deficit in ASAT technol-

---

ment and research and development spending by the National Reconnaissance Office and the air force, and reconnaissance spending by the navy and the CIA. As Burrows comments regarding this commitment of U.S. funds, "Allowing for inflation, that is almost twice what it cost to land men on the moon."

[98] Author's interview with Dr. Sally Ride, NASA, Washington, D.C., August 1987.

[99] For a discussion on the administration's perspective, see Hays, *United States Military Space*, p. 94.

[100] Burrows, *This New Ocean*, p. 542.

ogy, in light of Soviet developments with its co-orbital program. But experts recognized that the use of kinetic-kill interceptors would have negative effects for U.S. space security in terms of debris. Scientists from NASA and Teledyne-Brown Engineering "voiced concerns about the longevity of the [likely] debris" to the air force, causing the issue to be "vigorously debated" within U.S. Space Command in Colorado Springs.[101] Notably, these objections eventually won over NORAD's commander in chief. But the air force proceeded with a test anyway on September 13, 1985, launching a missile from an F-15 aircraft carrying a Miniature Home Vehicle (MHV) that successfully intercepted an aging Solwind satellite (P-78) at an altitude of 373 miles. The test created 285 pieces of trackable orbital debris,[102] and an unknown quantity of smaller particles. While some debris de-orbited quickly, trackable pieces remained in orbit until 2004, vindicating the predictions of the experts.

The administration's decision to overrule congressional opposition to the test (on the separate grounds of its strategic implications) started an even hotter political firestorm. Congress moved quickly to forbid additional tests of the ASAT system, unless Moscow resumed similar tests. The final 1986 appropriations bill forbade future U.S. ASAT testing against spacecraft for as long as the Soviet Union continued its own moratorium.[103] However, using a creative interpretation of the congressional language, the air force conducted two additional, nonexplosive ASAT tests in August and September of 1986, using background stars to test the heat-seeking sensor on the MHV interceptor.[104] Finally, after Congress put the same restrictions into the 1987 and 1988 bills, the air force canceled the ASAT test program in March 1988.[105] Similarly, rising anti-SDI sentiment led to congressional language insisting that the administration abide by the generally accepted understanding of the ABM Treaty used by the Nixon administration (the so-called narrow interpretation). Democrats now required, by this interpretation, that any testing of space-based ABM components should be banned. Despite lobbying by the administration for the "broad" interpretation of the ABM Treaty, the White House was unable to muster the necessary votes to block the new restrictions.

[101] Author's interview with Nicholas Johnson, now NASA's chief scientist for debris migration, Johnson Space Center, Houston, Texas, September 9, 2007 (conducted via e-mail). At the time of this 1985 test, Johnson worked on debris issues for Teledyne-Brown and participated directly in these debates in Colorado Springs with the air force brass.

[102] Ibid.

[103] Hays, *United States Military Space*, p. 111.

[104] Ibid.

[105] See Theresa M. Foley, "Slowdown in SDI Growth Delays Deployment Decisions," *Aviation Week and Space Technology*, February 22, 1988, p. 16.

More importantly, perhaps, the air force had begun to take seriously the threats to U.S. space assets and the broader orbital environment posed by kinetic-kill activities, particularly in highly trafficked areas of low–Earth orbit. In March 1986, the defense department formally acknowledged the problem of space debris and its potential impact on U.S. military operations. It also pledged itself to new debris mitigation guidelines, which stated: "Design and operations of DOD space tests, experiments and systems will strive to minimize or reduce accumulation of space debris consistent with mission requirements."[106] While the new policy did not ban debris-causing tests, it noted a new recognition of the problem and DOD's role in helping to mitigate it. Accordingly, the Strategic Defense Initiative Organization (SDIO) revised plans for a boost-phase interceptor test (Delta 180) on September 5, 1986, by conducting it at a "much lower altitude" (130 miles) than was originally planned, thus allowing debris from the test to de-orbit within six months.[107] This marked the beginning of the air force's adoption of debris mitigation guidelines as a matter of policy. As NASA debris expert Donald Kessler noted, "People are starting to see the dimensions of the problem, and do something about it."[108] At NASA, meetings began to be held with international launch companies to raise their awareness of the debris problem on the civilian side as well, where fuel-related explosions, exploding bolts used in stage separations, and other design and operational flaws continued to add steadily to the now-recognized orbital debris problem.

Despite this progress, safety had begun to be compromised in other areas of NASA owing to ongoing budget problems and political pressures to get the troubled shuttle program moving. After several launch delays, NASA site officials, facing subfreezing weather at the Kennedy Space Center in Florida on January 28, 1986, waived prelaunch requirements aimed at preventing a shuttle takeoff when critical O-ring booster seals might fail to do their job due to low temperatures. Unfortunately, this decision proved very costly. The *Challenger's* boosters exploded two minutes after takeoff, sending the space plane careening into the ocean, killing its crew of seven. Those on board included the first teacher (Christa McAuliffe) scheduled for a mission in space.[109] The disaster threw the nation into a period of shock as administration officials began to examine the web of political meddling in the space program that had led to

[106] DOD statement quoted in William J. Broad, "Orbiting Junk Threatens Space Missions," *New York Times*, August 4, 1987, p. C1.

[107] Author's interview with Nicholas Johnson, NASA, Johnson Space Center, September 9, 2007 (conducted via e-mail). The test caused two spent rocket stages to collide.

[108] Kessler quoted in Broad, "Orbiting Junk Threatens Space Missions."

[109] On the *Challenger* accident and the political, economic, and technical factors behind it, see Trento, *Prescription for Disaster*.

a mistaken decision under patently unsafe launch conditions.[110] While space activity had always been risky and NASA had often been used as a competitive political tool, this time its luck had run out. The shuttle would be grounded for two years.

The effect of the shuttle disaster proved to be not altogether negative for space security interests. For NASA, the refocusing of public attention on the organization led to the resignation of Acting Administrator Graham and a renewed concern for technical issues over political competition with Moscow.[111] In August 1986, the White House authorized approximately $2 billion to pay for a new shuttle to replace the *Challenger*, and within Congress there was a new interest in rebuilding the once respected U.S. space program. In the reevaluations that followed the disaster within NASA, changes could be seen in the agency's aims for the future. No longer would high-profile one-upmanship be the goal of manned space missions. No longer would politicians and foreign astronauts receive priority over scientists and researchers. The goal now was not to simply use space but to ensure that the United States remained focused on a new priority: developing space for commercial and scientific purposes. By 1989, the first U.S. commercial booster would lift a private materials processing experiment into space from the White Sands missile facility in New Mexico.

Gorbachev and Major Shifts in the Soviet Approach to Space

In the international realm, changing events also promoted reevaluations in American space policy. Most notably, in the Soviet Union, the death of General Secretary Chernenko in March 1985 had resulted in the Politburo's selection of Mikhail Gorbachev—the first Soviet leader to come from the post-Stalin generation. In the early 1980s, Reagan and his advisors had remained unconvinced of the sincerity of Soviet changes in space policy because of the enduring conservative nature of the Soviet leadership under first the ailing Brezhnev and then under the brief tenures of the old and sickly Andropov and Chernenko. But, as would become clear within a year after Gorbachev's assumption of power, U.S. hard-liners could no longer simply rest on past assumptions.

---

[110] For the results of this investigation, see NASA, *Report of the Presidential Commission on the Space Shuttle Challenger Accident*, in compliance with Executive Order 12546 of February 3, 1986, available on the NASA Web site at <http://science.ksc.nasa.gov/shuttle/missions/51-l/docs/rogers-commission/table-of-contents.html> (accessed October 3, 2007).

[111] Ironically, President Reagan brought in former NASA head James Fletcher to act as chief administrator during the reorganization.

Within months of entering office, the new Gorbachev leadership had begun a full-scale reevaluation of past policies. Recognizing the link between the secrecy surrounding Soviet space activities and Western assumptions about Soviet space militarism, Gorbachev and his advisors moved to break the stranglehold of the military on the space program. The creation of the commercial agency Glavkosmos in 1985 marked the first step in this diffusion of organizational power. Soviet information policy regarding space also became less restrictive, and cooperative overtures—to the West especially—were more forthcoming. Realizing that confrontation had neither stopped SDI nor won Moscow many supporters in the Western camp, the Gorbachev leadership took a more sophisticated tack in Soviet foreign affairs and space security efforts by moving to a policy of quiet negotiation.

What Gorbachev and his advisors were seeking from space now was much more significant than military capabilities. They were seeking a springboard for entry into the world political economy in the high-technology area, both to increase Soviet productive capabilities through the use of space technologies in domestic industry and to gain hard currency from sales of space equipment abroad. Although military systems already in the pipeline would move forward, the Gorbachev Politburo would no longer authorize new elements for an endless military space race. Soviet space-related technologies now constituted one of the few areas where the USSR was on par with—and, in some cases, even ahead of—world leaders. To capitalize on this advantage, the Soviets tried to create civilian agencies and production facilities that would make this asset usable in Soviet foreign trade policy. As one article explained the new philosophy behind these changes, "Never before has space played such a role in the business of the transformation of civilization as it does in our time."[112] Clearly, this commercial expansion was not the only aim of Soviet space policy, but it was an important new development that had a significant impact on the Soviet attitude toward international cooperation.

Taken together, changes in American and Soviet policy provided the groundwork for progress on the international front. Over the course of 1985, hostile rhetoric on both sides began to ease and concepts of détente slowly reemerged. Washington saw early signs of a possible shift in Soviet policy in the peaking of the virulent Soviet propaganda campaign against SDI, and began to respond. The framework for expert-level negotiations became the Nuclear and Space Talks in Geneva, which went through many rounds as the United States worked to convince the Soviets to allow a phased SDI deployment without responding

---

[112] V. M. Gavrilov and M. Yu. Sitnina, "Militarizatsiya kosmosa: novaya global'naya ugroza" [The militarization of space: a new global threat] *Voprosy Istorii*, No. 11, 1985, p. 94.

with a military buildup or breaking out of the arms control process.[113] At the same, the Soviet side worked to convince the American side that its willingness to engage in nuclear reductions was sincere—but would not go forward without U.S. restraint on SDI. In the end, the code words used by the Reagan administration to justify its gradual acceptance of nuclear arms control, while negotiations on SDI continued, would quote from the Russian phrase "Doveryai no proveryai" (Trust but verify).

In September 1985, a secret trip to Moscow by Director of the Jet Propulsion Laboratory Lew Allen opened a back channel for U.S.-Soviet communication aimed at the resumption of U.S.-Soviet space science cooperation.[114] Over the course of 1986, domestic follow-up talks ensued within both governments to work out positions for direct negotiation at the first Reagan-Gorbachev summit, planned for November.

The Reykjavik (Iceland) summit was remarkable for a nearly achieved accord for total elimination of nuclear weapons over ten years, but it was blocked by Gorbachev's concerns about the weaponization of space and his belief that the United States would try to use SDI for unilateral military gains. Or, viewed differently, the nuclear deal was blocked by Reagan's refusal to give up SDI. Although public records, including President Reagan's later speech to the nation, excluded mention of the details, Reagan's handwritten notes from the meeting (now at his presidential archive) and Gorbachev's later-published memoirs provide lively (and competing) descriptions of the course of the conversations. After covering a number of other topics, Reagan discusses his agreement with Gorbachev that intermediate-range nuclear-tipped missiles could be reduced and possibly eliminated in Europe and Asia. Reagan's notes continue:

> This then brought up the subject of S.D.I. and whether the deployment of such a system would violate ABM. I offered a proposal that we continue our present research and if [the technologies] reached the stage of testing we would sign *now* a new treaty that would permit Soviet observation of such tests and if the program was practical we would both eliminate our offensive missiles and then we would make available the S.D.I. system to the Soviets and others.[115]

---

[113] The space component of the Nuclear and Space Talks was referred to as the Defense and Space Talks. On the U.S. side, this delegation was headed by Ph.D. engineer, former air force official, and strong SDI supporter Ambassador Henry Cooper. Ambassador Max Kampelman, a lawyer, professor, and longtime U.N. and state department diplomat, headed the overall U.S. delegation to the Geneva talks. Kampelman, ironically, was a conscientious objector during World War II. He later embraced missile defense as a way out of the nuclear arms race and adopted a number of neoconservative stances.

[114] Robert C. Cowen, "Joint Program for US-Soviet Study of Mars Is on the Horizon," *Christian Science Monitor*, November 10, 1986, p. 9.

[115] Reagan's draft speech and notes from the Reykjavik meeting, Ronald Reagan Library, SP 1107, "Speeches: Iceland, Address to the Nation, D.C., 10/9/86," Folder 1, Document #439177, pp. 3–4.

Reagan then describes how he and Gorbachev adjourned, turning over the details of working out a specific formula to their respective summit staffs. The two sides wrestled until 2 A.M. to develop a concept presented to the leaders the following morning: a 50 percent reduction in strategic nuclear weapons within five years followed by their total elimination over the next five years. SDI research could continue during this time and then both sides would withdraw from the ABM Treaty and move to deploy SDI. However, according to Reagan, in the follow-on talks the next day, Gorbachev insisted that SDI research be halted during the arms reduction process, something he was not willing to support.[116]

Gorbachev's own account of the summit offers a contrasting interpretation. The former Soviet president states that the dispute centered not on SDI *research* but instead on the U.S. insistence on unlimited *testing*. The Soviet position was that testing before the two sides' mutual withdrawal would violate the ABM Treaty.[117] The Soviets also feared the development of a U.S. advantage. Notably, according to his own later remarks, Gorbachev did not yet know about the Soviet military's secret Skif and Kaskad programs,[118] which might have made him more confident in the possibility of Soviet countermeasures to SDI. Doubting U.S. intentions and believing SDI would lead to an arms race in space, he held firm against any compromise. The new Soviet leader remained puzzled about U.S. intentions, as he reflected later:

> Ronald Reagan's advocacy of the Strategic Defense Initiative struck me as bizarre. Was it science fiction, a trick to make the Soviet Union more forthcoming, or merely a crude attempt to lull us in order to carry out the mad enterprise—the creation of a shield which would allow a first strike without fear of retaliation?[119]

With their other work done and the arms control talks at an impasse, the two sides began to wrap up their meetings. As Gorbachev was departing, Reagan told the Soviet leader, "Meet me halfway and you'll feel the beneficial effects of American co-operation."[120] Gorbachev repeated to Reagan, "I'm ready to go right back into the house and sign a comprehensive document on all the issues agreed [to] if you drop your plans to militarize space."[121] But for Reagan, SDI had become too strong a political commitment to pull back on. The two leaders said good-bye and departed from Iceland.

[116] Ibid., p. 5.

[117] Mikhail Gorbachev, *Memoirs* (New York: Doubleday, 1995), p. 418.

[118] Konstantin Lantratov, "The 'Star Wars' That Never Happened: The True Story of the Soviet Union's Polyus (Skif-DM) Space-Based Laser Battle Stations" (Part II), *Quest*, Vol. 14, No. 2 (2007), p. 11. According to Lantratov, Gorbachev said to engineers at Baikonur in 1987, "It is a pity that I did not know all of this [about the Skif-DM station] before Reykjavik!"

[119] Gorbachev, *Memoirs*, p. 407.

[120] Ibid., p. 418.

[121] Ibid., p. 419.

Ironically, had Gorbachev known about his own possible hedge, he might have acted differently. With either the Skif or Kaskad, the Soviet Union could have acquired the capability to defeat U.S. interceptors in a time of war. Their task would have been considerably easier than that of the SDI interceptors, because all they needed to do was to punch a hole in the U.S. system. In any case, such space deployments by both sides would have constituted what Gorbachev opposed most: another round of the U.S.-Soviet arms race. Moreover, the SDI program would have required a massive arsenal of space-based systems, likely equipped with dangerous automatic-fire mechanisms (in order to function effectively in the boost phase). Such testing and deployments would have increased the chances of inadvertent war, vastly expanded space debris, and created conditions where civilian, commercial, and passive military missions would have fallen under the risk of mistaken attack.

Finally, had Gorbachev known about the Skif and Kaskad programs, he would also have recognized that both of these efforts still faced major technical hurdles.[122] Although their system requirements were far simpler than those of the SDI weapons, their development still faced at least several more years of research, testing, and development. The bottom line, as events would soon show, was that Gorbachev simply had no interest in tit-for-tat military competitions, which he believed had wasted Soviet funds in the past.[123]

Although arms control discussions had dominated the Reykjavik meetings, in the less controversial field of civilian space activity, the two leaders had reached agreement on establishing direct NASA–Soviet Academy of Sciences meetings for the resumption of bilateral space cooperation.[124] For scientists, this constituted a significant step forward.

While Reagan and Gorbachev had missed an opportunity to end the nuclear arms race, tensions continued to ease in U.S.-Soviet relations. In early 1987, Gorbachev offered a major, unilateral compromise in the arms control area by announcing that the disputed Krasnoyarsk radar station would be turned over to the Soviet Academy of Sciences for use as an international space research facility. This proposal, delivered to the United States by chief Soviet negotiator Yuli Vorontsov in Geneva, raised the idea of creating an international space monitoring agency with the power to make on-site inspections of all space objects prior to launch. Although Washington did not respond favorably to

---

[122] Lantratov, "'Star Wars' That Never Happened" (Part I), p. 8.

[123] On this point, see Gorbachev, *Memoirs*, p. 136. As he had recognized about Soviet finances during the late Brezhnev era, "In virtually all branches of the national economy, military expenditure [had] sapped the vital juices."

[124] Although not reported in the general press, the trade magazine *Aviation Week and Space Technology* published an apparently "leaked" account of the basic agreement, which matched in most details the actual pact signed the following spring.

the space proposal, the Soviet compromise on Krasnoyarsk did begin moving Reagan administration conservatives toward a recognition that the new Soviet leadership might be different in substantive ways from its predecessors.

Soviet "New Thinking" Regarding Space

In conjunction with the politics of *perestroika* (restructuring) in the Soviet economy, substantive changes in the conduct of Soviet space activity became evident under Gorbachev. His enunciated policy of "new thinking" had been accompanied by a willingness to pursue new types of cooperative ventures with the West. The new emphasis seemed to be on developing the Soviet space program not as a political tool but as an *economic asset*. Western scientific cooperation was sought to aid in mutually beneficial space research. But, more importantly, a wider array of marketable Soviet space assets began to be traded for Western hard currency. These changes indicated not simply tactical adjustments but fundamental goal change and learning. The United States, however, made no such changes, thus preventing what might have been a full-scale transformation of space security relations through the formation of a broad U.S.-Soviet cooperative framework and the use of this rapprochement to plug loopholes in existing space arms control agreements.

Before he could make progress on this agenda, Gorbachev found that he had housecleaning to do in the Soviet military space program. In May 1987, Gorbachev and a senior Politburo delegation visited Baikonur to learn about the state of the space program.[125] What they found surprised the new general secretary and at least some of his colleagues. After progressing since the late 1970s on secret funds within the military budget, a mock-up and unarmed version of the Skif laser battle station (Skif-DM) had been prepared for launch during Gorbachev's visit. On learning of the nature of the project, Gorbachev had told the engineers present, "The Politburo will not permit the launch of this rocket."[126] Clearly, the Skif-DM conflicted with everything that Gorbachev had been saying to the Americans about space arms control. In the end, since the system had no laser aboard and its main missions were scientific in nature (designed that way because of technological problems and in order to deceive the United States), Gorbachev and the Politburo allowed the flight two days later.[127] In the end, the Skif-DM module—launched from the huge Energiya booster—failed to achieve orbit. With this, the prototype Soviet response to SDI came to an end, as Gorbachev refused any further funding for the program.[128]

[125] On this visit, see Lantratov, "'Star Wars' That Never Happened" (Part II).
[126] Ibid., p. 11.
[127] Ibid., p. 12.
[128] Ibid., p. 13.

Gorbachev's commitment to greater transparency in the space program slowly began to be seen in unprecedented changes in Soviet informational policies, witnessed in a new attitude toward publicizing "negative" aspects of its past space activities. Gorbachev's policy of *glasnost* (openness) brought to light such previously taboo subjects as the death of a cosmonaut-trainee in the early 1960s, the suicide of a pilot who failed to make the cosmonaut team, the cause of the plane crash that killed Yuri Gagarin, and the timely admission of launch difficulties (such as the September 1988 loss of the *Phobos I* probe). Gorbachev also began unprecedented live television coverage of manned launches and landings, including the December 1987 return from the *Mir* station of Yuri Romanenko (who set a record for long-duration flight with 326 days in orbit) and the December 1988 touchdown of Vladimir Titov and Musa Manurov (who broke Romanenko's record with a flight of 365 days). Clearly, like his predecessors, Gorbachev was interested in using space to increase popular support for his leadership both at home and abroad. As part of this process, he continued and expanded the Soviet program of "guest cosmonauts," which included individuals from Afghanistan, Cuba, India, France, Mongolia, Syria, and Vietnam, as well as all of the East European countries.

Organizationally speaking, the increasing commercial orientation of the Soviet space program saw the Ministry of Foreign Economic Relations emerge as a new center of Soviet space cooperation.[129] The creation within the ministry of a "Central Administration for Space Activities" (Glavkosmos) in 1985 marked the formal establishing of an organization expressly designed to sell and market Soviet space services to the West, including Proton launchers, satellites, research space on Soviet spacecraft, and remote sensing services. Glavkosmos was not a catchall organization (like NASA in the United States), but within the commercial area it began to enjoy gradually increasing autonomy from higher authorities in forming contracts with Western space organizations and private companies.[130] Notably, the spacecraft marketed during this period by Glavkosmos were no longer produced by the Ministry of General Machine Building but instead in their own production facility.[131] This emerging separation between

[129] Besides the Ministry of Foreign Economic Relations, there were a number of other scattered state ministries also involved in space activities, usually as auxiliary functions. For example, within the Ministry of Health, the Institute for Biomedical Research—under the longtime direction of Academician Victor Gazenko—directed experiments for the Soviet Kosmos biosatellite program. Gazenko worked primarily with academy personnel and was involved in numerous cooperative missions with NASA and independent U.S. scientists investigating the biomedical effects of spaceflight. U.S. equipment and experiments were regularly featured on the Kosmos biosatellites.

[130] Author's interview with Glavkosmos official Dmitri Poletaev in Moscow, April 6, 1988.

[131] Much of the production for unmanned "scientific" launches also came from the Babakin Center.

military and civilian space production marked a sharp break from policies of the past.

Against the claims of some Western defense analysts that the primary aim of Glavkosmos was the acquisition of Western technology for the military, Glavkosmos officials argued that their real aim was to gain increasingly critical Western currency.[132] They pointed to the chronic problem of low-quality Soviet exports in other fields, as well as the January 1988 implementation of Gorbachev's new policy of *khozraschet* (self-financing), which forced Soviet organizations to begin considering previously ignored capitalist concepts such as profit and loss. In accordance with the requirements of *khozraschet*, Glavkosmos shocked outsiders by selling advertising space on its commercial rockets and spacecraft to both Western companies and Soviet enterprises. U.S. contacts, however, remained limited by then-current export control laws.[133] Meanwhile, also under Glavkosmos, a small organization called Soyuzkarta began to market Soviet satellite photos.[134] Soyuzkarta benefited from Soviet official decisions to override defense considerations and allow the sale of high-resolution photographs,[135] allowing it to undercut its American (Landsat) and French (Spot Image) competitors successfully during this period.

Another important change under Gorbachev was the elevation of space science within the Soviet space program.[136] In some areas, this came even at the expense of the interests of the military and manned programs. Gorbachev's

---

[132] Author's interview with Glavkosmos official Dmitri Poletaev in Moscow, April 6, 1988.

[133] U.S. export control banned the launching of any satellite containing U.S. technology by the Soviet Union. Therefore, Glavkosmos's extensive marketing efforts focused on longtime allies such as India and the more independently minded (yet inexperienced in space) countries of Western Europe, particularly for renting research space on Soviet orbital stations.

[134] This organization actually predated Glavkosmos, but now fell under its jurisdiction.

[135] Under Gorbachev, the allowable resolution dropped to 5 meters.

[136] The main organization dealing with space science was the Institute for Space Research (IKI) in Moscow. Its activities ranged from astronomy to plasma physics to the chemistry of crystal growth in near-zero gravity. The approximately two thousand staff workers at the institute were involved in a wide range of domestic, Soviet bloc-wide, and East-West research projects. Although it appeared that some military work may have been done by certain IKI scientists (the entrance to IKI in the late 1980s was controlled by uniformed military personnel and a large, adjacent building housed a special "Military-Technical Project"), the institute itself was primarily devoted to scientific advancement. It provided a visible civilian organization comparable in intent—though not in budget or size—to NASA in the United States. However, unlike NASA, IKI was not heavily involved in the manned space program, although it provided many scientific experiments that flew on the *Mir*. During the Soviet period, certain cooperative activities were also organized under the rubric of Interkosmos. This body encompassed both a program that coordinated research among Soviet-allied countries and a council that handled cooperative projects with the West. A number of smaller institutes in the Academy of Sciences structure also engaged in specialized research on particular aspects of spaceflight. For example, the Institute of Geochemistry and Analytical Chemistry conducted research on rocket fuels. (Author's weeklong trip to IKI in April 1988.)

selection of Space Research Institute Director and Academician Roald Sagdeev as a science and arms control advisor seemed to have shaped this policy
considerably.

Notably, in the new wave of Soviet political writings, space began to play
an increasingly important role. In the 1960s, the so-called scientific-technical
revolution had become a watchword for Soviet material and social progress.[137]
Faced with the increasingly sluggish Soviet economy and with domestic technologies that lagged well behind Western standards, progress in this area began
in the mid-1980s to be identified increasingly with Soviet space achievements.
As an article in the Soviet journal *International Affairs* argued on the issue of
social progress in the modern world, "Space exploration has . . . [a] major role
to play—its achievements serve as an effective locomotive for the scientific and
technological revolution."[138]

Thus, in the last gasp of Soviet-style Marxist-Leninism, the unique importance of technological advancement for the achievement of socialism created
a political environment where space technology had tremendous *symbolic* significance. Despite the economic problems in the late Soviet period, the relative
Soviet commitment to civil space, judged in spending as a percentage of gross
national product, roughly doubled that of the United States, as the Soviet economy was only about 60 percent as large as that of the United States and absolute
spending by the two programs was approximately equal. Ironically, this situation
marked a little-recognized reversal from the 1960s, when the United States spent
substantially more than the Soviet Union on human spaceflight.[139]

While U.S. experts generally scoffed at the economics of space manufacturing, Soviet research in this area was extensive. Soviet space analysts portrayed
this evolution as nothing short of inevitable. As Raushenbakh argued, "Space
factories and space plants play a central role in the development of the scientific-technical revolution."[140] According to recent Soviet writings, space technologies would help to mechanize future production, rationalize its distribution, and provide solutions to a variety of real-world problems: hunger, energy

[137] Beginning in the 1960s, the Soviets have celebrated April 12 as a "space" holiday called Cosmonautics Day.

[138] Vitaly Sevastyonov and Vladimir Pryakhin, "Space Exploration and New Thinking," *International Affairs* (Moscow), No. 5, May 1987, p. 25.

[139] Soviet writings unanimously propounded the goal of a permanent manned presence in
space, looking forward to the creation of actual "factories" in the near future, even fully automated
ones. See, for example, the discussion of space factories by Feoktistov and Bubnov in their book *O
Kosmoletakh* [On spaceflight] (Moscow: Molodaya Gvardiya, 1982).

[140] Boris V. Raushenbakh, "Byt' kosmosu mirnym" [Space should be peaceful] in Yevgeniy P. Velik-
hov, Raushenbakh, et al., *Uchenie protiv voyny* (Moscow: Molodaya Gvardiya, 1984), p. 179.

shortages, and even drudgery in the workplace. Under these conditions, both communism and fundamentally new, more "engaged" forms of international cooperation would be realized.

The New Politics of U.S.-Soviet Space Relations

Despite the growing evidence of changes in Moscow's political goals in space, a strong faction of "Soviet reform skeptics" in the Reagan administration held to a policy of talking cooperation and tension reduction but offering little in the way of substantive policy change. The two leaderships could hardly have been more mismatched in terms of their directions in space, leading to problems reaching concrete agreements.

On the eve of Secretary of State George Shultz's planned visit to Moscow in April 1987, the U.S. government finally decided to move ahead with a space cooperation agreement as a gesture to the new Soviet government, given its surprising willingness to compromise on a variety of important arms control matters. On April 14, Shultz and his Soviet counterpart—Foreign Minister Eduard Shevardnadze—signed a new set of space science accords, ending a drought of five years. The new agreement called for sixteen specific cooperation projects to set the stage for further cooperation on a larger scale later on, possibly to include manned missions.[141] The first three major projects would be in the unmanned area. Two Soviet missions—*Phobos* and *Vesta*—aimed at Mars would benefit from use of the U.S. Deep Space Tracking Network as well as U.S. data from its Viking missions. The American *Mars Observer*—to be launched later—would benefit from the findings of these missions and other cooperative research. Within NASA, five groups were set up to facilitate specific negotiations and detail work with the Soviet Union in particular areas of space research. During the summer, proposals from both sides expanded in scope. Communication increased significantly and included a four-hour "spacebridge" teleconference via satellite between Soviet scientists in Moscow and Americans in Boulder, Colorado, on the eve of a week-long conference of the Planetary Society focusing on Mars research.

Responding to U.S. criticisms over arms control issues relating to space, Gorbachev agreed to allow an unprecedented visit by selected U.S. congressmen to the Krasnoyarsk radar site. In September 1987, U.S. Representatives Bob Carr, Thomas J. Downey, and James Moody toured the facility and made a videotape to document their findings. While the visit did not silence administra-

---

[141] Robert C. Cowen, "Shared Space, US-Soviet Cooperation May Yield a Trip to Mars," *Christian Science Monitor*, June 2, 1987, pp. 18–19.

tion complaints—the investigation was nontechnical in nature—it did reduce tensions by showing a new Soviet openness toward previous military secrets.

Ironically, however, NASA itself was now emerging as one of the greatest opponents of space cooperation, as its officials recognized that a link existed between the intensity of U.S.-Soviet competition and the level of U.S. space spending, especially in the area of manned spaceflight. As one senior NASA official commented frankly, it was now the White House and state department that favored cooperation with the Soviet Union on a manned Mars trip, not top NASA managers.[142] From the point of view of these bureaucratic leaders, cooperation could exacerbate technical problems, mission planning, and spacecraft development, while adding costly delays spent on political matters. The feared result was that objectives would be reached later or not at all.[143] At lower levels in NASA, by contrast, the situation was quite different. Among the bulk of engineers and scientists, there was considerably more support for cooperation with the Soviet Union based on shared goals of furthering scientific knowledge about space.[144]

U.S.-Soviet Space Arms Control Discussions

Since 1957, Soviet writings had consistently argued the case that space should be a completely peaceful realm, a position that had evolved since the Outer Space (1967) and ABM (1972) treaties to mean that there should be no weapons stationed in space. Although unstated in Soviet pronouncements, the past unwillingness of Moscow to give up its current stockpiles of ASAT weapons on the ground, to reconsider its position on the Krasnoyarsk radar installation (believed to be part of an ABM system), or to limit its laser research proved to be important factors—in American eyes—preventing a new space treaty, which might include ASAT weapons. But by the late 1980s, Soviet space officials seemed genuinely committed to the fundamental changes in their military space policy.

[142] Author's interview at NASA headquarters (name withheld), Washington, D.C., August 1987.

[143] International programs are often the first to fall on the cutting block during periods of fiscal conservatism. For example, NASA had to make an embarrassing withdrawal from the international solar-polar research project when Congress removed it from the budget in the early 1980s.

[144] Former astronaut Dr. Sally Ride comments that there is an inverse relationship between the proximity to Earth of any potential joint mission and the level of political obstacles. In near-Earth space, cooperation missions frequently meet with political opposition and charges of technology giveaways, especially in the manned area. By contrast, deep space cooperation is often facilitated by the nonparticipation of cumbersome manned space bureaucracies (like NASA), the high cost of unilateral missions, and the relative absence of fears regarding military use of newfound information. (Author's interview with Dr. Sally Ride at Stanford University, Stanford, California, December 1987.)

Finally, significant movement began to occur in U.S. space politics. Whereas in the late 1950s and early 1960s members of Congress were a key driving force behind U.S. competitive programs, by the late 1980s very few members of Congress remained committed to a "race" in space. Indeed, during the Reagan and Bush years, Congress proved to be one of the strongest brakes on potential U.S.-Soviet space competition.

Within the U.S. military, an evolution in thinking had begun on the desirability of aggressive space competition with the Soviet Union. Although within the isolated basement offices of SDIO in the Pentagon there was still vigorous support for space-related research, the views of the particular services on space spending had become much more ambivalent.[145] Lieutenant Colonel Hays writes regarding the F-15 ASAT program that it "generally received its strongest support from the civilian defense officials of OSD [Office of the Secretary of Defense] and the Secretary of the Air Force rather than from the uniformed military."[146]

In the case of SDI—the program the Soviet Union most frequently singled out as an impediment to arms control—U.S. military support was mixed.[147] Indeed, excluding Admiral Watkins, others in the Joint Chiefs of Staff had tried unsuccessfully to kill the program in the early 1980s. From the perspective of career officers, SDI represented a potentially ineffective system that would draw off tens of billions of military dollars—money that could be better spent on raising service pay and maintaining high levels of training, armament, and readiness in proven defensive systems.[148] In addition, individuals questioned the desirability of abrogating existing space arms control agreements, if the result might be a more robust Soviet offensive nuclear capability and the jeop-

[145] Frequently, Congress has relegated a particular space project to a service (as with the National Aerospace Plane [NASP] and the air force) as an explicit cost-cutting measure. Members of Congress know that such programs will not be overfunded if increases translate directly into cuts in other service programs. Critics, however, have charged that this may slow the progress of needed weapons.

[146] Hays, *United States Military Space*, pp. 112–13. Hays observes, "Indeed, at times it would have been difficult determining that the MHV [miniature homing vehicle] was an Air Force program" (p. 113).

[147] Author's interviews with uniformed and civilian personnel in the Department of Defense, Washington, D.C., summer 1987.

[148] On the issue of instrumental service interests, one midlevel air force officer commented that the services were no longer willing to support vague research projects in space just to guarantee themselves a stake. The services realized—in this time of budget austerity—that the effect would be to strip funds from more central service missions. In the air force, internal forces tried to halt cuts in funding for planes and pilots, because promotions were tied to service in "wings," not research projects. (Author's interview with officers at Andrews Air Force Base, Maryland, August 1987.) The situation was mirrored in the other services as well. Thus, there were now *internal* forces within the military holding back competitive space programs.

ardizing of U.S. photo-reconnaissance and other current surveillance advantages.[149] As the vice commander of NORAD, Vice Admiral William E. Ramsey, stated candidly at a space and defense conference held in Washington in March 1988, "If we could outlaw weapons in space, it would be a damn worthy goal."[150] During the Reagan administration, it was mainly hard-line political appointees in the defense department and uniformed personnel within SDIO (whose jobs the program had become) who tended to favor full-scale development of SDI.[151]

Overall, far from predicting an inevitable showdown in space—with the Soviet Union in the role as the "evil empire"—the bulk of key actors in the U.S. space program by the late 1980s had now adopted a guardedly optimistic view of future space relations. While civilian manned cooperation with the Soviet Union was not widely supported, the existing arms control framework had considerable backing, as did an expansion of unmanned scientific cooperation. The primary theme emerging at that time was that continued competition with the Soviet Union in space was desirable but within safe and secure boundaries.

On the Soviet side, Gorbachev's words and deeds regarding space showed that he did not believe in waiting for a full-fledged U.S.-Soviet détente to begin meaningful cooperation in space.[152] As one Soviet space analyst observed at the time, Gorbachev was changing the prevailing Soviet "policy of confrontation" in space.[153] Gorbachev, to a much greater extent than his Soviet predecessors, had begun to look for *political* solutions to military problems in space. Previously slow to initiate cooperative missions, Soviet civilian space officials under Gorbachev now began to issue a plethora of proposals to the West, ranging from the exchange of scientific data to a joint U.S.-Soviet mission to Mars, and even to a high-level Soviet offer to fund an international space station. In 1987, Soviet Premier Nikolai Ryzhkov proposed the creation of a World Space Orga-

---

[149] On this issue, political appointees tended to focus on the defensive advantages that would be obtained by the United States in an early SDI deployment, while uniformed military personnel were more concerned with the reciprocal increases in Soviet offensive countermeasures.

[150] Vice-Admiral Ramsey quoted in Fitzgerald, *Way Out There in the Blue*, p. 447.

[151] Author's interviews with civilian and uniformed personnel in the Department of Defense, Washington, D.C., summer 1987.

[152] Beginning in 1986, long before the new U.S.-Soviet political relationship had been established, Soviet scientists and space officials—with government approval—had begun to call for a joint U.S.-Soviet manned mission to Mars as a possible forerunner to a general improvement in relations. The issue was raised by Gorbachev on the eve of the 1988 Moscow summit. However, the United States declined the proposal.

[153] Author's interview with Yuri Streltsov in Moscow, April 7, 1988.

nization to monitor a new set of space security treaties. By late 1988, Gorbachev had decided to make space a centerpiece of his address to the United Nations in New York. He elaborated on Ryzhkov's proposal by targeting, in particular, U.S. concern over the controversial Krasnoyarsk radar site, which many American conservatives saw as a violation of the ABM Treaty and an indicator of Soviet intentions to carry out a possible first strike on the United States: "We have put forward our proposal to establish [a world space body] on more than one occasion. We are prepared to incorporate within its system our Krasnoyarsk radar system. A decision has already been taken to place that radar under the authorities of the U.S.S.R. Academy of Sciences."[154] He also offered to cooperate with other space-faring states in establishing what he described as "an international space laboratory or manned orbital station designed exclusively for monitoring the state of the environment."[155]

Two reasons seemed to underlie the Soviet proposals. First was a recognition that racing with the Americans in space had done little to promote Soviet security. Moscow usually followed, rather than led, nuclear and military space developments, and such a race would strengthen the arguments of hard-liners at home seeking concrete military responses to SDI. Second, Gorbachev wanted to slow this international security competition so he could concentrate on what, for him, was a higher-placed priority: his domestic reform agenda. To do this, however, he needed to economize on the space program rather than enter a new space race. State funding for space projects that were once routinely approved slowed to a trickle. By the late 1980s, senior Soviet space officials were noting that cuts in the U.S. space science budget had begun to have a negative impact on Soviet space scientists, many of whom now depended on Western cooperation to bring down the cost of (and thereby enable) expensive scientific projects.

But a few remaining, high-cost military space projects continued to receive closeout funding from the cash-strapped Soviet government. In November 1988, the Soviets culminated over a decade of planning[156] with the successful launch of their *Buran* (blizzard) space shuttle from a huge Energiya booster.[157]

---

[154] Speech by Gorbachev to the United Nations, December 7, 1988.

[155] Ibid.

[156] According to Siddiqi, the Communist Party's Central Committee approved the project in February 1976, fearing that U.S. plans for a space shuttle might leave it with some advantage if there were no Soviet response. On the roots of the *Buran* program, see Siddiqi, *Soviet Space Race with Apollo*, pp. 835–38.

[157] Craig Covault, "Soviet Shuttle Launched on Energia Booster," *Aviation Week and Space Technology*, November 21, 1988, p. 18.

After two orbits, the unmanned *Buran* was landed flawlessly by automatic controls, displaying a capability more advanced than the U.S. space shuttle.[158] But the question Soviet scientists asked was *why?*[159] Gorbachev's science advisor Sagdeev, who in private had earlier criticized Soviet expenditures on manned spaceflight,[160] now offered this less-than-flattering public assessment of the *Buran* flight to the Western media: "It went up. It came down. But it had absolutely no scientific value."[161] The fact that nothing happened to Sagdeev said a great deal about how far Gorbachev was willing to go to apply his "new political thinking" to the previously highly propagandistic Soviet military space program.

Although *Buran* was originally designed out of fear of the possible military uses of the U.S. space shuttle (as a delivery system),[162] by 1988 the Soviets had little real fear of an imminent U.S. attack, and considerable information that the U.S. shuttles (now grounded) were intended for no such purposes. Yet Gorbachev's reforms still had limited power against the entrenched influence of the Soviet military-industrial complex, which could not be stopped on a dime. Thus, some expenditures continued for programs—like *Buran*—that now had little strategic purpose.

Another key target of critics of the Soviet space program was the manned space program, which, as in the United States, continued to command a large percentage of the funding allotted to space activity. The Soviet government had made effective use in the past of manned spaceflight for propaganda value by taking up selected foreign passengers from key allied states (such as East Germany, Czechoslovakia, and Poland) and by conducting high-profile, long-duration spaceflights, far surpassing U.S. capabilities. Now leading Soviet politicians—including former Moscow Communist Party chief and Gorbachev opponent Boris Yeltsin—joined the voices of scientists in calling for cuts in Soviet manned space activities, although their aim was to redirect this funding out of the space program altogether to address the increasingly desperate shortages in the domestic economy.

[158] On the 1988 *Buran* flight, see Harvey, *Russia in Space*, pp. 25–26.

[159] As Siddiqi notes, "The lion's share of the Soviet space budget during the 1980s was taken by the Energiya-Buran effort, the most expensive program in the history of the Soviet space program." Siddiqi, *Soviet Space Race with Apollo*, p. 840.

[160] Author's interview with Academician Roald Sagdeev at the Space Research Institute in Moscow, April 4, 1988.

[161] "Soviet Expert Says Shuttle Not Too Useful," Associated Press article reprinted in *San Francisco Chronicle*, November 22, 1988.

[162] Siddiqi, *Soviet Space Race with Apollo*, p. 835.

The George H. W. Bush Administration
and the Soviet Breakup (1989–1991)

With the election of George H.W. Bush, supporters of expanded U.S.-Soviet space security cooperation seemed to have gained an ally. As Bush laid out his priorities in an address to Congress on February 8, 1989, SDI moved to the back burner and there was a clear emphasis on civilian space progress: "We must have a manned space station, a vigorous, safe space shuttle program and more commercial development in space."[163] The speech also called for a $2.4 billion increase in NASA's budget. Further efforts to improve coordination among U.S. space actors and to reverse the drifting course of the U.S. civilian space effort came on March 2, when Bush officially announced the formation of a National Space Council. With Vice President Dan Quayle as its chairman, the new council would be in charge of coordinating the various areas of national space policy: military, scientific, and commercial. Others on the council would include representatives from the Departments of State, Defense, Commerce, and Transportation, as well as NASA, the Office of Management and Budget, the National Security Council, the CIA, and the White House staff.

The progress toward "normalizing" the situation at NASA also gained momentum during 1989, as the United States continued with a series of successful shuttle launches (after the first post-*Challenger* manned space launch in November 1988). Early into his administration, Bush announced the appointment of former-astronaut Rear Admiral Richard Truly to replace James Fletcher as NASA's chief administrator. His choice met with widespread approval from a number of political corners and helped the organization increase its influence in congressional budget battles. The success of the air force's Titan IV unmanned booster took further pressure off the shuttle program and allowed it to specialize more on its areas of strength: civilian manned missions and space research.

In the spring of 1989, contacts between U.S. and Soviet scientists expanded with joint planning of the various stages of the U.S. *Magellan* mission to Venus. Data from the more extensive Soviet research program concerning Venus, including primary magnetic tapes and processed radar images from the *Venera* 15 and 16 missions in 1983, aided in the prelaunch planning for the *Magellan* project. Symbolic of the new era of openness between the two space programs, a team of Soviet scientists observed the May launch of the spacecraft as it lift-

---

[163] President George H.W. Bush, quoted in *Aviation Week and Space Technology*, February 20, 1989, p. 24.

ed off from Cape Canaveral aboard the shuttle *Atlantis*. The Soviet scientists would play a closer role in the analysis of the eventual data from *Magellan* when the spacecraft reached Venus in 1990. A second U.S. planetary research mission to include Soviet participation was the *Voyager 2* flight, which would fly to Neptune.[164]

This resumption of direct contact and cooperation by government scientists marked a reversal of a nearly decade-long decline and was indicative of the change in space relations that had occurred since 1985. In this new political environment, there were hopes on both sides for a further expansion of U.S.-Soviet space security cooperation. The rapid pace of Moscow's increasing openness in space (as well as in its broader political system) led to disbelief in many quarters of the U.S. government. This was clearly not the behavior of an "evil empire."

In another important area of scientific cooperation—orbital debris management—the Bush administration pushed forward in discussing the issue with Soviet space officials, as well as with others. Under the Reagan administration, NASA had already begun discussions with the European Space Agency in October 1987.[165] But these efforts gained new momentum early in President Bush's term following the completion by U.S. government scientists and officials of a "Report on Orbital Debris by Interagency Group (Space)" to the U.S. National Security Council in February 1989. On the strength of this study and recognizing the international character of the problem, the new Bush administration adopted a policy of seeking to promote debris mitigation "with other nations and relevant international organizations."[166] Accordingly, NASA held debris-related discussions with Japan in May 1989. In December 1989, NASA put together the first meetings in Moscow with Soviet scientists in creating a new U.S.-Soviet Orbital Debris Working Group.[167] After years of study by, in particular, U.S. scientists (aided by NORAD space tracking data), this increasingly important environmental issue had finally emerged onto the international space agenda.

With the dramatic decrease in U.S.-Soviet hostility, the Bush administration's new emphasis on civilian space included a more critical view of the role of the SDI program. As the administration enunciated a more positive political line toward the Soviet Union, policies of accommodation and even support for Mikhail Gorbachev began to be enunciated. In the context of this new détente,

[164] John Noble Wilford, "U.S. and Soviets Quietly Team Up to Explore Space," *New York Times*, May 8, 1989, pp. 1, 9.

[165] Author's interview with Nicholas Johnson, NASA, Johnson Space Center, October 3, 2007 (conducted via e-mail).

[166] Ibid., quoting this report.

[167] Ibid.

space competition became a lower priority, and the administration began a policy of seeking *political* solutions to space security problems. Space science cooperation programs increased in number, and the government for the first time allowed a private U.S. company to contract with the Soviet Union to fly a commercial experiment on a Soviet spacecraft. A sharp reduction in the hostile rhetoric previously directed at the Soviet Union's space activities provided further evidence of a new American space policy, returning to precedents set by the cooperative policies adopted under the Nixon administration in the early 1970s. This warming of relations carried over to the military space area as well—as an answer to alleviating problems in the federal budget.

In 1989, Chairman of the Soviet Council of Ministers Nikolai Ryzhkov released the first accurate figures for the Soviet space budget.[168] These data put overall space spending at $10.5 billion (6.9 billion rubles), of which slightly more than half ($5.93 billion, or 3.9 billion rubles)[169] went into military programs. Despite lower Soviet labor costs, Soviet military space spending was boosted by the disproportionately high number of launches the Soviets had to undertake in order to keep up enough of their relatively short-lived photo-reconnaissance satellites to maintain a functioning network. Thus, the bulk of this figure undoubtedly went into passive military systems. [170] While this new evidence showed that not all of past Soviet commentary on the purely civilian nature of its space program had been a ruse, the figures finally put a concrete measure to the Soviet military space efforts, allowing Washington to adjust its threat perceptions accordingly.

Moscow continued to open its space program, overwhelming the ability of most Americans even to comprehend the speed of these changes. The availability of research space on the orbiting *Salyut* and *Mir* space stations provided new means of cooperating with Western scientists and even—for the first time—private corporations. The trend toward commercialization had reached such a scale that the Soviets began to offer foreigners with enough money access to the *Mir* space station. Indeed, in May 1991, the Soviets would launch their first civilian, fee-paying passenger, a British chemist named Helen Sharman, sponsored by a consortium of major British companies. Clearly, the days of the highly secretive and inaccessible Soviet space program had ended.

One major impediment to concrete progress in U.S.-Soviet space arms control, however, was differences over the ABM Treaty and missile defenses. Although far more tepid in its commitment than the Reagan team, the Bush

---

[168] See Bill Keller, "Soviet Premier Says Cutbacks Could Reach 33% for Military," *New York Times*, June 8, 1989, pp. 1, 11. (Figures for space program cited on p. 11.)

[169] The figures are based on the August 1989 exchange rate of 1 ruble/1.52 dollars.

[170] In the United States, similar systems had long been more sophisticated and had lasted years rather than months.

administration continued to adhere to the goal of testing and eventually deploying SDI. Nonetheless, congressional insistence on the narrow interpretation of the ABM Treaty, further cuts in the SDI research budget, and the less than wholehearted commitment of the president himself (compared to his predecessor) suggested that compromises with the Soviets over this issue might still be possible. The Bush team also proved very cautious in the area of defense for fear of disrupting Soviet reforms and causing a hard-line backlash. Overall, although the U.S. commitment to some form of SDI continued to block a new agreement with Moscow to ban ASATs, missile defenses in space remained largely a back-burner item.

In the spring of 1989, President Bush announced that the Reagan administration's original SDI budget request for fiscal year 1990 would be scaled back from $5.6 billion to $4.9 billion. Part of the reason had to do with the increasing difficulty of proving the need for such a system to the American people and to Congress, part with problems in meeting the Nitze cost-effectiveness criteria, especially in laser weapons. As the Strategic Defense Initiative Organization's Acting Deputy Director Major General Eugene Fox commented on the decision to cut back on laser research, "I don't know whether I would call lasers a disappointment, but they have not come along as fast as I thought they would."[171] Furthermore, despite cost and launch obstacles, SDIO announced a shift in the emphasis of the missile defense program to the Brilliant Pebbles concept—the massive deployment of small, single-shot rockets using highly developed sensors.

In response to these changes, Congress took an increasingly wary attitude toward SDI in the summer 1989 budget hearings. Passive sensor technologies received support, but directed energy weapons suffered new cuts, as did research in the enormous power generators needed to run them. New funding was directed to the consideration of more limited, ground-based defensive systems. In the end, the House cut SDI's 1990 budget to $3.1 billion, and the Senate to $4.3 billion.[172] The final conference committee agreed to a $3.6 billion level, representing a significant decline in inflation-adjusted dollars.[173] The new consensus among American political leaders was that SDI—at least in the original Reagan sense—represented a system whose purpose and time had largely passed. Former senior Kennedy administration official McGeorge Bundy summarized the prevailing wisdom in the United States regarding SDI in a long

[171] Quoted in Joseph Romm, "Pseudo-Science and SDI," *Arms Control Today*, Vol. 19, No. 8 (October 1989), p. 15.

[172] See Missile Defense Agency (MDA), "Historical Funding for MDA FY85–07" chart, MDA Web site, at <http://www.mda.mil/mdalink/pdf/histfunds.pdf> (accessed July 3, 2006).

[173] Ibid.

article published soon after the budget debates: "It is too soon to announce the end of SDI, but it is not too soon to say that it is unlikely to prosper during the next several years. The program is likely to be less and less pretentious, and more and more devoted to the prudent research that has always been needed as insurance against surprise."[174]

The following year's budget debates (for 1991) brought a further decline in SDI spending. Despite a presidential request of $4.5 billion, the Democratic Congress approved a mere $2.9 billion.[175] Clearly, the United States no longer considered space defenses a critical national priority. The dramatic events of late 1989 and early 1990, as the countries of Eastern Europe followed one another in staging largely peaceful and successful rebellions against their communist leaderships—as Moscow sat by passively—convinced even the most conservative members of the Bush team that Gorbachev wanted an end to U.S.-Soviet confrontation. Given this setting, negotiations with Moscow for strategic arms reduction took center stage over the course of 1990. The administration also began to rethink its need for space-based defenses. In January 1991, with the initial bombing campaign for the allied attack on Iraqi forces already begun, the Bush administration rechristened SDI with a new and more limited mission: Global Protection Against Limited Strikes (GPALS). The idea of a massive, umbrella-like defense had been put aside as too expensive and too difficult, and the United States had new defense priorities.

This war would be the first to benefit from an operational GPS system. Two advanced Block II satellites had been launched into 12,000-mile orbits shortly before the war to provide additional coverage for the nearly completed network.[176] This technology allowed those U.S. platforms with GPS receivers to navigate and deliver weapons with improved accuracy, reducing casualties among both civilians and U.S. forces. But only a small percentage of weapons in the 1991 Gulf War against Iraq used devoted precision guidance (laser, infrared, terrain-mapped); later, new GPS applications during the 1990s meant that a majority of heavy bombs dropped or delivered by ships and submarines in the fall 2002 Afghan conflict and the 2003 Iraq War would employ onboard GPS-aided packages. These developments only further increased the military

---

[174] McGeorge Bundy, "Ending a Common Danger," *New York Times Magazine*, August 20, 1989, p. 56. In the immediately preceding lines, Bundy comments, "Given this severe restriction on the Reagan dream, the public appeal of the program, in a new Administration that includes no dreamers in its first team, is bound to decline year by year. Very large expenditures for untested and fractional defenses of missile launching sites and command centers are unlikely to be attractive to military leaders or indeed to ordinary citizens."

[175] MDA, "Historical Funding for MDA FY85–07."

[176] Friedman, *Seapower and Space*, p. 267.

support value of space and the continued importance of rights of safe passage and noninterference.

In terms of missile defense policy, the effect of the Gulf War—and Iraq's use of Scud missiles against Israel and U.S. forces in Saudi Arabia—shifted Congress's focus from the Soviet threat to rogue state missile proliferation. The main emphasis in funding shifted accordingly to *theater* missile defense missions and upgrading the Patriot system, the seeming "hero" of the coalition missile campaign (although later data revealed that the Patriot failed in most if not almost all of its encounters with Iraqi Scuds).[177] As for Moscow, it had evolved into more of an onlooker than a threat. Its role in the Gulf War had been positive from Washington's perspective—Gorbachev had supported critical U.N. resolutions and restrained his defense and security forces from providing significant aid to Saddam Hussein.[178] Suddenly, the concept of "collective security" seemed to be a reality, as U.S.-led U.N. forces routed Iraqi troops and set back a major challenge to peace in the Middle East. U.S. leaders saw new opportunities for reducing defense costs and for building a more stable footing for relations with Moscow. Accordingly, in July 1991, Bush and Gorbachev announced a major new arms control agreement, the Strategic Arms Reduction Treaty (START). The pact called for specific limits on launchers and the first actual reductions in U.S.-Soviet strategic nuclear weapons ever agreed to, with planned, verifiable cuts to six thousand warheads within ten years. Worried about the course of SDI, however, Soviet negotiators attached a unilateral declaration that U.S. failure to abide by the ABM Treaty could lead to a Soviet withdrawal.[179]

The domestic momentum of Gorbachev's reforms had finally surpassed his ability to control events, especially in the fifteen Soviet republics (including Russia itself) that now had their own formal presidents. On the eve of the signing of a new union treaty in August 1991, Communist Party hard-liners staged a coup and seized power, with the aim of restoring Soviet power. Aided by the powerful force of international media covering these events, Russian President Yeltsin and thousands of pro-democracy supporters faced off against Soviet

---

[177] See, for example, Theodore A. Postol, "Lessons of the Gulf War Experience with Patriot," *International Security*, Vol. 16, No. 3 (Winter 1991/92). Postol subsequently conducted further assessments that eventually succeeded in getting the Pentagon to admit these failures, but not until late in the Clinton administration.

[178] Only some years after the Gulf War did information surface about Soviet satellite imagery of allied forces having been provided to the Iraqis. Clearly, it made little difference in the final outcome of the conflict.

[179] Fitzgerald, *Way Out There in the Blue*, p. 487.

tanks sent into Moscow by the anti-Gorbachev coup plotters. When the army wavered and finally retreated rather than kill Russians in their own capital, the coup collapsed and its participants were arrested.

Gorbachev's return to power would not last. Through the fall of 1991, more and more republics declared their full independence from Moscow. Although a number of states in Central Asia clung to Moscow until the end, the trend of republican nationalism built an unstoppable momentum through the end of the year. Finally, in early December, a meeting in Minsk among the presidents of Russia, Ukraine, and Belarus declared a new, noncommunist commonwealth of independent states, formally ending their allegiance to the Soviet Union. Within a few days, Gorbachev acceded to the division and collapse of the powerful empire he had done so much to reform—but which now slipped through his fingers.

As for SDI, holdovers from the Reagan administration continued to push for development of the program, despite its recent demotion and the emphasis on shorter-range missiles. The new head of SDIO, former Reagan administration arms control negotiator Dr. Henry Cooper, argued for deployment of a mixed system with one thousand Brilliant Pebbles and some five hundred to one thousand ground-based interceptors.[180] Buoyed by post–Gulf War fervor (and mistaken views of the Patriot's effectiveness), Congress rushed through the Missile Defense Act of 1991, which called for deployment of a limited ground-based system within five years (even though the Patriot was not part of the SDI program). With support from respected Senate Democrat Sam Nunn (Georgia) and Senate Republican John Warner (Virginia), the legislation cleared the Senate without debate and, with backing from Congressman Les Aspin (Dem., Wisconsin), the bill passed the House as well.[181] Debate then ensued on the bill's meaning, as the deployment contradicted SDIO's plans and the legislation called for the program to be ABM Treaty compliant, leading the missile defense agency to argue that this could be treated simply as stage one rather than constituting an alternative. To sweeten the offer, Congress had boosted missile defense spending from $2.9 billion in fiscal year 1991 to $4.1 billion for FY 1992.[182] Politics had clearly taken hold, although not exactly in the direction that space-based SDI supporters had wanted.

[180] Ibid., p. 484.

[181] Ibid., p. 487. Senator Albert Gore, an early supporter of SDI on the Democratic side, voted against the measure, as did several other leading Democrats.

[182] "Historical Funding for MDA FY85–07," Web site of the Missile Defense Agency, U.S. Department of Defense, at <http://www.mda.mil/mdalink/pdf/histfunds.pdf> (accessed October 6, 2007).

Conclusion

The Soviet Union's dissolution in December 1991 marked the official end of the Cold War in space. While U.S.-Soviet tensions had been decreasing steadily for some time, it was still possible right up until the end for U.S. critics of space cooperation to refer to the Soviet "bogeyman" as a means of raising instinctive domestic opposition. In the eyes of the arms control community, however, much time for the expansion of space security had been wasted by overly conservative White House policies. An opportunity for a transformation of space security from a series of tentative agreements to an all-inclusive regime, with strong bilateral verification mechanisms, had seemingly been missed in the chimerical pursuit of a "defensive" space weapons revolution. These events showed the importance of *timing* to space security and its problems.

The Soviet government under Gorbachev had reached an epiphany near its end: weapons could not guarantee its security. The United States was in the opposite mode, at least in terms of spending and rhetoric if not yet in actual deployments. Had a Nixon-Kissinger team been in office, the outcome conceivably might have been very different. In the end, a strong reluctance to pursue new treaties (seen later in the George W. Bush administration) prevailed. What kept space from a possible arms race—the likely outcome had SDI testing moved forward in space—was a combination of the "institutional memory" of collective space security within other parts of government (especially the Congress), technical problems with SDI, and Gorbachev's own unwillingness to participate in such a race. Despite the change in U.S. (and Soviet) leaderships, the PTBT, the Outer Space Treaty, the ABM Treaty, and norms against harmful acts continued to restrain *actual* military space policies, as opposed to research and rhetoric, where they had little restraining effect. U.S. SDI theatrics had failed to promote space security, despite President Reagan's hopes of transforming the international security system. Ironically, if anything "broke" the Soviet space effort, it was the cost of Energiya and the *Buran* shuttle, programs the Soviets had inflicted on themselves. The mistaken Soviet belief that the U.S. space shuttle represented a dangerous military technology caused them to waste the equivalent of billions of dollars on a system that would never be used. Before eventually cooperating, the two sides muddled through with brinkmanship, continued overestimation of each other's military capabilities, and a little luck. They both found themselves poorer from the experience—with excessive military spending and the strain of an intense global competition—but still resilient enough to patch things up in the late 1980s.

Collective approaches to security in space had survived, albeit barely, the decade and a half following the informal end of U.S.-Soviet détente in 1975.

The two sides had emerged from these years of intense military space competition somewhat the wiser. The Soviets would no longer engage in the kind of action-reaction projects that had characterized much of the Cold War. In the United States, a different lesson had been learned. The air force had gained an appreciation for an emerging threat on the horizon—orbital debris—through one of its own tests. As with Starfish Prime in 1962, the experience of having "junked up" its own front yard had led to new understanding on the dangers of weapons testing in space.

By the end of this period, it was clear that the Gorbachev regime had failed in its efforts to convince Washington to build a more expansive security regime for space. At a time when it needed a cooperative partner, it faced instead a highly nationalistic and distrustful administration under President Reagan. The Bush-Gorbachev relationship eventually proved to be the most cordial, but the weight of the Soviet past made U.S. leaders remain standoffish and wary. The two governments eventually made major strides in nuclear arms control, but President Bush showed no such vision for transforming space, relying instead largely on inherited policies, albeit with some leavening. By the end of the period, the status of space security remained largely where it had started.

The United States now faced a new, quasi-democratic successor state in Russia, albeit one run by a former communist, Boris Yeltsin. Washington's thrill at its unexpected victory in the Cold War was tempered by its concern over the new Russian government's extensive financial and political problems. U.S. leaders faced a number of critical questions: Would it survive? Was a weak Russia—decaying from within (and threatening to implode further)—actually better for the United States than a cooperative Soviet Union? What would happen to the Soviet Union's missile and space expertise, particularly as the new Russian government found itself unable to pay salaries to a large array of scientists and engineers with know-how in ballistic missiles, nuclear re-entry systems, ASAT weapons, and other sensitive technologies? Finally, how should the United States react to its newfound space dominance? Was this the time to race forward with SDI to seize the high ground in space, or was it more appropriate to craft new cooperative structures to try to bring Russia closer into the Western space community? Would Russia's weakened and ailing command and control system and space-based early-warning network put the United States at risk of an *accidental* nuclear war? These challenges would bring both cautious responses and bold new initiatives from U.S. space and national security decision makers over the next decade.

CHAPTER SIX

Post–Cold War Space Uncertainty

1992–2000

With the Soviet successor states—including the Russian Federation—scrambling to establish viable governments, worries about space security drifted far from the main radar screen of U.S. military planners compared to the 1980s. President Boris Yeltsin now used every opportunity to emphasize his close ties to Washington as he sought to root out vestiges of the old communist system, making the concept of a Russian ballistic missile attack on the United States highly unlikely. Instead, defense planners in Washington began to worry about the possible contribution of impoverished Russian and Ukrainian scientists to missile programs in countries like Libya, Syria, Iraq, Iran, and North Korea. The Chinese space program remained in its infancy and not on Washington's list of potential concerns, particularly given the generally pro-Chinese direction of policies under Presidents George H.W. Bush and (until the Tiananmen Square massacre) Bill Clinton. More worrisome, by the end of the 1990s, was Beijing's apparent success in attracting former Soviet engineers to help modernize its strategic delivery systems.

In many respects, Washington now held all of the cards in space. Although Russia retained know-how, extensive launch capabilities, and unparalleled experience in manned spaceflight, it lacked a key commodity for an active space program: money. The United States faced three basic options for maintaining space security: (1) a focused strategy to dominate space and force Russia into a subservient position (aggressive space nationalism); (2) a transformational strategy to expand collective security mechanisms by using America's newfound hegemony to introduce stricter forms of space arms control (forward-leaning global institutionalism); or (3) a muddling-through approach aimed at reducing costs, preserving weapons options, and dampening Moscow's incentive to revert to military responses to U.S. space activity (a mixture of slowed techno-logical determinism and tepid social interactionism). For a president elected on the mantra "It's the economy, stupid," it is not wholly surprising that President Clinton opted for the third strategy. But neither space power enthusiasts nor critics of missile defense found the new U.S. policy very satisfying.

By the end of his second term, President Clinton lost control of even this rather limited space security strategy. Three factors emerged over the course of the 1990s to change the balance of domestic forces in the space weapons/missile defense debate. First, the dramatic failure of the Democratic Party in the 1994 midterm elections brought Republican control of both houses of Congress for the first time since 1955, reshaping the U.S. political map. While the new Re-publican majority stressed lower taxes, deficit reduction, and small government as its main priorities, the party's senior leadership retained President Reagan's perception of the need for active U.S. missile defenses, including in space. Un-der these pressures, Clinton administration funding for missile defense re-search increased significantly after 1994, despite the president's lack of serious interest in or support for the program. Second, in 1998, a special congressio-nal commission tasked with examining global missile proliferation concluded that a rogue state might be able to develop an intercontinental-range ballistic missile (ICBM) within five years if it had foreign assistance. This upped the ante for space-based defenses and led to new legislation, signed reluctantly by President Clinton, which mandated deployment of such a system as soon as the technology became available. Finally, the reemergence of national security as a core political issue—given the Republican claim of a lack of attention to U.S. military needs under the Clinton administration—made the 2000 presidential election a referendum, in part, on missile defense. In this context, support for robust missile defense became a matter of concern to both of the major-party presidential candidates. Ironically, Vice President Albert Gore, a strong sup-porter of the Strategic Defense Initiative (SDI) during the 1980s and a Vietnam veteran, found himself painted as weak on defense, while far less-experienced Texas Governor George W. Bush effectively labeled Clinton-era policies of de-fense budget-trimming as years of military "neglect." Gore moved further to the right as the 2000 presidential campaign wore on and voiced his support for certain types of missile defense. In the end, however, his less-than-effective campaign style, failure to define his issues or his successes, and linkage to the perceived moral turpitude of President Clinton caused Gore to lose the elec-tion. President-elect Bush would bring a very different perspective in regard to space security.

Bush-Yeltsin Relations, Russia's Decline, and New Opportunities

As Russian President Yeltsin sought to reap the "rewards" of his revolution over communism, he sought out U.S. assistance for his ailing economy, including its space program. He saw private U.S. investment in Russian space technology as the most reliable means of ensuring the industry's survival rather than simply waiting for U.S. government handouts, which he doubted would ever appear. Washington, for its part, sought both to benefit from some of Russia's unique assets—especially in the space-launch field—and to ensure that its own national security would not be harmed by the sudden and uncontrolled exodus of scientists from this critical industry. Accordingly, Congressman Les Aspin and Senators Sam Nunn and Richard Lugar had pushed a revolutionary piece of legislation (known as the Cooperative Threat Reduction or "Nunn-Lugar" program) through the Congress in 1991 to help the ailing pieces of the Soviet Union deal safely with their inheritances of weapons of mass destruction (WMD) and delivery vehicles. When the Soviet breakup occurred a few months later, these tasks became even more urgent. While the Nunn-Lugar program focused initially on providing technology and funding to ensure the compliance of the four Soviet nuclear successor states (Russia, Ukraine, Kazakhstan, and Belarus) with the 1991 Strategic Arms Reduction Talks (START I) agreement, the effort soon expanded to cover other WMD, troubling personnel-related problems (such as housing for military officers mustered out from eliminated strategic forces units), defense conversion, and civilian research funding for scientists with WMD know-how. Eventually, missile scientists would be included in this last program through the Western- and Japanese-funded International Science and Technology Center in Moscow and the Science and Technology Center in Kyiv, Ukraine. Other government-encouraged outreach efforts by U.S. industry sought to retain the skilled labor forces at such former Soviet space/missile enterprises as Energomash, Khrunichev, Lavochkin, and Yuzhmash (in Ukraine). For U.S. companies, the opening of these facilities created a fascinating new "gold mine" of technology, allowing them to add skills and capacity at minimal cost. Such cooperation also satisfied the professional curiosity of many U.S. technicians to see what it had been like on the other side of the Iron Curtain during the long years of U.S.-Soviet space competition.

To maintain its existing space capabilities, Moscow now had to scramble to reach agreements with a welter of new states that it formerly controlled. A particular ignominy for Russia was maintaining access to its main space launch

facility at Baikonur, now inconveniently located in the new country of Kazakhstan. An intergovernmental commission cobbled together a working arrangement in 1992, but Kazakhstan naturally began to assert more and more rights,
while also demanding payment for use of the facility. Kazakhstani environmen-
talists had long complained about the harmful effects of toxic fuels used in
Soviet space launches, which routinely polluted locations in the ground track
of popular Soviet routes to space. Now, the Kazakhstani government took up
these charges as a nationalist issue and began to press claims for compensation
against Moscow. The one brake on President Nursultan Nazarbayev's assertive-
ness was the fragility of Kazakhstan's independence, in light of the large number of Russians still living in the republic. Speculation continued throughout
the next year about possible attempts by Russia to seize the northern part of
Kazakhstan, eventually causing Nazarbayev to move his capital from the beau-
tiful mountains of Almaty to the insect-infested steppe of Astana. Still, neither
side could afford a major conflict.

On the Russian side, space now seemed like an expensive Soviet-era play-
thing that the struggling government could ill afford, somewhat like its previously coddled Olympic team. The end of Soviet state planning and the privatization of many industries meant that thousands of Russian government officials, military personnel, and scientists now faced the nonpayment of salaries
and the ravaging of their savings due to rampant inflation of the Russian ruble.[1]
Over the course of 1992, inflation surged to 2,323 percent,[2] resulting in the shutdown of many formerly state-supported facilities and extensive layoffs, work
slowdowns, and the loss of valuable staff. Still, many scientists stayed at their
enterprises for various reasons: loyalty to their profession and their colleagues,
health care and other remaining social benefits, and simply a lack of other opportunities, particularly for older employees.

Despite these very serious problems, the Russian Duma and President Yeltsin himself proved relatively unsympathetic with previously well-funded space
facilities, in the context of other budgetary priorities and socioeconomic crises
going on around them. As a result, struggling Russian scientists and enterprises
began selling off pieces of the Soviet space program—equipment, clothing,
technical manuals, and designs—at bargain-basement prices.[3] Even highly sen-

---

[1] On these conditions, see Andrew Lawler, "Financial Woes May Cripple Russian Space," *Space
News*, August 1–7, 1994, p. 1.

[2] "Russia: Commanding Heights," on the WGBH (Boston) Web site at <http://www.pbs.org/
wgbh/commandingheights/lo/countries/ru/ru_money.html> (accessed July 11, 2006).

[3] On one such sale at Sotheby's in New York in late 1993, see William E. Burrows, *This New
Ocean: The Story of the First Space Age* (New York: Random House, 1998), p. 587.

sitive military space hardware was now marketed to the West. In 1992, for example, the U.S. Department of Defense (DOD) purchased two Soviet-designed Enisy (Topaz II) space-nuclear reactors for $13 million from the Moscow-based Kurchatov Institute to study their design and possible relevance to U.S. missile defense projects.[4] DOD purchased an additional four reactors in 1993. In addition, the U.S. Air Force worked through third parties to purchase a large number of photo-reconnaissance images on the United States—just to check their resolution.[5] Clearly, there were no secrets now that Moscow was not willing to part with for a price. A major concern in Washington, however, was that the United States was probably not the only buyer.

While some equipment and designs leaked out to countries like China, Libya, and perhaps North Korea, Moscow hoped mainly to enter Western markets, which were far more profitable. Thus, its behavior began to change in order to build new bridges with the United States and Western Europe. A key development in the process of opening up the Russian space industry to possible U.S. investment and expanded state-to-state cooperation was the creation of a new civilian entity—the Russian Space Agency (RSA)—in April 1992.[6] Prior to this reorganization, the bulk of space enterprises fell under the military-controlled Ministry of General Machine Building, which both blocked its activities from public view and placed them under strict defense controls. While the RSA's new director, Yuri Koptev, came directly out of the old ministry, he had a mandate to create a serious counterpart to NASA and to open up long-closed facilities to Western companies and national space agencies.

Of particular importance regarding space cooperation with Russia was a new policy adopted after the Soviet breakup by the states involved in the U.S.-led Coordinating Committee for Multilateral Export Controls (CoCom), an organization of Western countries formed in 1949 to prevent the diffusion of sensitive military technology to the Soviet Union. On June 1, 1992, the group had agreed to start a dialogue with Russia and the other Soviet successor states to assist these countries in forming effective export control systems and facilitate their ability to receive sensitive Western technology.[7] The progress of this

[4] Brian Harvey, *Russia in Space: The Failed Frontier?* (Chichester, England: Praxis, 2001), p. 262.
[5] Ibid., p. 113.
[6] John M. Logsdon and James R. Millar, eds., "U.S.-Russian Cooperation in Human Space Flight: Assessing the Impacts," Space Policy Institute and the Institute for European, Russian and Eurasian Studies, George Washington University, February 2001, Appendix B, p. 2, available at <http://www.gwu.edu/~spi/usrusappb.html> (accessed July 5, 2006).
[7] See White House Press Release, "COCOM Issues," June 17, 1992, posted on the Web site of the Federation of American Scientists at <http://www.fas.org/spp/starwars/offdocs/b920617e.htm> (accessed July 5, 2006).

process would eventually create a two-way street, allowing U.S. companies to interact commercially with Russian space enterprises without fear of U.S. government reprisal.

In mid-June, the soon-to-retire President Bush met with President Yeltsin in Washington for their second meeting since the Soviet breakup. The centerpiece of the summit was an agreement to reduce their nuclear arsenals by two-thirds in a forthcoming START II agreement, which would also eliminate multiple-warhead, land-based ICBMs, particularly the huge, ten-warhead SS-18s that had worried U.S. military planners since the late 1970s. In addition, the United States announced its provision of $4.5 billion in financial assistance to Russia for currency stabilization and other macroeconomic reforms, $400 million for Nunn-Lugar program WMD dismantlement activities under the DOD, and an array of material and financial assistance projects to alleviate critical food and medical supply shortages. The summit yielded important progress in terms of space as well. The Agreement Concerning Cooperation in the Exploration and Use of Outer Space for Peaceful Purposes created a framework for future space relations, including the possibility of direct funding by the United States for projects in Russia.[8] The new pact called for the first flight of a Russian cosmonaut aboard the U.S. space shuttle, the first flight of U.S. astronauts to the *Mir*, and future *Mir*-shuttle missions, formally ending seventeen years of separation between the two manned space efforts.[9] In the commercial sector, the new framework facilitated a July visit by forty-five U.S. corporate and NASA representatives to some forty Russian enterprises. These meetings resulted in a number of new deals.[10] Other trips followed. Russian facilities now welcomed increasing U.S. contacts as a source of critically needed funds. Space expert Brian Harvey notes that within a period of just a few years the Russian space program would undergo a dramatic transformation from being one of the world's most closed to being one of the "most commercialized."[11]

Official NASA-RSA cooperation stepped onto firmer ground in October 1991, when the two sides signed an agreement for cooperation in human spaceflight, as well as a $2 million contract for Russian contributions to a future joint Mars undertaking.[12] NASA also began to consider how Russian help

---

[8] On this agreement and specific follow-on activities, see Matthew J. Von Bencke, *The Politics of Space: A History of U.S.-Soviet/Russian Competition and Cooperation in Space* (Boulder, Colo.: Westview Press, 1997), pp. 100–110.

[9] Ibid., p. 104.

[10] Ibid., p. 101.

[11] Harvey, *Russia in Space*, p. 300.

[12] See Von Bencke, *Politics of Space*, p. 101; and Logsdon and Millar, "U.S.-Russian Cooperation in Human Space Flight," p. 3.

might reduce costs for the extremely expensive *Freedom* space station.[13] With the NASA budget facing a steady decline (relative to inflation), it had become clear to U.S. space officials that a partner would be needed—even at the cost of compromises in this still highly nationalistic program. But U.S. and Russian supporters of greater cooperation remained tethered by enduring conservative biases within the two governments and their militaries. Some U.S. statements on space, such as those by Vice President (and Chairman of the National Space Council) Dan Quayle continued to emphasize the need to deploy SDI and establish U.S. "space control."[14] Meanwhile, Russian critics accused Yeltsin of selling the Soviet "crown jewels" in space to Moscow's erstwhile capitalist enemy. In August 1992, the Ministry of Defense (still staffed with Soviet hard-liners) announced the formation of the Russian Space Forces,[15] a wing of the military devoted to managing Russia's military space assets and countering a perceived threat from U.S. plans for missile defense and eventual space dominance.

The November 1992 presidential election brought victory to Democratic candidate Clinton. His transition team brought a new perspective to U.S.-Russian space relations, one less colored by the Cold War, less committed to the SDI program, and more optimistic about prospects for space cooperation with Russia. Although it played little role in the election, a further blow to space-based missile defense supporters was the September 1992 revelation from the U.S. General Accounting Office that the results of at least four of the seven major SDI flight tests from January 1990 to March 1992 had been exaggerated by the Strategic Defense Initiative Organization (SDIO).[16]

As the final accomplishment of his administration, President Bush signed the START II accord with Russian President Yeltsin in January 1993. This landmark agreement codified Russia's willingness to destroy its multiple-warhead, land-based missiles in return for American willingness to engage in deep reductions in deployed strategic warheads—which Russia could no longer afford to maintain at existing levels—to between 3,000 and 3,500 warheads, including limiting submarine-launched ballistic missiles to no more than 1,700 to 1,750 warheads.

---

[13] The station's name would soon change to the more neutral *Alpha* and finally, given the addition of new partners, the *International Space Station*.

[14] January 1993 report headed by Quayle, quoted in Frances Fitzgerald, *Way Out There in the Blue: Reagan, Star Wars and the End of the Cold War* (New York: Simon & Schuster, 2000), p. 490.

[15] Burrows, *This New Ocean*, p. 586.

[16] Fitzgerald, *Way Out There in the Blue*, p. 488.

The Clinton Administration's Space Security Policy

Soon after taking office in January 1993, the Clinton administration made a number of changes affecting U.S. policy on space security. First, as part of a major review of defense priorities, Secretary of Defense Les Aspin downgraded SDIO in May. Instead of reporting directly to the secretary, the renamed Ballistic Missile Defense Organization (BMDO) would now answer to the assistant secretary of defense for acquisitions and technology.[17] Secretary Aspin also reoriented the program, consistent with the threats exposed by Iraqi Scud use in the 1991 Gulf War, to focus on ground-based theater missile defense (TMD) instead of space-based, national missile defense (NMD). Along with a slight decline in the first BMDO budget returned by Congress, the less proven and more futuristic space elements were now put on the back burner. Instead, systems like the advanced Patriot missile, the sea-based Aegis program, and the Theater High Altitude Area Defense (THAAD) would receive the lion's share of funding. Finally, the Clinton administration halted tentative efforts under the Bush administration and by old SDI supporters to build cooperative bridges with Russia for the construction of joint missile defenses.

The new Clinton team also sought to downgrade the link between national security and civilian space by disbanding the National Space Council. The hawkish direction of the council under the Reagan and Bush administrations—and the focus of its final report on SDI and eventual goals of U.S. space weapons—moved in the wrong direction, according to the new administration. The nexus of U.S. space decision making now returned to NASA and the Office of Science and Technology Policy.[18] Given President Clinton's focus on cost cutting, however, the administration also quickly put together a Station Redesign Team for *Freedom* to examine means of economizing on what was threatening to become an unaffordable, Cold War-inspired white elephant. Already, the space station development program had cost $11 billion, with nothing yet in orbit. A letter from RSA head Koptev outlined the possible role of a planned Russian *Mir 2* as a module for the *Freedom* station, influencing the work of the study group. Soon, both sides embraced the idea of a merger of the two elements to enable a joint effort, since parallel national programs no longer seemed to make either political or economic sense. One hurdle, however, stood in the way. With the Cold War only just having ended and with new fears of rogue missile threats on

---

[17] Ibid., p. 491.

[18] On this shift, see Dwayne A. Day, "A New Space Council?" *Space Review*, June 21, 2004, online at <http://www.thespacereview.com/article/163/2> (accessed July 10, 2006).

the horizon, Washington insisted that the extensive cooperation envisaged under the new space station design meant that Russia would have to agree to follow the guidelines of the Missile Technology Control Regime (MTCR). These strictures would require Moscow to refrain from future sales of missiles with a range in excess of 300 kilometers and with a payload capacity of 500 kilograms, except to MTCR member states already possessing such capabilities. Otherwise, administration officials feared a situation where NASA funds and advanced space technology were being shared with Russian space enterprises that might then be selling missiles out the back door to regimes that could threaten the United States.

Traditional supporters of NASA and the U.S. space industry in Congress proved initially skeptical to the idea of including the Russians in the planned *Freedom* space station. NASA, when all was said and done, had always been popular for two reasons: (1) nationalistic pride; and (2) good, high-paying domestic jobs. Some members also worried about the reliability of any Russian promises to restrict sales of missile technology. However, NASA Chief Administrator Dan Goldin, Vice President Gore, and President Clinton lobbied members of Congress heavily with arguments about the benefits of cooperation for NASA in terms of cost and technology. The bill also included $400 million in cooperative space station projects aimed to keep Russian scientists and technicians employed as a means of preventing the proliferation of their missile know-how.[19] In late June, the House passed (by one vote) a budget that included the revised plan after defeating two amendments to scuttle the project entirely.[20] The bill went forward for Senate consideration late in the summer.

Meanwhile, a debate brewed over the first test case of administration policy, namely, Russia's proposed deal to sell advanced cryogenic engines and manufacturing technology to India—a member of neither the nuclear Nonproliferation Treaty nor the MTCR. The sale pitted Russia's commercial space goals against U.S. nonproliferation aims. The Bush administration had already in March 1992 placed sanctions on the Russian state body responsible for the proposed sale (the commercial agency Glavkosmos), barring it from receiving funds from the United States or its companies.[21] Finally, after considerable internal delibera-

---

[19] Susan Eisenhower, ed., *Partners in Space: US-Russian Cooperation After the Cold War* (Washington, D.C.: Eisenhower Institute, 2004), p. 43.

[20] Marcia S. Smith, "NASA's Space Station Program: Evolution and Current Status," Testimony before the House Science Committee, April 4, 2001, available on the NASA Web site at <http://history.nasa.gov/smith.htm> (accessed July 11, 2006).

[21] The sanctions also included the proposed recipient, the Indian Space Research Organization. See M. Lucy Stojak, "Recent Development in Space Law," in J. Marshall Beier and Steven Mataija,

tions, the Yeltsin administration decided in July to reverse its planned transfer of production technology to India, moving forward only with the sale of the actual launchers.[22] Russia's decision indicated a new willingness to accept international nonproliferation norms in exchange for acceptance in a broader commercial and military space community. This clear Russian preference for negotiated forms of space security impressed members of the Senate. In September, the Senate approved the appropriations bill that included the space station, rejecting an amendment to cancel the project altogether.[23] The newly renamed *International Space Station* (*ISS*) marked a major turning point for Russia in space, now rejecting the competitive politics of the Cold War and instead engaging itself in a major cooperative venture with its erstwhile enemy. For the United States, the move allowed NASA to rein in a ballooning $30 billion station budget to a more manageable $17.4 billion, while providing new opportunities to benefit from the technology of former Soviet space enterprises.[24]

The Clinton administration's focus on the U.S. economy and erasing the Reagan-era budget deficit led to a pro-business attitude in the commercial space field as well. Given the growth of international (including Russian) services in remote sensing, the government issued new guidelines in March 1994 to improve the competitiveness of U.S. commercial satellites, allowing sales of images with a resolution of up to one meter.[25] The president's new National Space Transportation Policy, issued in August, instructed the Departments of Commerce and Transportation to encourage the development of commercial space launch vehicles.[26]

In the field of debris mitigation, the Clinton administration continued past U.S. policies but also moved them up to the next level. Whereas the Bush administration had pursued bilateral talks with the European Space Agency (ESA), Japan, and the Russian Federation, the new U.S. team merged these efforts in 1993 into a single Inter-Agency Space Debris Coordination Committee (IADC). By now, these efforts included participation by ten space agencies and by the multiple countries represented in ESA.[27] Gradually, what had been pri-

---

eds., *Arms Control and the Rule of Law: A Framework for Peace and Security in Outer Space* (Toronto: Center for International and Security Studies, York University, 1998), p. 58.

[22] Eisenhower, *Partners in Space*, pp. 44–45.

[23] Smith, "NASA's Space Station Program," p. 3.

[24] Ibid.

[25] Steven Lambakis, *On the Edge of the Earth: The Future of American Space Power* (Lexington: University Press of Kentucky, 2001), p. 231.

[26] Ibid., p. 232.

[27] Author's interview with Nicholas Johnson, NASA's chief scientist for debris migration, Johnson Space Center, Houston, Texas, September 9, 2007 (conducted via e-mail).

marily a U.S. concern was now becoming increasingly international. Moreover, the institutionalization of these talks and the growing exchange of research helped move this process toward practical cooperation and discussion of best practices.

Space and the Changing Political Balance in the U.S. Congress

The November 1994 midterm elections marked a major political setback for the Clinton administration. Instead of being able to work with Democratic majorities in the House and Senate, it now faced powerful opposition after the Republican success in capturing both bodies. The results affected space security policy by pushing the political pendulum back toward the Reagan administration in terms of prevailing attitudes on missile and space defenses. The Republicans' Contract with America called for rewriting the rules of congressional committees to increase the relative power of individual (and newly elected) members. It also listed as high priorities cuts in discretionary domestic spending, enactment of balanced budget legislation, efforts to redress a perceived decline in national defense, and minimization of the role of international organizations in U.S. national security decision making.[28] The new majority's National Security Restoration Act proposed specifically that Congress should "rapidly provide the American people, United States forces, and United States allies with a capable defense against missile attacks."[29] Notably, the draft legislation failed to make any reference to existing limitations against NMD deployments under the ABM Treaty.

By 1995, the fruits of the first Bush administration's and early Clinton team's space agreements with the RSA (Russian Space Agency) had begun to bear fruit. In March, U.S. astronaut Norman Thagard rode a Russian Soyuz rocket into orbit for what would be a U.S. record-setting flight of 115 days (far short of the Russian mark) and the first by any American on the *Mir* space station.[30] In late June, the U.S. shuttle *Atlantis* conducted the first U.S.-Russian docking since the Apollo-Soyuz flight twenty years earlier.[31]

With this burgeoning cooperation, President Clinton resisted efforts to shift

[28] See the 1994 "Republican Contract with America," posted on the U.S. House of Representatives Web site at <http://www.house.gov/house/Contract/CONTRACT.html> (accessed July 10, 2006).

[29] See text of the proposed "National Security Restoration Act," U.S. House of Representatives Web site at <http://www.house.gov/house/Contract/defenseb.txt> (accessed July 10, 2006).

[30] Eisenhower, *Partners in Space*, pp. 69–70.

[31] Ibid., p. 70.

U.S. policy from a reliance on treaties to an emphasis on military means. As the Pentagon's official history of SDI notes, the Clinton administration "dismissed the broad interpretation of the treaty in 1993 and focused its energies on 'strengthening' the ABM Treaty" with efforts to demarcate allowed TMD from banned NMD efforts.[32] At the Moscow summit in May 1995, Presidents Clinton and Yeltsin issued a joint statement outlining their shared interpretation of the ABM Treaty, which would allow development of slower TMD systems (such as the Patriot) but disallow faster, NMD-capable interceptors that might threaten the viability of Russia's strategic deterrent.[33] The Republican-dominated Congress lacked the power to overrule this joint presidential statement, which fell short of treaty status. Thus, the Clinton administration was—for now—able to maintain its preferred policy.

Congressional legislation passed in late 1995 attempted to force the White House to deploy an NMD system by 2003. However, President Clinton used his veto power to reject the bill and Republicans lacked the votes to override him. Nevertheless, Republican members of Congress succeeded in doubling the Pentagon's NMD request for research and development.[34] Their goal of deploying a broad, multilayered NMD network soon became part of the proposed 1996 Defend America Act, although high costs aired by the Congressional Budget Office for the proposed deployment eventually caused deficit-conscious members to withdraw the legislation.[35] But Republicans continued to pressure for changes in the administration's missile defense priorities.

Despite the lack of interest by the air force in continuing to test kinetic-energy systems in space, the Republican-controlled Congress believed that the United States needed such systems in order to have a capability to attack hostile spacecraft. In 1996, a multiyear effort began under the leadership of Senator Bob Smith (Rep., New Hampshire) to force the administration to restart a terminated army program called the Kinetic Energy Anti-Satellite (KEASAT) system. The KEASAT program involved use of a ground-based interceptor to destroy satellites in low–Earth orbit.[36] With no request from the Pentagon, Re-

[32] See Department of Defense, "The ABM Treaty and the President's Decision Not to Deploy NMD," from "Fact Sheet: History of the Missile Defense Organization," online at <http://www.defenselink.mil/specials/missiledefense/history4.html> (accessed July 15, 2006).

[33] Armed Forces Newswire Service, "U.S. and Russia Issue 'Clarification' to ABM Treaty," May 11, 1995.

[34] Fitzgerald, *Way Out There in the Blue*, p. 492.

[35] Ibid., p. 493.

[36] Laura Grego, "A History of Anti-Satellite Weapons Programs," study available on the Web site of the Union of Concerned Scientists at <http://www.ucsusa.org/global_security/space_weapons/a-history-of-asat-programs.html> (accessed October 27, 2006).

publican supporters added $30 million in funding for the project to the FY 1996 budget.[37]

Although the congressional majority hoped for changes in the White House, the November 1996 elections returned Bill Clinton to a second presidential term, thus ensuring a continued battle over space and missile defense issues.

Further Russian Space Decline and Reorganization

For those who had invested their lives in the glorious space accomplishments of the Soviet period, the mid-1990s marked an ignominious end and the start of nearly a decade of struggle for survival. State orders had largely dried up, and those enterprises that did win state orders often found well into the work that the government had reneged on its financial pledges. The civilian space budget had plummeted to around $700 million in 1996, putting at risk Russia's pledge to contribute $3.3 billion to the *ISS* (compared to commitments of $17 billion by the United States and $6 billion by other Western allies).[38] In fact, Russia could no longer continue to supply its own *Mir* space station. These conditions created a vise that only tightened as time went on: Russia could not afford to continue both its own national space program and its now extensive international commitments to the *ISS*. For the time being, however, the fee of $400 million paid by NASA to allow its astronauts—beginning with Thagard in 1995—to train for future *ISS* activities aboard the *Mir* plugged part of this gap.

Meanwhile, Russia's conflict with Kazakhstan over Baikonur had reached a head. Recognizing that Baikonur could not easily be replaced—given its massive infrastructure, on-site personnel, and favorable location for geostationary launches—Moscow agreed to pay Kazakhstan for continued use of the facility and its operation by Russian troops. Although the deal reached in March 1994 failed to end Kazakhstani claims for various usage fees and environmental cleanup costs, the two sides reached a basic yearly lease agreement of $115 million (paid in hard currency) for continued Russian use of Baikonur until 2024.[39] However, Russia hedged its bets by also announcing construction of a major new spaceport in the Russian Far East (at Svobodniy, an old Soviet missile base), aiming at the eventual replacement of Baikonur.[40] Where the ex-

---

[37] Ibid.

[38] Burrows, *This New Ocean*, p. 601.

[39] Harvey, *Russia in Space*, p. 196.

[40] Dmitriy Frolov, "Strategic Missile Base Will Become Cosmodrome," *Segodnya* (Moscow), June 17, 1994, p. 1 (FBIS-SOV-94-117). The report listed a conversion cost of four trillion rubles over ten years.

tensive funding required for such a shift would come from—in the midst of Russia's financial crisis—remained far from clear.

Over the course of the 1990s, Russia's *military* space program entered a period of even more precipitous decline. Deprived of state funding and with no opportunity to attract outside support—unlike the civilian-oriented RSA—the newly created Russian Space Forces saw their once-plentiful spigot of Cold War activities slow to a trickle. As one analyst reported, given these financial conditions, "it is remarkable that Russia [was] able to maintain a military space programme at all."[41] However, some enterprises had leftover monetary and material reserves built up under the Soviet system, partly a result of the past padding of inventory to guard against production shortfalls in case of a failure to meet the yearly state plan. Such resources, for those facilities that had them, came in handy as cash and materials disappeared from the Russian economy in the early 1990s. Russian military space launches remained above twenty per year through 1994, but then plummeted sharply to nine a year in 1996 and finally to only four during 1999.[42] Further, Moscow's previously extensive space tracking network began to fall apart, research and design bureaus collapsed or moved in new directions, and access to other important facilities, inputs, and sites now had to be negotiated and paid for in advance, rather than simply assumed. Under these conditions, gaps in satellite coverage were bound to occur, some with direct military consequences.

Of greatest international concern was the stability of Russia's early-warning network for detection and identification of foreign missile launches. The two-tiered Soviet system had been based on satellites in highly elliptical orbits (HEO) and, after 1984, geostationary orbits (GEO).[43] Over the course of the 1990s, the Russian government struggled to keep this network functioning. While coverage of the globe narrowed, the HEO satellites continued to cover the main missile sites in at least the United States.[44] The GEO network, however, began to fall apart by the end of the decade.

The toll of the deterioration of Russia's network of early-warning satellites and radars became evident in January 1995 when a Norwegian scientific rocket to study the northern lights was mistaken for a U.S. Trident sea-launched ballistic missile.[45] Although other parts of the Russian government had been no-

[41] Harvey, *Russia in Space*, p. 140.

[42] Ibid., p. 140.

[43] Pavel Podvig, ed., *Russian Strategic Nuclear Forces* (Cambridge, Mass.: MIT Press, 2001), pp. 431–32.

[44] Ibid., p. 432.

[45] On this incident, see Geoffrey Forden, Pavel Podvig, and Theodore A. Postol, "False Alarm, Nuclear Danger," *IEEE Spectrum*, Vol. 37, No. 3 (March 2000).

tified of the test, the radar operators and parts of the military had not. After several tense minutes, the Russian government decided that this was not the start of World War III and halted before any counterattack orders were given. The incident highlighted serious problems in the Russian system and the need for greater cooperation in early warning.

To prevent future accidents, support Russian scientists, and appease domestic supporters of missile defense, the Clinton administration proposed a joint satellite project to share early-warning data from missile launches. Known as the Russian-American Observation Satellite (RAMOS) program, the effort envisaged development of U.S. and Russian spacecraft and sharing of information as a confidence-building and threat reduction measure.[46] The two sides formalized the agreement at a Clinton-Yeltsin summit in 1997, allowing U.S. funds to be shared by the Missile Defense Agency with the Kometa Central Scientific-Production Organization in Russia.[47]

In the area of photo-reconnaissance, the short lifespan of Soviet-designed spy satellites (measured in months rather than years) exacerbated the problem of maintaining a reliable surveillance network, because each had to be replaced more frequently than their U.S. counterparts. As a result of its worsening financial situation, the Russian military experienced its first "blind" period in space reconnaissance coverage from late September 1996 to May 1997.[48]

Russia's financial problems also put the safety of its ships and planes at risk. The Soviet-developed GLONASS[49] satellite network—an analogue to the U.S. Global Positioning System launched in 1978—had only been completed in 1995.[50] Its original aim had been to provide highly accurate navigational signals for Soviet navy ships and freighters, as well as military and civilian aircraft. Launched by Proton boosters, the large GLONASS spacecraft were designed to function in a twenty-four-satellite constellation, with each satellite lasting approximately three years and with replenishment launches coming at two per year.[51] However, the technology employed by the Russians lagged far behind

---

[46] G. Wayne Glass, "U.S. and Russian Cooperation on Missile Defense: How Likely? The Troubling Story of the Russian American Observation Satellite (RAMOS) Program," Center for Defense Information, Washington, D.C., May 29, 2002.

[47] Pavel Podvig, "U.S.-Russian Cooperation in Missile Defense: Is It Really Possible?" PONARS Policy Memo 316, November 2003, available on the Web site of Stanford University's Institute of International Studies at <http://iis-db.stanford.edu/pubs/20733/podvig-pm_0316.pdf> (accessed October 27, 2006).

[48] Harvey, *Russia in Space*, p. 120.

[49] The Russian name is *Globalnaya navigatzionnaya sputnikovaya systema* (literally, global navigational satellite system).

[50] Harvey, *Russia in Space*, p. 134.

[51] Ibid., p. 135.

that of the U.S. counterpart, and by 1996, financing to maintain even this second-tier system had already begun to disappear, commensurate with the sharp decline in defense ministry resources and the absence of any significant commercial users. No launches took place in 1997, and serious gaps began to appear in the constellation by the following year. Given these problems, the Russian government opted to shift responsibility for the system from the military to the RSA with the aim of eventual commercialization.[52] But progress in this direction proved slow, leading to functional limitations on military (and civilian) navigational capabilities.

The Way Out for Russia: Space Cooperation and Joint Ventures

Russia's sharp economic decline required that its space industry radically transform itself or face imminent collapse. Fortunately, it had technologies that Western companies valued, providing the basis for a series of important new joint ventures that had emerged over the course of the mid-1990s.

The first major beneficiary of the new Russian relationship with the United States was the scientific production association (NPO, in Russian) Energiya, producer of a number of Soviet-era launch vehicles, spacecraft, and space stations, including the *Mir*.[53] This was the exact kind of expertise the United States needed as it attempted to move forward with the *ISS*, thus making Energiya a valuable partner and the prime contractor under the first space station agreement between NASA and the RSA (part of the $400 million mentioned in the space bill). This relationship continued and expanded during the course of the 1990s.

The first truly commercial venture involved a bold initiative between Lockheed and Khrunichev NPO to market the highly successful Soviet-designed Proton rocket. Conceived in December 1992, the new company survived a series of mergers and acquisitions (involving Martin Marietta and Energiya) to emerge in 1995 as International Launch Services (ILS). By offering rates below those charged by Western commercial launchers and possessing a long and reliable track record, ILS quickly succeeded in capturing 15 percent of the world commercial space-launch market by 1995.[54]

Another innovative and successful four-way partnership aimed at marketing Ukrainian Zenit boosters—one of the components of the Energiya launch

[52] Ibid.

[53] On Energiya's new ventures, see Von Bencke, *Politics of Space,* pp. 160–61.

[54] Harvey, *Russia in Space,* p. 289.

system that had boosted *Buran* in 1988. The concept involved use of a mobile, sea-based launch platform to enable it to access popular orbits. The team that formed the so-called Sea Launch partnership in 1995 included the Dnepropetrovsk-based Yuzhnoe design bureau (Ukraine), Boeing (United States), Energiya (Russia), and Kvaerner Maritime Group (Norway).[55] Putting together the deal proved complicated, but by 1999, the new company had finally begun launch operations, winning a modest market share.

A more lucrative partnership involved some of the most modern Russian space hardware, specifically, the highly advanced RD-180 engine. This joint venture emerged out of talks between Pratt and Whitney (responsible for the U.S. Atlas program) and the Russian manufacturing enterprise Energomash. By the end of the decade, Energomash would amass $1 billion in orders, ironically, largely for service contracts with the U.S. military.[56] This unique cooperative venture showed how much U.S. and Russian perspectives on space security had changed within the past decade.

Other major international deals included the 1995 agreement between France's Arianespace and the Central Specialized Design Bureau in Samara, Russia, called Starsem. This effort marketed the Soyuz launcher. Another joint venture linked Khrunichev with Germany's Daimler-Benz to market the small Russian Rockot booster, a converted SS-19 missile, renamed Eurorockot.[57] Despite setbacks, these swords-into-ploughshares concepts would be well under way by the late 1990s. These changes marked a formal end to the atomized nature of space activity, which before had been dominated by competing national programs and industries. The emergence of growing internationalization within the space industry gradually picked up speed and, by the end of this period, became firmly established.

One initial limitation, however, on Russia's ability to enter the world space market was the problem of U.S.-imposed launch quotas.[58] Before 1991, U.S. national security legislation banned U.S.-made technology in any spacecraft launched by the Soviets. The main fear was that Soviet space technicians would attempt to steal and reverse-engineer U.S. systems and allow Moscow to divert them to hostile military purposes. Given the dominant U.S. role in the space industry, such restrictions sharply limited the Soviets' ability to enter the commercial launch market. However, with the decline of the Soviet threat, the Unit-

---

[55] Eisenhower, *Partners in Space*, p. 161.

[56] Harvey, *Russia in Space*, p. 289.

[57] Ibid., pp. 290–91; and Von Benke, *Politics of Space*, p. 168.

[58] On the quota system, see Harvey, *Russia in Space*, pp. 294–95.

ed States shifted to more practical concerns. The delicate balancing act involved several contradictory aims: continued U.S. technological control (which could now be achieved by the continuous presence of U.S. technicians), prevention of too-severe Russian price undercutting of U.S. commercial launchers, and on the other hand, Washington's provision of *enough* launches to Russian firms so that they would not collapse, possibly causing their workers to enter the employment of rogue-state missile programs. The first Russian boosting of a satellite with U.S. technology, a Proton launch of an Inmarsat satellite in 1992, took place through a special export control waiver under President Bush. By 1993, a new system had emerged under which Washington granted Russian enterprises nine commercial launches through 2001, but with prices tied to within 7 percent of established international space-launch prices.[59] Worried about increasing threats of Russia's space collapse and the implications for proliferation, Vice President Gore and Russian Prime Minister Victor Chernomyrdin amended the deal in January 1996 to boost the number of Russian launches to twenty and the cost window to a more flexible window of within 15 percent of world prices.[60]

Moscow's simultaneous efforts to open the Russian space program, save on defense expenditures, and facilitate a more cooperative international environment led the Russian Federation to increase its calls for new restrictions against possible space weapons. Indeed, in August 1993, the Russian Duma had passed a Law of the Russian Federation on Space Activities, which called upon the Yeltsin government to oppose the emergence of an arms race in outer space.[61] At the international level, Moscow had continued to support efforts to negotiate a treaty within the Conference on Disarmament (CD) until the mandate of the Ad Hoc Committee on the Prevention of an Arms Race in Outer Space (PAROS) expired in 1994.

In the fall of 1996, the White House issued a new space policy, the first revision since the end of the Cold War. The new document recognized the changed political context of its space activities and also the new economic opportunities presented by the improved international environment. In the space security

[59] Ibid., p. 295.

[60] Ibid. This arrangement more closely coincided with the agreement reached with China in 1989, which granted nine launches if China ensured an open market, and a subsequent eleven-launch deal with prices pegged to within 15 percent of world prices. However, export control issues remained problematic with each launch. On this issue, see Brian Harvey, *China's Space Program: From Conception to Manned Spaceflight* (Chichester, England: Praxis, 2004), pp. 116–17.

[61] Detlev Wolter, *Common Security in Outer Space and International Law* (Geneva: U.N. Institute for Disarmament Research, 2006), p. 64.

realm, the new space policy noted under "key priorities" the critical need to use space to "monitor arms control and non-proliferation agreements."[62] Concerning the issue of treaties, U.S. space policy under President Clinton remained open to possible additional controls, stating that Washington would consider entering into new agreements "if they are equitable, effectively verifiable, and enhance the security of the United States and our allies."[63] Emphasizing the importance of new understandings on the need to keep space safe and accessible, the official policy stressed the U.S. government's continued priority of taking "a leadership role in international fora to adopt policies and practices aimed at debris minimization."[64] Indeed, through the IADC (Inter-Agency Space Debris Coordination Committee), the United States would lead an effort—beginning in the late 1990s—to begin drafting formal Space Debris Mitigation Guidelines for adoption at the international level.[65]

At the CD in Geneva, however, little evidence of the official U.S. open-mindedness on space arms control measures emerged, despite the fervent efforts of a number of countries, including Russia, Australia, and Canada, to reconstitute the PAROS Ad Hoc Committee.[66] Instead, the United States continued a policy begun under the Bush administration of seeking no new space treaties. In search of a compromise, India and Brazil proposed in 1998 that satellite safety and a ban on anti-satellite weapons form the mandate for the new CD committee, instead of the broader subject of complete nonweaponization, which could include missile defenses. Again, Washington opposed even the new, more restricted formula for discussions, thus blocking action (given the CD's rule of consensus). The United States believed instead that nuclear nonproliferation and progress toward a Fissile Material Cut-Off Treaty (FMCT) should constitute the main focus of the CD. Debate continued through the end of Clinton's second term on the possibility of multiple ad hoc committees but without a consensus on forming a program for actual work. Around this time, China emerged as the main U.S. adversary by pushing for renewed space negotiations.

At the United Nations in New York, meanwhile, the United States continued to fight a lonely battle against the yearly resolution on PAROS, usually voting with Israel and occasionally one or two other states to abstain from the oth-

[62] "Fact Sheet: National Space Policy," Office of Science and Technology Policy, September 19, 1996, p. 4, on the Federation of American Scientists' Web site at <http://www.fas.org/spp/military/docops/national/nstc-8.htm> (accessed October 21, 2006).

[63] Ibid., p. 11.

[64] Ibid.

[65] Author's interview with Nicholas Johnson, NASA's chief scientist for debris migration, Johnson Space Center, Houston, Texas, September 9, 2007 (conducted via e-mail). Johnson led the U.S. IADC team in these negotiations from 1999 to 2002.

[66] On the CD debates at this time, see Wolter, *Common Security in Outer Space*, pp. 67–69.

erwise unanimous international vote of support for the measure. The United States remained unmoved by strong arguments in support of PAROS even from close allies such as Canada and France. Fearing the implications for strategic stability of a foreseen, unilateral U.S. deployment of nationwide missile defenses, a wide majority of countries (including France) voted 80 to 4 in the fall 1999 U.N. session to recognize the ABM Treaty as a "cornerstone" of international security, and called upon treaty members (the United States and Russian Federation) to refrain from deploying NMD systems and to maintain "full and strict compliance."[67] The highly polemical tone of the resolution increased to sixty-eight the number of countries that abstained from the measure, but the United States remained isolated with Albania, Israel, and Micronesia in opposing it. Clearly, despite its own ambivalence in the pursuit of NMD, the Clinton administration had adopted a position opposed to international attempts to restrict the U.S. right to deploy space defenses.

The Democrats' Gradual Relinquishment of the U.S. Space Security Agenda

In an effort to head off mounting Republican opposition to perceived administration foot-dragging on missile defense, President Clinton decided to strike a compromise and move toward gradual deployment, but with an escape route. The concept became known as "three-plus-three," or three years of research and development that could lead to a decision to deploy a system within the following three years.[68] The focus of the program was considerably more modest than even the Bush administration's GPALS framework, envisaging an initial deployment of only some twenty missiles to guard against a rogue-state attack. Still, administration critics largely hailed the move as a slippery slope bound to benefit their preferred outcome sooner or later. However, the White House continued to insist on a negotiated—rather than unilateral—approach to the ABM Treaty. Russia remained opposed to NMD but faced increasing U.S. pressure to allow testing of high-altitude theater systems, such as THAAD. In 1997, Russia agreed to "grandfather" THAAD into the ABM Treaty but balked at further compromises. Republican supporters of NMD sharply criticized the new agreement and continued to demand U.S. withdrawal from the ABM Treaty.

In 1998, the internal U.S. debate only intensified further. Following a CIA

[67] U.N. General Assembly A/RES/54/54, "Preservation of and Compliance with the Treaty on the Limitation of Anti-Ballistic Missile Systems," January 10, 2000, United Nations, *Treaty Series*, Vol. 944, No. 13446.

[68] See Fitzgerald, *Way Out There in the Blue*, p. 494.

report that concluded that it was unlikely that a rogue state—such as Iran, Iraq, or North Korea—could develop a long-range ballistic missile capable of striking the United States within the next decade, conservative critics in Congress commissioned a follow-on study under former Secretary of Defense Donald Rumsfeld. The Republican-leaning though nominally bipartisan commission reported very different findings based on a lower standard of evidence. Their study—issued in July 1998—argued that it was *possible* that a rogue state could develop an ICBM within five years if it had assistance from a foreign country.[69] More troubling was their conclusion that the United States would likely not know about the decision to develop an ICBM until it was too late to build defenses. The authors argued that prudence dictated an *advance* response, before the new missiles could be launched. While critics noted the very limited number of hostile states with even medium-range missile capabilities, world events soon intervened to effectively weaken these arguments. North Korea's launch of a three-stage, medium-range Taepodong over Japan in August seemed to "prove" exactly what the Rumsfeld commission had warned against. Although the third stage of the missile failed and the range proved to be a mere 828 miles,[70] those concerned about the missile threat concluded that the *potential* range of the missile might be enough to reach the Aleutian Islands—had everything worked. This claim soon led to assertions in leading conservative newspapers that North Korea was readying itself to hit the United States with a ballistic missile.[71] Supporters of missile defense ran with this threat to try to outflank the defense policy of the Clinton administration and, more importantly, the 2000 presidential campaign of Vice President Gore.

With former Republican Senator William Cohen leading the Clinton defense department, supporters of NMD had a natural ally in the Pentagon. In January 1999, Secretary Cohen added $6.6 billion to the long-term NMD planning budget to facilitate the system's deployment if a decision to go ahead were made in the coming year. The secretary pushed out the deployment date, however, from 2003 to 2005 because of the scale of work required and continued slowness of the test program in achieving its milestones.

Senator Smith and his Republican allies in Congress, meanwhile, continued to add unrequested funding to the Pentagon budget for the controversial KE-

[69] See "Executive Summary of the Report of the Commission to Assess the Ballistic Missile Threat to the United States," July 15, 1998, Pursuant to Public Law 201, 104th Congress, posted on the Web site of the Federation of American Scientists, at <http://www.fas.org/irp/threat/bm-threat.htm> (accessed October 22, 2006).

[70] Sonni Effron and Norman Kempster, "N. Korea's Latest Provocation May Be Last Straw," *Los Angeles Times*, September 1, 1998, p. 1.

[71] See, for example, "A Line in the Sand?" (Editorial), *Wall Street Journal*, July 14, 1999, p. A22.

ASAT program. Given the system's relatively primitive, kinetic-kill technology and the likely damage it would cause to other U.S. space assets, the air force had been "openly critical" of the effort.[72] Nevertheless, by the end of the Clinton years, a total of $125 million would be appropriated for research and development costs.[73]

The main problem for supporters of missile defense again proved to be technological. Even under conditions favorable to the interceptor (such as launch prenotification, fixed speeds, and sharp limits on decoys), the elements of the proposed system, including a series of THAAD missiles and midcourse, ground-based interceptors, registered repeated failures. U.S. Undersecretary of Defense Paul Kaminski admitted that hit-to-kill tests had failed in 70 percent of attempts even under these non-battlefield conditions.[74] Nevertheless, the weight of support behind NMD and the fears of opponents of being branded as "weak on defense" (particularly in the wake of the 1999 Monica Lewinsky impeachment scandal that had hurt not only the president but also the Democrats in Congress who had supported him) led to Republican momentum for passage of the National Missile Defense Act of 1999. Introduced by Senator Thad Cochran (Rep., Mississippi) and sponsored in the House by Curt Weldon (Rep., Pennsylvania), the bill read: "It is the policy of the United States to deploy as soon as is technologically possible an effective National Missile Defense system capable of defending the territory of the United States against limited ballistic missile attack (whether accidental, unauthorized, or deliberate)."[75] Democrats had mixed feelings about the bill but seemed assured by its language about "limited ballistic missile attack" and its pledge that yearly appropriations votes would be required on the system. Notably, the bill failed to specify what kind of defenses would be built, and unlike in 1991, the legislation made no specific mention of the ABM Treaty. Still, most Democrats saw it as a "Mom and apple pie" bill that committed them to little and burnished their defense credentials, making it hard to vote against. Although initially opposed to the legislation, the Clinton administration decided reluctantly to support it when it became clear that the bill would pass with a veto-proof majority.[76] Freed from White House pressure, more Democrats opted to support the bill. The Senate version passed

[72] Grego, "History of Anti-Satellite Weapons Programs."

[73] U.S. General Accounting Office, chart entitled "Summary of KE-ASAT Funding," in "KE-ASAT Program Review," December 2000, posted on the Web site of GlobalSecurity.org at <http://www.globalsecurity.org/space/library/report/gao/d01228r.pdf> (accessed October 27, 2006).

[74] Fitzgerald, Way Out There in the Blue, p. 564, n. 61.

[75] Text of Senate Bill 257 (as passed), on the Web site of the Library of Congress at <http://thomas.loc.gov/cgi-bin/query/D?c106:3:./temp/~c106XOwrI2::> (accessed July 12, 2006).

[76] On this point, see Fitzgerald, Way Out There in the Blue, p. 496.

on March 17, 1999, with 97 in favor and only 3 opposed (Democrats Durbin, Wellstone, and Leahy). The House version of the bill passed the following day 317 to 105, with 103 Democrats crossing party lines to support the legislation.

The Continuing Russian Economic Crisis

During the course of the late 1990s, space security developments continued to be affected by declining conditions in the Russian economy, leading to heightened concerns about brain drain, lax export controls, and the proliferation of WMD technologies. In general, the impact of U.S. funding helped curb tendencies to proliferate, for Russian manufacturers faced a serious trade-off: play by the rules and continue to receive U.S. contracts or violate the rules and risk sanctions and the end of Western cooperation. For larger enterprises with more to lose, compliance generally improved. For smaller companies, the temptation (and need) to export proved more difficult to overcome, particularly as the Russian economy entered a new phase of its ongoing crisis.

In the early 1990s, the Russian economy had been opened up by Yeltsin to what many Russians (and observers) called "wild capitalism." The government had failed to keep up with legislative and legal protections for consumers and would-be Russian shareholders. Plans for the privatization of large Russian companies had led to massive corruption, as workers and others who were supposed to benefit found themselves swindled by shady dealers and unscrupulous entrepreneurs. Moreover, more than a few banks continued to play Ponzi schemes, promising absurdly high returns but actually moving money from one set of accounts to another to stave off sudden withdrawals. In addition, a number of financial institutions and importers made money by playing off the gap in Russia's periodically adjusted exchange rate and the real street rate, leading to speculation, inadequately supported loan operations, and declining price competitiveness of Russian-made products. With dropping oil revenues, a mounting current account debt, and an inability to maintain the inflated ruble and associated high interest rates, the Russian government finally devalued its currency, causing the bottom to fall out of the economy.[77] Inflation soared and the stock market lost 90 percent of its value. Again, Russia's survival as a state appeared to be in jeopardy. However, by pricing imports out of the reach of ordinary Russians, the crisis led eventually to the revival of domestic industry.

In the commercial space sector, the inability of firms to pay for component parts created an expanding ripple of delays in the fulfillment of contracts. Sala-

---

[77] On these developments, see Padma Desai, "Why Did the Ruble Collapse in August 1998?" *American Economic Review*, Vol. 90, No. 2 (May 2000).

ries began to lag again, and rampant inflation stripped workers of purchasing power even when wages were obtained. Russia's reliability as a commercial partner began to be questioned.[78] For NASA, its heavy investment in the RSA for critical deliveries on the *ISS* project now began to seem like a serious mistake.

The impact of Russia's financial collapse on space proved even more severe in the military sector. After completing ten military launches in 1998, the Russian military could only afford to carry out four in each of the next two years.[79] As tensions grew due to ethnic cleansing by Serb forces in Kosovo, and the United States and NATO began their attack on Serbia in March 1999, Russia—Serbia's erstwhile protector—found itself at a severe intelligence disadvantage.[80] With only one photo-reconnaissance satellite and one electronic intelligence satellite in orbit, Russia received only periodic updates from the battlefield. Clearly, this gap worked to the U.S./NATO advantage. Despite recognition of the problem by the Russian General Staff, the systemic problem of financing produced difficult trade-offs, particularly given the high cost of Russia's renewed war in Chechnya in August. These conditions contributed to Russia's second "blind" period in space-based photo-reconnaissance from December 1999 to May 2000.[81]

In the more critical area of early-warning satellites, Russia also experienced problems. The basic system of early warning established in the 1990s consisted of geostationary Prognoz satellites and Oko spacecraft in elliptical orbits. By 1999, Russia found itself too strapped to replace the *Kosmos 2350* launched into GEO April 1998, which malfunctioned and ceased transmitting data after only two months.[82] Thus, Russia had to rely on a constellation of Oko satellites in lower orbit, which offered solid views only of Western Europe and the east-central portion of the United States, supplemented by early-warning radars in former Soviet states with which it still maintained military relations. While Russia was not technically "blind" to possible ICBM attack, its range of vision was now dangerously limited. This increased risks of late or faulty detection and raised serious uncertainties about the ability of decision makers to make reliable decisions in what could be a very short amount of time.

Despite these difficulties, the RSA managed to fulfill critical pledges to the United States in establishing itself as a reliable, and indeed essential, partner in the construction of the *ISS*. In November 1998, a Russian Proton rocket launched the first system element for the *ISS* from Baikonur: the *Zarya/FGB* (or

---

[78] On politics of these problems, see Eisenhower, *Partners in Space*, pp. 84–85.

[79] Harvey, *Russia in Space*, p. 140.

[80] For a good description of Russia's gap at this time, see Harvey, *Russia in Space*, p. 109.

[81] Ibid., p. 120.

[82] Ibid., p. 138.

functional cargo block).[83] The module represented the first significant product under Khrunichev's subcontract for the *ISS* with the Boeing corporation.[84] A U.S. shuttle *Endeavor* flight in December added the U.S. *Unity* module to the Russian spacecraft to complete the first segment of the *ISS*, thus beginning this unprecedented U.S.-Russian construction project in space.[85] The Cold War in space finally seemed to be over: old policies of defending space had now turned to joint development.

End of the Clinton Years and Increasing Tensions over NMD

Despite this progress, space security relations between the United States and Russia in the last years of the Clinton administration witnessed increased hostility. The White House felt pressed by domestic political forces to try to wrest further amendments in the ABM Treaty from an increasingly reluctant Russian side. The treaty-designated U.S. ABM site remained Grand Forks, North Dakota, where the lone deployment of interceptors had once stood in the mid-1970s. Now, the administration's plans, which included an X-band radar site on Shemya Island, Alaska, to track hostile missiles coming from Asia, required moving the designated site.[86] No precedents existed for such a shift, and Moscow protested vehemently, seeing that it could not benefit from these changes. Instead, this path seemed to be leading toward a U.S. breakout from the treaty. To prevent such a development, incoming Russian President Vladimir Putin pushed the Russian Duma to ratify the long-delayed START II Treaty (which the U.S. Senate had approved in 1996) in April 2000 and the Comprehensive Test Ban Treaty (which the U.S. Senate had rejected in the fall of 1999) in June. But President Putin emphasized that Russia's START II adherence would be linked to continued U.S. compliance with the traditional, narrow version of the ABM Treaty that rejected testing of new NMD technologies and related space

[83] In Russian, *funtsional'niy gruzovoi blok*. On this module, see the Web site of the Khrunichev Space-Missile State Production Center at <http://www.khrunichev.ru/khrunichev/live/mod_mks.asp > (accessed October 24, 2006); also, Anatoly Zak, "Milestones of Space Exploration in the 20th Century," on Russianspaceweb.com, at <http://www.russianspaceweb.com/chronology_XX.html> (accessed July 16, 2006).

[84] NASA, "The Zarya Control Module: The First International Space Station Component to Launch," *NASA Facts*, January 1999 (IS-1999-01-ISS014JCS), online at <http://spaceflight.nasa.gov/spacenews/factsheets/pdfs/zarya.pdf> (accessed July 16, 2006).

[85] Zak, "Milestones of Space Exploration."

[86] Department of Defense, "The ABM Treaty and the President's Decision Not to Deploy NMD."

systems. He further warned that U.S. violation of the ABM Treaty would lead to Russia's cessation of arms reductions and its possible deployment of new multiple-warhead missiles.[87]

In recognition of the importance of U.S.-Russian relations and the significant progress being achieved in arms reductions, Vice President (and now Democratic presidential candidate) Gore argued that it would be a mistake for the United States to insist on sea- and space-based missile defenses.[88] However, his campaign did not rule out a limited land-based system, such as that being considered under the Clinton administration, which would intercept missiles in space. But he also called for further discussions with Russia on modifying the ABM Treaty.

For the administration, the politics of missile defense had become a delicate tightrope. On October 4, 1999, the Pentagon announced a successful flight test of the ground-based interceptor's exo-atmospheric kill vehicle (EKV),[89] although GPS data and an exaggerated infrared signal had been provided to assist the EKV in reaching its target. In the next test in January 2000, however, the EKV missed the incoming missile when its cryogenic system failed and interrupted signals from its infrared sensors. Meanwhile, close allies remained strongly opposed to a unilateral U.S. abrogation or withdrawal from the ABM Treaty. French President Jacques Chirac argued that U.S. moves to break out of the treaty would "retrigger a proliferation of weapons, notably nuclear missiles."[90] The British government urged the United States to "seek other ways of reducing the threats it perceives,"[91] rather than turning to missile defense. German Chancellor Gerhard Schroeder cautioned about the possible destabilizing division of NATO itself if the United States created a "two-class security system," where some countries were protected by missile defense and others not.[92]

---

[87] Michael R. Gordon, "In a New Era, U.S. and Russia Bicker over an Old Issue," *New York Times*, April 25, 2000, p. A1.

[88] Frank Wolfe, "Gore Campaign: Sea-, Space- NMD Options Bad Idea," *Defense Daily*, March 27, 2000, p. 1.

[89] Department of Defense, "Evolution: NMD," from "Fact Sheet: History of the Missile Defense Organization," online at <http://www.defenselink.mil/specials/missiledefense/history3.html> (accessed July 16, 2006).

[90] Chirac quoted in *New York Times*, July 22, 2000, as cited in "Quotations on Anti-Ballistic Missile (ABM) Treaty/National Missile Defense (NMD)," on the Web site of the British American Security Information Center at <http://www.basicint.org/nuclear/NMD/quotes_EU.htm> (accessed October 27, 2006).

[91] See Tom Buerkle, "U.K. Panel Questions U.S. Missile Shield Plans," *International Herald Tribune*, August 3, 2000.

[92] "Clinton, Schroeder Discord over U.S. Missile Shield Plan," *Deutsche Press-Agentur*, June 1, 2000.

In this difficult environment, Clinton's closest advisors believed that a "go slow" strategy for missile defense made the most sense. One key reason was concern over restarting the arms race with Russia or, perhaps even worse, a reversal of Russia's progress toward democracy. The Clinton team also continued to worry about cost escalation within the NMD program, given its focus on achieving a balanced budget. Finally, its estimate of the near-term threat remained low, and the program was struggling to overcome a number of nagging technological hurdles.

But missile defense continued to enjoy tremendous support among influential Republicans in Congress, who controlled both houses and therefore the power of the purse. Most supported the demise of the ABM Treaty because it stood in the way of developing an NMD system for the United States. As chairman of the Senate Foreign Relations Committee Jesse Helms (Rep., North Carolina) warned the Clinton administration in the spring of 2000, "Any modified ABM treaty negotiated by this administration will be dead-on-arrival at the Senate Foreign Relations Committee."[93] Therefore, any efforts by the Clinton administration at space security through negotiation had to be accomplished *around* Congress rather than in consultation with it.

Meanwhile, Governor Bush's presidential campaign sensed the salience of the issue and began to hit harder on this topic, although not yet pledging itself to ABM Treaty withdrawal. Still, Bush's speech to the National Press Club on May 23 laid down a gauntlet to the Gore campaign. As Governor Bush argued, "The Clinton administration at first denied the need for a national missile defense system. Then it delayed. Now the approach it proposes is flawed—a system initially based on a single site, when experts say more is needed."[94]

Bush called for thorough review and revision of the ABM Treaty to allow for a system that would "protect all 50 states—and our friends and allies and deployed forces overseas." He concluded, "No decision would be better than a flawed agreement that . . . prevents America from defending itself."[95]

Despite this pressure, President Clinton decided to move forward with a special Joint Declaration of Principles at the Moscow summit in June with President Putin, again reiterating the importance of the ABM Treaty as "a cornerstone of strategic stability." To promote the stability of early-warning mechanisms in Russia, the two sides agreed to establish an unprecedented Joint Data Exchange Center (JDEC) in Moscow for coordinating missile flight data

---

[93] Helms, quoted in editorial, "Take No for an Answer," *Wall Street Journal*, May 1, 2000, p. A34.

[94] George W. Bush, quoted in Frank Wolfe, "Bush Pushes Missile Defense, Rebuilding Military," *Defense Daily*, May 24, 2000, p. 1.

[95] Ibid.

and preventing accidental war.[96] The new center would follow on progress in exchanging information and plans for possible problems caused by the international changeover of computers to the year 2000. But critics proved unmoved by these attempts to push weapons issues to the back burner of space and missile security efforts.

Following the summit, ardent missile defense supporter Frank Gaffney lambasted the Clinton administration for having "liquidated the Strategic Defense Initiative." Gaffney described the president's continued commitment to the ABM Treaty in the post–Cold War environment as "simply bizarre."[97]

In the summer, a number of problems in the NMD program emerged. In June, a report to Secretary of Defense William Cohen noted likely delays in the development of the new booster for the system.[98] On July 8, these problems seemed to be confirmed by a highly embarrassing test of the NMD interceptor in which the kill vehicle failed to separate from the booster itself, a fundamental operational mistake that reflected poorly on the system's overall readiness. In early August, Assistant Secretary of Defense and Director of Operational Test and Evaluation Phillip Coyle issued an internal report, which concluded that the current NMD system could not be recommended for deployment by 2005 because of its lack of proven technology.[99] In late August, Defense Secretary Cohen—a supporter of the program—suggested in his recommendation to the president that the administration announce a "limited green light" that would begin to construct key system elements, like the X-band radar, but defer a decision on the complete system.[100] After several days of deliberation, the president delivered a speech at Georgetown University on September 2 at which he announced his decision not to proceed with NMD deployment, stating, "I simply cannot conclude with the information I have today that we have enough confidence in the technology and the operational effectiveness of the entire N.M.D. system to move forward to deployment."[101]

[96] See "Memorandum of Agreement Between the United States of America and the Russian Federation on the Establishment of a Joint Center for the Exchange of Data from Early Warning Systems and Notifications of Missile Launches," U.S. State Department, Bureau of Verification, Compliance, and Implementation, June 4, 2000, on the state department Web site at <http://www.state.gov/t/ac/trt/4799.htm> (accessed October 27, 2006).

[97] Frank Gaffney, "Clinton and Gore Scuttled SDI: Bush Offers Real Choice with Pledge to Deploy National Missile Defense," *Human Events* (Washington, D.C.), June 16, 2000, p. 1.

[98] Paul Richter, *Los Angeles Times*, August 9, 2000, p. 5.

[99] Roberto Suro, "Woes Undermined Missile Defense Cause; Clinton Weighed Test Failures, Development Delays in Addition to Diplomatic Costs," *Washington Post*, September 3, 2000, p. A4.

[100] Ibid.

[101] "A Call for Realism and Prudence: Excerpts from President Clinton's Speech," *New York Times*, September 2, 2000 (late East-coast edition), p. A10.

Conclusion

The hostility of the fall 2000 U.S. presidential debates over NMD provided a dichotomous background to the unprecedented progress that had been made in U.S.-Russian space cooperation in the decade since the Soviet breakup in 1991. While not rejecting cooperation with Russia in space, the Bush campaign portrayed space as an arena primarily for military defense, given what it perceived as the rise of missile threats to the United States and the growing irrelevance of Cold War arms control. Under these conditions, Governor Bush argued that the United States should make the most of its current window of opportunity to develop and deploy a missile defense system in space before such threats emerged. The Gore campaign voiced its support for missile defense as well, including possible space-based elements, but emphasized the need for collaboration both with allies and with the Russian Federation to avoid negative political and military repercussions, which might prove worse than a few rogue missiles.

The October 31 launch of the first *ISS* crew from Baikonur—with one American and two Russians—marked yet another milestone in moving beyond the Cold War to a new footing for broader space security relations. However, as events would soon show, cooperation in human spaceflight seemed to have little impact on military relations, at least on the American side, where concerns about missile and space vulnerabilities would dominate the agenda of the next administration.

In one of the last acts of the outgoing Congress in January 2001, a commission on U.S. management of its space assets outlined the major U.S. dependence on space for commercial and military activities, and identified a range of possible vulnerabilities.[102] The language of the report reflected its authors' belief in a hostile form of technological determinism in space and the likely need for U.S. space weapons to protect American assets from emerging threats. Again, the stage had been set for renewed space nationalism and a challenge to norm- and treaty-based approaches.

[102] "Report of the Commission to Assess United States National Security Space Management and Organization," Pursuant to Public Law 106-65, Committee on Armed Services, U.S. House of Representatives, January 11, 2001.

PART III

Considering Twenty-First-Century Space Security

CHAPTER SEVEN

A New U.S. Perspective on Space
and Its Opponents

2001–2007

The period after 2001 brought a fundamental challenge to the norms of space security established and followed since the early 1960s. In many respects, the rhetoric and direction turned back to the Reagan administration, but in other respects it turned back to 1950s' assumptions about space competition. Changes in U.S. policy caused this confrontation with existing approaches: the result of a dramatic shift in power and perspective in the White House and single-party control over both houses of Congress. The November 2000 presidential election proved to be among the more controversial in American history. As the Supreme Court ended a contentious recount in the last state still in play (Florida), a team dominated by leading neoconservative thinkers and former officials entered office with a self-perceived mandate for dramatic change. In space, they identified shortcomings of simultaneous U.S. dependence and vulnerability. Instead of continuing Clinton-era and Cold War space security policies and co-operation with Russia, the George W. Bush administration promoted concepts of space security aimed largely at military means, stating that the Cold War had ended and that U.S. security could no longer be held hostage to Moscow or the Anti-Ballistic Missile (ABM) Treaty. It also pointed at potential future threats to its space assets. Not surprisingly, this perspective pitted U.S. policies directly against the attitudes of nearly all states active in space security discussions—and particularly those of the two leading players, Russia and China. These countries continued to promote new forms of space arms control and restraint-based approaches, although with no real progress. By the end of this period, both sides would begin brandishing space weapons options.

The Bush administration endured one of the most difficult first years of any president in office. The terrorist attacks in the United States on September 11, 2001, reoriented the security agenda and required immediate American attention and response. But the 9/11 attacks provided one benefit to the administration: they emasculated Democratic opposition to its defense budget requests, missile defense policies, and vision for space for the next six years. These dynamics ushered in a period of untrammeled "America first" thinking in regard to space security, leading to Washington's withdrawal from the ABM Treaty, the doubling of missile defense spending, and the enunciation of broad plans for space weapons for both defensive and offensive (Global Strike) purposes.

In President Bush's second term, however, the continued limited nature of the threats to U.S. space assets from major space powers and the high costs of the administration's simultaneous wars in Iraq, Afghanistan, and against global terrorism more generally, limited funding for the deployment of space defenses. The military services and Congress felt increasing pressure to pay greater attention to near-term needs. An expanded debate by the middle of Bush's second term—as during the 1980s—brought forth new concerns about possible negative implications of space weapons, even within the military. This did not necessarily dampen military, industrial, or congressional willingness to support research funding for space-based defenses, but it did begin to limit the range of realistic options. Moreover, a number of voices—including within the military—began to question the sagacity of possibly spurring opponents and generating orbital debris through the testing and deployment of kinetic-kill defenses. But the administration moved toward voluntary guidelines rather than outright bans on their use in order to keep its options open.

At the international level, this period witnessed a major change in the strategic dynamics of space relations. With the Chinese launch of its first "taikonauts" into orbit, Beijing established itself as a major space player, quickly eclipsing Russia in terms of its dynamics, its commitment to expanded space activity, and its possible collision course with the United States. On the one hand, China's limited military activities before January 2007 and its focus on human spaceflight postponed conflicts over space, and Beijing continued to voice its strong support for a new treaty banning space weapons. On the other, China's apparent laser weapons research—revealed in 2006—made U.S. officials wary, if not skeptical, of China's intentions. Beijing's ultimate goals in space remained unclear, thus undercutting arguments about the "inevitability" of space conflict but also leaving critics of arms control to doubt the viability of new space treaties.

The trend toward a more nationalistic U.S. military space policy after 2001 conflicted with growing U.S. commercial engagement with foreign aerospace manufacturers in joint ventures, which had begun to blur the lines separating domestic space industries. Russian engines launched U.S. military space payloads; Arianespace opened its launch facilities to Russian boosters; and Russian hardware, designs, and know-how facilitated China's first manned spaceflights. By the end of this period, small companies like SpaceX had emerged with plans to offer low-cost, private access to space. These changes threatened the dominance of state-sponsored, heavy-lift rockets in space activity and opened the prospect of small companies and ordinary civilians becoming influential space actors, instead of governments solely. The increasingly international direction of the space industry seemed to support negotiated forms of space security, yet even companies with significant overseas partnerships proved interested in seeking contracts for proposed space-based defenses.

Finally, China's test of an anti-satellite (ASAT) weapon in January 2007 marked the first violation of a tacit norm of no destructive ASAT testing in place since the U.S. test in 1985. Not only did this event threaten a now much more crowded space environment with considerable new debris (adding almost 10 percent to the amount of trackable objects), but it risked starting a cascade of testing by others, including the United States. Thus, as space activity completed its first fifty years, it remained unclear in what direction space security would eventually move: into efforts to forge new forms of international cooperation for the commercial development of space or into a new era of space nationalism, replete with space-based offensive and defensive weapons.

The Bush Administration's Return to a Modified SDI

Although Governor Bush's initial plans for space seemed somewhat inchoate during his election campaign, the staffing of his new administration in early 2001 began to clarify and harden its direction. By appointing Donald Rumsfeld as secretary of defense, President Bush signaled his strong support for a more military-oriented approach to space security. Rumsfeld had chaired the congressionally appointed Commission to Assess United States National Security Space Management and Organization, which had issued its report on January 11. The report described de facto norms regarding military activities in space by saying, "The U.S. and most other nations interpret 'peaceful' to mean 'non-aggressive.'"[1] But the report went further. It warned of the need to "take seriously the possibility of an attack on U.S. space systems" to avoid what it called a "Space Pearl Harbor."[2] It repeated assumptions from leading space weapons

[1] "Report of the Commission to Assess United States National Security Space Management and Organization," Pursuant to Public Law 106-65, Committee on Armed Services, U.S. House of Representatives, January 11, 2001, p. 17.

[2] Ibid., pp. 8–9.

advocates that space would inevitably become a zone of military conflict, just as the sea and air had before it, and that the United States must therefore "develop the means both to deter and to defend against hostile acts in and from space."[3] In regard to arms control, it noted that "there is no blanket prohibition in international law on placing or using weapons in space, applying force from space to earth or conducting military operations in and through space."[4] Consistent with Clinton policies, the commission stated that Washington should be "cautious" of any agreements that might restrict future U.S. military space activities.[5] In regard to missile defense—a topic outside the purview of the Rumsfeld commission—President Bush's appointment of a combination of past Strategic Defense Initiative (SDI) supporters (Stephen Cambone and Douglas Feith) and leading neoconservatives (John Bolton, Stephen Hadley, and Paul Wolfowitz) to senior positions in the Departments of Defense and State and to the National Security Council presaged an activist approach to deployment and a hostile view of the ABM Treaty.[6]

Within the U.S. Air Force, a gradual shift had already begun toward a more active approach on space defenses and offenses. Given the air force's mission of defending U.S. space assets, the service had seized on the opportunity to move beyond past domestic constraints. Its old vision of a manned space bomber now evolved into the concept of an unmanned Common Aero Vehicle (CAV) for possible space engagements and quick-descent bombing of targets on Earth. The air force's plan for the next two decades envisaged a seamless web of space-based, air-based, sea-based, and land-based systems that would allow the United States to achieve "full spectrum dominance."[7] Other documents foresaw the opportunity to use space for more aggressive force projection. A broad-strokes planning document issued by the air force spoke of developing the "ability to control space when need be" and of the capacity to "strike effectively wherever and whenever necessary with minimal collateral damage."[8] A more closely defined concept that would eventually gain a considerable audience within the Bush administration was the notion of using space for so-called Global Strike missions. In other words, with the emergence of new threats, technological

---

[3] Ibid., p. 10.

[4] Ibid., p. 17.

[5] Ibid.

[6] Notably, a number of these individuals had worked previously for major defense contractors involved in missile defense work.

[7] Air Force, "America's Air Force: Vision 2020," available on the Web site of the Defense Department's Defense Technical Information Center (DTIC) at <http://www.dtic.mil/jointvision/jvpub2.htm> (accessed August 9, 2006).

[8] "America's Air Force: Vision 2020," available on the Web site of the U.S. Army, at <http://www.army.mil/thewayahead/afvision.pdf> (accessed August 9, 2006).

developments, and the broadening of the target geography since the end of the Cold War, air force thinkers like General John P. Jumper began to discuss the viability of using space to support a constantly available fleet of advanced weapons platforms (manned aircraft, unmanned aerial vehicles, and airborne lasers) to provide the United States with the ability to hit any target on Earth at very short notice. As General Jumper argued in an influential article published in spring 2001, "Today, we stand on the brink of technological advances that can prompt a new concept of aerospace power employment."[9] While neither Jumper nor these official vision statements specifically mentioned space-based weapons, members of the new administration proved only too eager to append such systems to the evolving air force defense framework.

Steven Lambakis, a leading thinker on space at the neoconservative National Institute for Public Policy (NIPP) who soon entered the Bush Pentagon, had set the stage by arguing, in a book that appeared in early 2001, "We can state with certainty that we will be challenged in space . . . simply because it makes military sense to do so."[10] His critical views of the history of U.S. military space policy and the need for space weapons summed up well the political thinking of senior members of the Bush administration: "Rather than preparing adequately to meet new dangers to national security, policy makers have spent the past forty years arbitrarily, and with astonishing irregularity, constraining military activities in the realm above the atmosphere without clear public justification."[11] From this perspective, the United States needed to worry particularly about the emergence of Chinese military space power. Lambakis discussed China's possible use of nuclear weapons in space as a likely asymmetric response to U.S. space dominance.[12] To meet this purported threat, he called for early, preventive deployment of space weapons to deny the Chinese the ability to carry out such an attack.

Writing in 2001, military strategist Everett Dolman took this argument one step further, linking the need for U.S. space dominance to U.S. goals of promoting liberal democracy and capitalism. These ideas played well with evolving administration doctrines of neoconservatism, which would later be adopted as the U.S. vision for Iraq and the broader Middle East. As Dolman outlined his prescription for space:

[9] Gen. (USAF) John P. Jumper, "Global Strike Task Force: A Transforming Concept, Forged by Experience," *Aerospace Power Journal*, Vol., 15, No. 1 (Spring 2001), at <http://www.airpower.au.af.mil/airchronicles/apj/apjo1/spro1/jumper.htm> (accessed August 9, 2006).

[10] Steven Lambakis, *On the Edge of the Earth: The Future of American Space Power* (Lexington: University Press of Kentucky, 2001), p. 137.

[11] Ibid., p. 2.

[12] Ibid., pp. 183–84.

> By following the three-part *Astropolitik* strategy—immediately renouncing the OST [Outer Space Treaty] and acting to structure a property-based free-market regime in its place; deploying a space-based BMD [ballistic missile defense] system which would eliminate missile-borne threats and guarantee domination of space; and [by] establishing a proper, cabinet or ministry level space coordination agency to encourage space efforts . . . a dominant liberal democracy like the United States can usher in a new era of peace and prosperity.[13]

Dolman's work soon became very influential in air force circles and was taught widely in its professional training courses.

Critics of these ideas questioned the rationale, the threat (given the strong international *opposition* to space weapons), and the implications of moving forward unilaterally with a strategy of space dominance. As Michael Krepon argued in *Foreign Affairs*, "The repercussions will include new international competition to put weapons in space, further strains in alliance relations, closer strategic cooperation between Russia and China, deeper partisan division at home, weakened nonproliferation treaties, and, ironically, greater difficulties in developing . . . missile defense."[14] Krepon concluded that "because of the threat of asymmetrical warfare, [space] dominance would be very hard to achieve and would have many adverse effects."[15] Other critics of U.S. investment in large-scale space weapons for national missile defense (NMD) noted that deploying extremely costly space weapons could be counter-productive, as states might simply avoid using missiles as a delivery system for weapons of mass destruction. Lieutenant Colonel (USAF) Charles E. Costanzo, writing in the pages of the journal *Aerospace Power*, cited the failed Sentinel program of the 1960s and 1970s as a negative example for space-based missile defense. Instead, he warned, "Alternative delivery methods such as aircraft, cruise missiles, ships, trucks, special-operations troops, or terrorists would be less expensive than developing and deploying ICBMs—and more reliable than long-range missiles. Moreover, nonmissile delivery would circumvent the NMD."

While the U.S. debate raged, the Russian government sponsored a major international conference in Moscow in April 2001 aimed at rallying support for a verifiable new treaty to prohibit the placement of weapons in space.[16] Despite the publicity the meeting generated, in part due to the personal participation

---

[13] Everett C. Dolman's *Astropolitik: Classical Geopolitics in the Space Age* (London: Frank Cass, 2002), p. 165.

[14] Michael Krepon, "Lost in Space: The Misguided Drive Toward Antisatellite Weapons," *Foreign Affairs*, Vol. 80, No. 3 (May/June 2001), p. 3.

[15] Ibid., p. 7.

[16] On the meeting, see Detlev Wolter, *Common Security in Outer Space and International Law* (Geneva: U.N. Institute for Disarmament Research, 2006), p. 75.

of President Vladimir Putin, the results remained scanty, as U.S. opposition to opening space treaty talks blocked work at the Geneva-based U.N. Conference on Disarmament (CD). However, in contrast to U.S. support for a wide array of space defenses, the incoming Bush administration announced no similar plans for an expansion of NASA or civilian space activities. Instead, it seemed to take a "steady as she goes" attitude, consistent with its initial priority on entering government of reducing the size of the public sector. President Bush waited until longtime NASA Chief Administrator Dan Goldin resigned in October 2001 to name a new NASA director. His eventual choice was telling: Sean O'Keefe, a man with no space experience and a reputation as a budget-cutter from his time as the comptroller of the defense department from 1989–1992 and, during 2001, as the deputy director of the Office of Management and Budget.[17] O'Keefe would continue to be known as a cost-cutter rather than emerging as someone with a broader vision for U.S. civilian space activities.

Meanwhile, in Russia, the space program continued to deteriorate due to an even more severe lack of funding. In the absence of government support for new research, the Russian Space Agency (RSA) had to de-orbit the *Mir* space station after fifteen years. Although the spacecraft had exceeded its planned term of operation, the ignominy of having to shut it down for lack of funding struck a chord of humiliation within the Russian space community. World news agencies tracked its meteor-like re-entry in some fifteen hundred trackable pieces over the western Pacific in late March.[18] To the relief of all, no one was hurt. Over vigorous U.S. objections, the RSA instead moved ahead with an effort to remove the state's former role in the manned space program by using paying customers. The first of Russia's self-financed space tourists, American businessman Dennis Tito, took off from Baikonur on April 28 for the *International Space Station* (*ISS*). He would soon be followed by a small but steady list of wealthy customers paying $20 million for the privilege. NASA would eventually get over these unexpected commercial visitors, which it deemed the cost of keeping in business the RSA, whose help it need to complete the *ISS*.

As under President Reagan and consistent with the second President Bush's campaign pledges, funding for missile defense rose sharply in the administration's first budget. Until fiscal year (FY) 2001, NASA's budget had exceeded the

[17] Notably, O'Keefe's biography on the NASA Web site listed no space experience prior to his appointment, although it did list broader experience in defense affairs and service as President George W. Bush's Secretary of the Navy. See <http://www.nasa.gov/about/highlights/AN_Feature_Administrator.html> (accessed July 31, 2006).

[18] Richard Stenger, "Mir Destroyed in Fiery Descent," CNN, March 23, 2001, available on the CNN Web site at <http://archives.cnn.com/2001/TECH/space/03/23/mir.descent/> (accessed August 2, 2006).

Pentagon's published space budget throughout the late 1990s. Under Bush, the military now pulled even in these figures at slightly over $14 billion. In FY 2002, the Pentagon surged ahead by nearly $1 billion, highlighting the change in the national direction in space. As during the SDI days, Pentagon planners quickly resumed consideration of an array of space-based weapons previously seen as too exotic, too expensive, or simply unnecessary. Now, they were described and pursued as critical systems for the tasks of missile defense and satellite protection. To provide it with greater autonomy, the administration upgraded the Ballistic Missile Defense Organization into the Missile Defense Agency,[19] while also dropping the distinction between theater and national missile defense systems. This new emphasis sought to create a seamless web of defenses to combat all threats, whatever their range.

Unlike under the Reagan administration, however, missile defenses were no longer seen as having a revolutionary strategic role. That is, as no peer space competitor to the United States existed any longer, the administration did not link its missile and space defense plans to broader disarmament goals like the replacement of an offensive with a defensive emphasis or the total elimination of nuclear weapons. Thus, many countries—especially those that had signed the Nonproliferation Treaty as nonnuclear weapon states—criticized the administration for what seemed to them to be defense "overkill" and for fomenting an "arms race" in outer space. The Bush administration rejected both claims and reiterated in its official statements that no decision had been made to deploy space weapons. Instead, the United States intended to investigate all options—without Cold War limitations—and then determine what systems would be most useful.

Longtime supporters of missile defenses now came out with arguments calling for rapid deployment of SDI-type, space-based systems. Former U.S. Ambassador to the 1980s' U.S.-Soviet Defense and Space Talks and ex-SDIO Director Henry F. Cooper made an impassioned case in the op-ed pages of the *Wall Street Journal* for the Bush administration to commit itself to early launch of a constellation of space-based Brilliant Pebbles interceptors, arguing, "It would be a travesty for President Bush to fail to revive this important program."[20] The administration, though, remained cautious and refused to adopt a policy of rapid deployment, despite voicing its firm support for studying such concepts.

---

[19] On this organizational change, see Missile Defense Agency, "Missile Defense Timeline: 1944–2004," on the Web site of the MDA at <http://www.mda.mil/mdalink/html/milstone.html> (accessed October 16, 2006).

[20] Henry F. Cooper, "Why Not Space-Based Missile Defense?" (Op-Ed), *Wall Street Journal*, May 7, 2001, p. A22.

In July, the missile defense program registered its first success in some time. A kill vehicle launched from Kwajalein Atoll in the Marshall Islands successfully intercepted a missile launched from Vandenberg Air Force Base in California,[21] although sensors failed to register the hit because of a system overload. A single decoy with a larger-than-normal infrared signature was avoided. Despite the limited parameters of the test, its overall success put the president's plans for early deployment back on track. During meetings with Putin advisor and former defense minister Igor Sergeyev, U.S. National Security Advisor Condoleezza Rice spoke of the urgency of deploying missile defenses, saying, "The new threats that we face . . . won't wait."[22] However, the Russian government reiterated the need for negotiations on the ABM Treaty, expressing its willingness to "work night and day" to reach an agreement but also its frustration with the administration's lack of a clear plan for missile defense and a description of its technology.[23]

The 9/11 Attacks and U.S. Withdrawal from the ABM Treaty

While the horrific events of September 11, 2001, did not affect space activity directly—and indeed over time distracted attention from the new administration's priority for missile and space defenses—it revived the waning credibility of a key administration player, Secretary of Defense Rumsfeld. In the summer of 2001, critical commentary appearing even in the pro-Republican *Wall Street Journal* had criticized Secretary Rumsfeld for his aloofness and failure to push needed goals of "transformation" at the Pentagon, including downsizing Cold War weapons and modernizing the rest of the military.[24] For many Republicans, these defense goals, which ranked high on their agenda, would provide both the funds and a compelling rationale for missile defenses. They were also widely stated incoming goals of the secretary of defense. Within two months of this critical commentary, however, everything had changed.

As the nation reeled from the shock of the September terrorist attacks on

---

[21] Greg Jaffe and Carla Anne Robbins, "Bush's Hopes for Missile System Get Boost With Successful Test," *Wall Street Journal*, July 16, 2001, p. A24.

[22] Carla Anna Robbins and Andrew Higgins, "Moscow's Message to U.S. on Missiles: Details, Please—Putin's Top Military Adviser Stands by ABM Treaty Until New Pact Is Forged," *Wall Street Journal*, July 27, 2001, p. A7.

[23] Ibid.

[24] See, for example, Ken Adelman, "Stop Reviewing; Start Reforming" (Op-Ed), *Wall Street Journal*, July 13, 2001, p. A10; also, Capitol Journal (column), "Military Reform: Is the Opening Slipping Away?" July 11, 2001, *Wall Street Journal*, p. A18.

New York and Washington, Secretary Rumsfeld restored his stature by leading the charge for retaliation against Al Qaeda's safe havens in Afghanistan and for removing the Taliban government. The rapid success of American and coalition forces in the conflict caused Rumsfeld's—and the president's—popularity to soar, giving them tremendous clout in budget debates on Capitol Hill, where they pushed through truly massive increases in defense spending. Military space programs, including those for missile defense, proved to be among the key beneficiaries, as Democrats quietly shelved a pre-9/11 plan to protest expected boosts in missile defense funding. Although these programs had nothing to do with protection against future terrorist attacks, Democrats feared that opposing any defense appropriations would make them vulnerable to charges of being divisive, disloyal, and unpatriotic at this time of unprecedented national turmoil.

The one exception to the otherwise quiet debate over the Pentagon's missile defense request came from longtime Senator Robert Byrd (Dem., West Virginia), who gave a long floor speech on September 26 against the Bush administration's budget request. Some of his remarks responded specifically to a recent front-page story in the *New York Times Magazine* called "The Coming War in Space." The article had outlined in detail the status of a bewildering variety of U.S. military research programs on space weapons, including the space-based laser, hyper-velocity rods, and unmanned space bombers. It had described the administration's plans to boost funding for a number of these systems in the 2002 budget.[25] Senator Byrd's extended floor comments reflected on what he saw as a rush to deploy these weapons. He noted that he was "troubled . . . because we are spending huge sums on them without being sure in our own minds that the weaponization of space is the best course of action to ensure our security."[26] He cited President Eisenhower's Farewell Address in January 17, 1961, which had cautioned Americans about the military-industrial complex and its "unwarranted influence."[27] He concluded by reiterating that the nation now faced new threats of greater priority and by stating his belief that "it would be both wise and prudent to back off just a little bit on the accelerator that is driving us in a headlong and fiscally spendthrift rush to deploy national missile defense and to invest billions into putting weapons in space. . . . That heavy foot on the accelerator is merely the stamp and roar of rhetoric. The threat does

---

[25] See Jack Hitt, "The Coming Space War," *New York Times Magazine*, August 5, 2001.

[26] "Space Wars," Statement by Senator Robert Byrd, Congressional Record, September 26, 2001, available online at <http://thomas.loc.gov/cgi-bin/query/z?r107:S26SE1–0011:> (accessed July 31, 2006).

[27] Ibid.

not justify the pace."[28] Despite Byrd's plea, Congress approved $7.8 billion for missile defense. The new appropriation represented a 62.5 percent increase over the 2001 figure of $4.8 billion. A successful December 3 test of the ground-based interceptor—aided by a radio beacon from the target missile and other data received prior to the test—buoyed Pentagon optimism that the NMD program was beginning to achieve its goals.

The administration quickly capitalized on its newfound momentum to move ahead with its planned but still highly controversial withdrawal from the ABM Treaty. After a series of unsuccessful negotiations with the Russian government during the fall, including a summit meeting with President Putin at the Bush ranch in Texas in November, the administration issued its formal six-months' notice of its planned withdrawal from the ABM Treaty on December 13. Despite the unlikelihood of terrorist use of ballistic missiles, President Bush keyed on the then-strong 9/11-inspired, antiterrorist feelings of the country in stating his rationale for the withdrawal: "I have concluded the ABM Treaty hinders our government's ability to develop ways to protect our people from future terrorist or rogue-state missile attacks."[29] Russian President Putin's response to the U.S. decision noted his government's recent efforts to save the ABM Treaty and concluded, "We consider this decision a mistake."[30]

The U.S. withdrawal from the 1970 treaty also met with strong allied opposition. The usually staid *Financial Times* of London referred to the U.S. withdrawal as "bad for the future of arms control."[31] Its editorial board highlighted NATO's opposition to the move and predicted Russian efforts to get out of the START II agreement, concluding, "Moscow is [now] free to act unilaterally, just as Mr. Bush has done." The paper criticized this go-it-alone U.S. policy in the face of the need for the NATO alliance to act in unison in order to win the new war on terrorism. Other commentators feared that the move might push Russia back toward communism. In regard to space, the decision marked a clear break from the Cold War collective restraint framework. Washington countered that the dissolution of the Soviet Union made bilateral space and missile defense agreements no longer relevant and the logic of ABM Treaty restraint strategically anachronistic.

A key, unstated, motivation for supporters of missile defenses, however, was

[28] Ibid.

[29] Gopal Ratnam, "Missile Defense Hurdles Removed," *Defense News*, December 17–23, 2001, p. 40.

[30] Ibid.

[31] "Risks of Ripping Up a Treaty" (Editorial), *Financial Times* (London), December 13, 2001, p. 12.

concern over the perceived rising threat to the United States from China. One example of such views came in the form of an impassioned public letter in early December from Congressman Bob Schaffer (Rep., Colorado) to Secretary of Defense Rumsfeld. Citing a CIA report, Representative Schaffer stated that China was developing laser and electromagnetic pulse weapons for use against U.S. space assets.[32] Combined with what he described as highly accurate Chinese ballistic missiles, Schaffer used the Rumsfeld Commission's words to warn of the possibility of a "surprise attack like the Japanese at Pearl Harbor" and complained that U.S. "ballistic missile defenses are non-existent except for the short-range Patriot."[33] In conclusion, Representative Schaffer charged China with having a "strategic alliance with Saddam Hussein"—although he offered no evidence for this claim—and stated that Beijing was "preparing for direct military confrontation with the United States on its own terms."[34] The obvious answer for Schaffer and other like-minded Republicans was near-term deployment of a robust missile defense. In March, the Pentagon announced yet another successful test of the ground-based interceptor, which identified a target missile from six large decoys, although under less-than-battlefield conditions (given the test's initial use of a target beacon to cue the interceptor).

With this background, congressional budget debates began in spring 2002 on missile and space defenses in the FY 2003 budget. Although Congress was ready to provide massive increases in defense spending to counter threats raised by the 9/11 attacks, one specific aspect of the administration's missile defense proposal caught the eye of both Democrat and Republican legislators: a planned study by the Defense Science Board on possible use of low-yield nuclear weapons in space for missile defense purposes.[35] Pentagon planners wanted to examine options for "near hits" using nuclear weapons in case hit-to-kill technologies proved too difficult to master. To the surprise of the White House, Senator Ted Stevens (Rep., Alaska) led an effort, along with Senator Diane Feinstein (Dem., California) to block funding for the plan. Senator Stevens, a longtime supporter of missile defenses, declared that news of the nuclear study made him "mad."[36]

---

[32] "Letter to Secretary of Defense—Hon. Bob Schaffer," *Congressional Record*, December 14, 2001, online at <http://thomas.loc.gov/cgi-bin/query/D?r107:12:./temp/~r107ktUkbG::> (accessed July 31, 2006).

[33] Ibid.

[34] Ibid.

[35] The issue came to light when a newspaper report quoted William Schneider, Jr., chairman of the Defense Science Board, as saying that Secretary of Defense Rumsfeld had raised these issues and expressed an interest in such a study. See Bradley Graham, "Nuclear-Tipped Interceptors Studied; Rumsfeld Revives Rejected Missile Defense Concept," *Washington Post*, April 11, 2002, p. A2.

[36] Sen. Stevens, quoted in Wade Boese, "U.S. Reportedly to Study Nuclear Warheads for Missile Defense," *Arms Control Today*, Vol. 32, No. 4 (May 2002), p. 27.

He sharply criticized the administration for threatening to disrupt the "fragile balance" in Congress over missile defense funding by introducing the dubious and highly controversial concept of nuclear-tipped missile defenses, which most members of Congress and the informed public had long seen as discredited. While a far cry from the 5-megaton Spartan missile deployed in the now-dismantled Grand Forks, North Dakota, ABM system of the mid-1970s, the concept revived concerns about electromagnetic pulse radiation and possible fallout on humans. The House of Representatives, however, passed the measure, and language for the study was included by the joint House-Senate conferees who met to reconcile differences over the defense appropriations bill. When the bills went back to their respective chambers, the House version included the study, but the Senate deleted the funding. The study ended there. Notably, both sides' versions had forbidden development of any weapons, showing shared recognition of the potentially destabilizing effects of such systems.

During the spring, Washington began to act on widespread concerns that its planned ABM Treaty withdrawal would damage arms control progress with Russia. The Bush administration early on had made a policy decision to oppose any formal arms control treaties with Moscow, stating that such agreements were time-consuming anachronisms since the new Russian Federation was now a friend of the United States. In this context, however, Washington eventually succumbed to President Putin's demand for a treaty to codify planned reductions to lower levels than those elaborated in START II. In a summit meeting in late May in Moscow, the two sides signed the Strategic Offensive Reductions Treaty (SORT)—also known as the Moscow Treaty—which codified planned, bilateral reductions of deployed strategic weapons to numbers not to exceed 1,700 to 2,200 by 2012. However, in keeping with its own predilections (and in a reversal of recent U.S. practices), the Bush administration refused to include any on-site verification measures or requirements that the two sides eliminate warheads taken out of service. The treaty, like its Cold War predecessors, would be verified from space. Finally, the agreement would be fully reversible once the day of implementation had been reached. Critics called it a "non-treaty." They also noted that, given the risk of a terrorist seizure of nuclear materials in a still-unstable and crime-plagued Russia, the United States should seek to eliminate its nuclear stockpile as fast as possible.[37] Supporters called the treaty a realistic means of providing the United States with a nuclear "hedge" capability against possible future uncertainties in Russia or threats from some other potential adversary. Thus, for President Putin, the Moscow Treaty came as only a small

[37] Senator Joseph Biden, "Beyond the Moscow Treaty" (Op-Ed), *Washington Post*, May 28, 2002, p. A17.

victory following his failure to keep controls on future U.S. development of space-based missile defenses.

U.S. withdrawal from the ABM Treaty became effective on June 13, 2002. Russian officials promptly announced that they would no longer abide by the START II agreement, which would have forced them to dismantle all of their land-based, multiple-warhead missiles.[38] As this had been the very type of missile that had led to the Reagan military buildup in the early 1980s, getting out of the START II treaty provided a moral victory for Russian conservatives, even though its primary motivation was financial, not strategic. Put simply, it was cheaper for Moscow to keep old missiles than to build new single-warhead Topol Ms. Russia also began to raise additional barriers to cooperation on the already ailing Joint Data Exchange Center (JDEC) and the Russian American Satellite (RAMOS) project. Unlike under Yeltsin, President Putin began to see such projects as offering little benefit to Moscow for the cost of opening up national defense procedures to what he saw as prying American eyes.[39]

The Bush administration responded to Russia's exit from START II by arguing that the action was expected and that the United States did not need to fear Russian forces any longer.[40] With its ABM withdrawal completed, the Bush adminstration moved forward with a renewed effort to pursue all azimuths of missile defense. These programs proved to be wide and varied: the Space-Based Infrared System (SBIRS) High and Low for early warning; an X-band radar for missile tracking; the Patriot Advanced Capability (PAC) III, the Theater High Altitude Area Defense (THAAD), the sea-based Aegis system, and the Airborne Laser for short- and medium-range missiles; and the midcourse Ground-Based Interceptor, Space-Based Laser, and space-based kinetic-kill interceptors for long-range missiles. Research funds flowed into companies and facilities around the country and longtime supporters believed their hopes of an early deployment would soon be realized.

But the missile defense program's lack of focus and the loose *technical* criteria for inclusion of systems for near-term deployment seemed to bear out Byrd's predictions. Soon, engineers and systems analysts following the program—on

---

[38] Michael Wines, "After U.S. Scraps ABM Treaty, Russia Rejects Curbs of START II," *New York Times*, June 15, 2002, p. A2.

[39] On Russian views of U.S. nuclear and military space efforts under the Bush administration, see Victor Mizin, "Russian Perspectives of Space Security," in John M. Logsdon, James Clay Moltz, and Emma S. Hinds, *Collective Security in Space: European Perspectives* (Washington, D.C.: Space Policy Institute, George Washington University, January 2007).

[40] Moreover, it argued, the prior signing of the Moscow Treaty already bound the Russian side to further reductions in the total number of its deployed strategic nuclear forces, which, Washington argued, was what counted most.

both sides of the debate—began to identify problems with the administration's "scatter shot" approach. Philip Coyle, a recent defense department assistant secretary and the former director of operational tests and evaluation, stated in a review he conducted of the status of missile defense programs in 2002, "During the first year of the Bush administration, all U.S. missile defense programs—both theater and national—have slipped."[41] He emphasized the long-term nature of the necessary research and development program and concluded about the optimism exuded by the administration, "Many decision-makers in Washington—and . . . the president himself—seem to be misinformed about the prospects for near-term success."[42] In the fall of 2002, the president's own Defense Science Board issued a report with much the same conclusion. It called on the White House and the Pentagon to focus their missile defense attentions on at most two of some eight current systems, if it wanted to succeed in deployment of even the limited national missile defense that President Bush had promised for his first term.[43] But administration officials countered these assessments by arguing that cost concerns for the broad research being undertaken could be overcome with the assistance of U.S. allies, who it presumed would pick up the costs for certain layers of the proposed system.[44] As time wore on, however, U.S. allies proved interested mostly in receiving lucrative U.S. missile defense contracts from the Pentagon rather than in making their own contributions. The one exception, because of the increasing reality of North Korean missile capabilities, was Japan, but its contribution remained tiny compared to overall system costs and it remained limited by a national ban on the export of weapons technology to any country—even the United States.

Other countries viewed U.S. missile defense developments and emerging support for space weapons with greater concern. Although the Bush administration largely continued an existing Clinton policy at the Geneva-based CD and at the United Nations in New York opposing a new space arms control treaty, the tone changed to a harder line and prospects for compromise evaporated. From an outside perspective, Washington was now the primary player standing in the way of new agreements to limit space weapons. Nevertheless, as meetings about how to deal with the deadlock opened in Geneva in late May 2001, the Russian and Chinese delegations both submitted new proposals for moving

[41] Philip Coyle, "Rhetoric or Reality? Missile Defense Under Bush," *Arms Control Today*, Vol. 32, No. 4 (May 2002), p. 9.

[42] Ibid.

[43] Bradley Graham, "Missile Defense Choices Sought; Panel Urges Focus on 2 Approaches," *Washington Post*, September 3, 2002, p. A1.

[44] Sharon Weinberger, *Aerospace Daily*, September 6, 2002, p. 7.

toward negotiations on a new treaty.[45] The Bush administration remained unresponsive. The following year, China and Russia submitted a joint paper to the CD on banning space weapons, with the compromise that tests of missile defenses that only passed through space would be allowed.[46] Their plan, however, would ban the stationing of weapons of any kind in space and the use of force (or the *threat* of its use) against space objects. But these concessions brought no change from the American side, which continued to insist that China first agree to open negotiations on a Fissile Material Cut-off Treaty before any discussions on space could begin. Regarding space, U.S. Ambassador to the CD Eric M. Javits rejected Russo-Chinese claims about space insecurity and argued instead for continuing the status quo: "We believe that the existing multilateral arms control regime adequately protects states' interests in outer space and does not require augmentation. There simply is no problem in outer space for arms control to solve."[47] In the view of other states, U.S. plans for space defenses constituted a highly destabilizing development. Ambassador Javits countered that the United States had decided only to move forward with a "limited" missile defense system. He observed that "missile defense has not upset strategic stability or led to a new arms race"[48] and reaffirmed the U.S. commitment to the 1967 Outer Space Treaty. But foreign observers remained skeptical of U.S. claims of restraint and wary of loopholes for the placement of dual-use defensive systems, which might be converted to offensive purposes (such as anti-satellite use). Chinese Foreign Ministry spokesperson Cheng Jingye cautioned in response that "the weaponization of space seems closer than ever" and that the continuation of U.S. policies could cause human society to "lose access to space."[49]

One inadvertent step for greater space restraint emerged in September 2002 when the one-man crusader for the Kinetic Energy ASAT (KEASAT)—Senator Bob Smith (Rep., New Hampshire)—was defeated in the Republican primary by moderate John Sununu, the son of the former Reagan administration official. While Smith's defeat had little to do with his position on the KEASAT, the

[45] On these negotiations, see Wolter, *Common Security in Outer Space*, p. 73.

[46] Ibid.

[47] Amb. Eric M. Javits, "A U.S. Perspective on Space," in James Clay Moltz, ed., *Future Security in Space: Commercial, Military and Arms Control Trade-Offs*, Center for Nonproliferation Studies, Monterey Institute of International Studies, Occasional Paper No. 10, July 2002, p. 52, available online at <http://www.cns.miis.edu/pubs/opapers/op10/op10.pdf> (accessed August 8, 2006). This paper was based on remarks given the previous month at the CD and approved by the Department of State.

[48] Ibid.

[49] Cheng Jingye, "Treaties as an Approach to Reducing Space Vulnerabilities," in Moltz, *Future Security in Space*, p. 48.

continued efforts he had made to add money not requested by the Pentagon for the system to successive defense budgets ended. The Bush administration, recognizing the problem of the KEASAT as a debris-producer and the lack of Pentagon interest in the system, made no effort to revive it.

China's Rise in Space and Its Implications

Up to this time, China's role as a major player in the international space arms control debate had not been matched by any significant operational space capability. To U.S. critics, Beijing's diplomatic efforts constituted mostly an effort to prevent Washington from gaining too much of a lead in an arena where China could not yet compete. However, this began to change in the early part of the Bush administration as a series of successful Chinese space launches and growing evidence of technological cooperation with Russia began to attract U.S. attention.

China's emergence as a genuine space power now came at a sensitive time for the United States. On February 1, 2003, the U.S. space shuttle *Columbia* experienced a massive structural breakup as it re-entered the atmosphere, killing all seven astronauts aboard, including the first Israeli in space. As after the *Challenger* incident, NASA immediately shut down the shuttle program and ordered a full-scale investigation. Quickly, suspicion focused on the problem of foam insulation falling from the shuttle's boosters at the time of takeoff, which had damaged the shuttle's delicate system of protective tiles. This disaster would cause soul-searching within NASA and a long delay in the planned U.S. flight schedule for serving the *ISS*. In the meantime, NASA rushed to sign an agreement with the RSA to fill in the gaps opened by the U.S. flight moratorium. Ironically, Russian hardware now became the sole means of maintaining contact with the multibillion dollar (and largely U.S.-funded) *ISS*, putting into perspective how international manned space activity had become.

The perceived momentum in human spaceflight now passed to China. In March, the Chinese successfully orbited *Shenzhou 3* and, in December, *Shenzhou 4*—both with dummies aboard, indicating a strong likelihood that a manned spaceflight would be the next Chinese mission. Finally, in October 2003, China fulfilled years of planning and flight testing by launching its first taikonaut—Yang Liwei—in a fourteen-orbit mission on *Shenzhou 5*. Despite the lack of surprise, China's success in becoming only the third country to cross the manned-space-launch threshold created a stir in the U.S. Congress, as representatives publicly bemoaned the grounded status of the shuttle fleet and the country's lack of vision for future human spaceflight. As during the *Sputnik* period, it seemed to many that the United States had lost ground in comparison to a major international rival and was threatened with a serious loss of prestige if it failed to compete more effectively in space. Naysayers pointed out that the Chinese had merely borrowed Russian *Soyuz* technology to carry out a mission that the United States had accomplished in 1962, but the perception of Chinese technological momentum could not be overcome and soon appeared in heated congressional debates. Some members of Congress called for a vigorous response by NASA; others called for a focus on the Pentagon's planned space weapons.

In its official policy, the Bush administration studiously avoided any discussion of space cooperation with China. Several factors influenced this policy, which had deep roots. First, memories of the Tiananmen Square massacre in 1989 remained a potent symbol of China's attitude toward human rights, forging a coalition of liberal Democrats and conservative Republicans in opposing funding for any such opening. Second, the congressionally mandated Cox Commission in 1999 had concluded that the Chinese military had gained technology relevant to nuclear delivery systems from the Loral Corporation in the early 1990s.[50] Loral had sought to correct a problem that had cost it the loss of an expensive satellite when a Chinese booster rocket failed to deploy it properly. But it had crossed the line of dual-use export controls, incurring steep fines, as had Hughes Space and Communications International in separate cases. Third, the Wen Ho Lee nuclear spying case—in which a Taiwanese-born scientist working at Los Alamos National Laboratory had been arrested in 1999 for allegedly supplying China with classified nuclear information[51]—still resonated with members of Congress, particularly after 9/11 when fears of foreign spying in the United States reached new heights. Finally, China's downing of a U.S. EP-3 intelligence-gathering aircraft in April 2001 and its temporary detention of the crew had set a sour tone for U.S.-Chinese relations soon after the administration had entered office. These obstacles would take time to overcome.

Although the administration denied any connection between the *Shenzhou* 5 launch and President Bush's first major speech on space activity, delivered at NASA on January 14, 2004, the timing of the address seemed at least serendipitous. The administration had "lost face" in space and needed to do something about it. With the *Columbia*-induced launch moratorium still in force,

[50] See Select Committee of the House of Representatives, *Final Report of the Select Committee on U.S. National Security and Military/Commercial Concerns with the People's Republic of China*, May 25, 1999 (unclassified version), "Overview," available via the Web site of the House of Representatives at <http://www.house.gov/coxreport/pdf/overv.pdf> (accessed October 8, 2007).

[51] Lee was released after nine months in prison. Despite the ferocity of his initial prosecution, he was exonerated on all charges of spying, although he pled guilty to one charge of mishandling classified information.

the administration sought to redress its damaged space reputation and flagging NASA morale by issuing a bold plan for returning to the Moon and carrying out a manned mission to Mars. The vision called for reallocation of funding in the existing five-year NASA budget, the abandonment of the *ISS* (and the shuttle) at an earlier date than planned, and a concentration of NASA resources on developing a new spacecraft. NASA would first seek to establish a Moon base, and then push off for a Mars mission from the Moon's favorable low-gravity environment. Coming on the heels of the successful landing of NASA's two Mars Rovers on the surface of the red planet, the speech rallied NASA employees and supporters and painted a bright and exciting future, trying to evoke the heady days of the Moon race. But the president offered a pledge of only $1 billion in new funding, compared to the $400 to $500 billion estimated cost of the Moon/Mars program. As with missile defenses, funding would clearly have to come from other sources than just the NASA budget. In this regard, the president's speech indicated a willingness to cooperate with other states in this endeavor, explaining, "We'll invite other nations to share the challenges and opportunities of this new era of discovery. The vision I outline today is a journey, not a race, and I call on other nations to join us on this journey, in a spirit of cooperation and friendship."[52] After the speech, however, NASA Chief Administrator Sean O'Keefe dampened hopes in regard to China, noting about Beijing's possible participation, "I wouldn't want to speculate."[53] Skeptics suggested that the speech was a direct response to China's space mission—a charge the administration denied—and questioned the sincerity of President Bush's commitment in light of the low level of funding promised. The *Christian Science Monitor* commented that the financing for the program "has a car-salesman ring to it," since the vast majority of the U.S. costs would come under future presidents.[54] The administration also provided no evidence of prior consultations with other governments regarding their possible financial pledges or technical contributions. Conservative columnist Charles Krauthammer, however, praised the president's vision to move beyond the aged shuttle system and called the president's plan for a Moon base "the most glorious human adventure since the Age of Exploration five centuries ago."[55] But tepid public reception to the plan in the weeks following the speech and its high expected costs

[52] Text of President Bush's speech at NASA, January 14, 2004, White House Web site, at <http://www.whitehouse.gov/news/releases/2004/01/20040114-3.html> (accessed August 1, 2006).

[53] O'Keefe, quoted in David E. Sanger and Richard W. Stevenson, "Bush Backs Goal of Flight to Moon to Establish Base," *New York Times*, January 15, 2004, p. A19.

[54] "A Countdown to Mars; Bush's Space Mission Needs a 10-Point Reality Check" (Editorial), *Christian Science Monitor*, January 16, 2004, p. 10.

[55] Charles Krauthammer, "A Modest Proposal" (Op-Ed), *Washington Post*, January 16, 2004, p. A19.

caused the president to drop mention of the proposal entirely from his State of the Union address in late January. It seemed that this would not be a second Apollo after all.

Technical and Political Problems for Space Defenses

By 2004, the administration's early enthusiasm for space defense had translated into relatively little concrete progress toward its bold vision for a new space security framework based on U.S. military strength. While spending had continued to increase, with a record $9 billion budget for missile defense passed by Congress in the fall of 2004, too many programs and too few successes brought back memories of the SDI program. To both critics and supporters, the lack of flight tests, combined with an administration decision to deploy the ground-based, midcourse interceptor—a system known to have serious technical problems—left the administration open to criticism. Other parts of the program faced an even more withering attack.

The American Physical Society, for example, came out with a detailed technical report in July 2003 on the existing capabilities of proposed boost-phase interceptors and the parameters of making such a system effective. It concluded that foreseeable U.S. defenses had poor chances of intercepting fast-burning, solid-propellant ICBMs, that there were serious range and reload problems with the proposed Airborne Laser, and that space-based elements of the system would be extremely costly due to the large number of interceptors needed in orbit to ensure timely coverage of targets.[56] The authors argued that only a few of the components needed to justify the planned deployment of boost-phase defenses within the next five years even existed and that "we see no means for deploying an effective boost-phase defense against ICBMs within 10 years."[57]

By the fall of 2004, the Pentagon had begun to deploy a half-dozen interceptors at Ft. Greely, Alaska, along with a handful of additional missiles at Vandenberg Air Force Base in California. Although the critical sensors required to guide the missiles were still missing, the administration used these deployments in President Bush's successful reelection campaign against Democratic Senator John Kerry to claim fulfillment of his pledge to defend the American people against ballistic missiles.

In January 2005, with the coming end of President Bush's first term, the

---

[56] American Physical Society, "Boost-Phase Intercept Systems for National Missile Defense: Scientific and Technical Issues," July 2003, available on the APS Web site at <http://www.aps.org/media/pressreleases/loader.cfm?url=/commonspot/security/getfile.cfm&PageID=57862> (accessed July 31, 2006).

[57] Ibid.

White House attempted to get the Pentagon to certify the operational capability of its NMD systems in Alaska and California. Given the failure of recent tests and the unavailability of the X-band radar, the SBIRS-high system, or the SBIRS-low system—required for effective, long-range tracking of missiles, warheads, and decoys—the defense department declined to confirm its operational status. Pentagon spokesperson Larry Di Rita walked a difficult political tightrope in accurately describing the NMD system as having a "nascent operational capability."[58]

As part of its more active approach toward space defenses, the Bush administration had by now largely abandoned efforts to settle technical disputes with Moscow over establishing cooperation in missile early warning. After spending some $120 million on RAMOS, the White House deleted the program from the FY 2005 budget.[59] For the Russians, the approximately $20–$30 million per year they had been scheduled to receive was not enough to overcome Moscow's political and security concerns about the project. Moreover, the satellite agency Kometa had proven to be a weak and inappropriate partner,[60] particularly given its required task of *pushing* the Russian government into a cooperative arrangement with a part of the U.S. government the Putin administration now had obvious contempt for: the Missile Defense Agency (MDA). As for the JDEC, an empty building constructed with U.S. funds in Moscow remained the only symbol of an idea whose fruition required a level of political trust and cooperation no longer present in the U.S.-Russian relationship.

New Collective Security Approaches for Space

Meanwhile, the international community continued its efforts to draw the United States into negotiations on space security. By late summer 2003, China had changed its position at the CD from one of requiring that the Ad Hoc Committee on the Prevention of an Arms Race in Outer Space (PAROS) meet for talks aimed at signing a new space *treaty* to one requiring only discussions, given current U.S. opposition to a formal agreement. In hopes of stemming a seemingly inexorable U.S. drive toward deployment of space weapons, Russian President Putin made a pledge at the United Nations in September 2003 that

[58] "Di Rita: Developmental Missile Shield Provides 'Nascent' Defense," *Defense Daily*, January 18, 2005, p. 1.

[59] Jeremy Singer and Warren Ferster, "MDA Request Cancels RAMOS, Slows Work on Space Interceptors," *Space News*, February 24, 2004.

[60] Pavel Podvig, "U.S.-Russian Cooperation in Missile Defense: Is It Really Possible?" PONARS Policy Memo 316, November 2003, available on the Web site of Stanford University's Institute of International Studies at <http://iis-db.stanford.edu/pubs/20733/podvig-pm_0316.pdf> (accessed October 27, 2006).

Moscow would follow a unilateral "no-first-deployment" policy on offensive space weapons, in effect, challenging the United States not to break this taboo.[61]

Given the ineffectiveness of their efforts thus far, critics of U.S. space policy within the U.S. and international arms control community began to shift strategies. Instead of calling for a treaty against future arms, some analysts began to examine approaches that might be consistent with the Bush administration's skeptical view of formal pacts and yet still push forward collective security in space. One approach that received considerable attention in various meetings and discussions held in Washington, including on Capitol Hill, was the notion of establishing international "rules of the road" for space, possibly as a way station until states could agree on how to move ahead with more formal controls. Michael Krepon of the Henry L. Stimson Center took the lead in initiating a series of related proposals based loosely on the U.S.-Soviet Incidents at Sea model, calling on states to: register their spacecraft promptly, undertake actions to avoid dangerous maneuvers that might lead to collisions in space, and refrain from close approaches or simulated attacks on satellites (including remotely with lasers).[62] Theresa Hitchens of the Center for Defense Information argued that such mechanisms were not only desirable but "necessary" in order to solve emerging space management problems (traffic control, orbital debris proliferation, and others) by creating a voluntary "international norm" of enhanced cooperation among major international space actors.[63]

In his statement to the CD in August 2004, Chinese Ambassador Hu Xiaodi also lowered the threshold for a new international space agreement by arguing that verification mechanisms (another U.S. hurdle) could be put aside indefinitely. As he noted, "For the time being a future legal instrument in outer space can be formulated without a verification mechanism; as science and technology progress, the addition of a verification mechanism may be discussed again when conditions are ripe."[64] Again, however, Washington's position diverged so

[61] However, the pledge did not state specifically that Russia would refrain from placing all possible weapons in space, thus leaving Putin's seemingly bold pledge open to interpretation. For the text of President Putin's speech on September 25, 2003, to the United Nations, see <http://www.un.org/webcast/ga/58/statements/russeng030925.htm> (accessed August 11, 2006).

[62] Michael Krepon (with Christopher Clary), *Space Assurance or Space Dominance? The Case Against Weaponizing Space* (Washington, D.C.: Henry L. Stimson Center, 2003).

[63] Theresa Hitchens, *Future Security in Space: Charting a Cooperative Course* (Washington, D.C.: Center for Defense Information, September 2004), pp. 73–74.

[64] For the full text of Amb. Hu's remarks on August 26, 2004, see the "Current and Future Space Security" Web site of the Center for Nonproliferation Studies at the Monterey Institute of International Studies at <http://cns.miis.edu/research/space/pdf/cdpv966.pdf> (accessed August 11, 2006).

far from international opinion that these compromises failed to register on its radar screen and resulted in no perceptible change.

In a subsequent June 2005 speech to the CD, Ambassador Hu continued the assault on U.S. plans for space-based missile defenses and possible anti-satellite weapons. He argued that space-based defenses would "undermine international security," damage the current arms control environment, and trigger a "new arms race."[65] Finally, the Chinese representative raised new environmental concerns regarding the testing and deployment of space weapons, arguing that they risked "harming the biosphere of the earth" and would "exacerbate the already serious problem of space debris."[66]

At the CD, the Russians and Chinese had now joined their campaigns and had together become the most active promoters of space arms control. Frequently, the Canadian government sought to act as a mediator between the United States and other countries, sponsoring conferences, workshops, and receptions in hopes of beginning at least informal dialogue. Russian Ambassador to the CD Leonid Skotnikov tried to reason with his U.S. counterparts in June 2005 by explaining at one session, "Hopes to achieve domination in space with the use of force are illusory, and ultimately such ambitions would weaken rather than strengthen the security of all States."[67] Ambassador Skotnikov reiterated Moscow's support for a treaty or other international legal instrument that would "reliably block attempts to place weapons of any type in outer space or to use or threaten to use force against space objects."[68] This phraseology sought in part to appeal to U.S. military demands about the issue of electronic jamming, which had earlier been included in definitions of space weapons that had banned *any* type of interference. The evolving Russian concept now implicitly allowed jamming of satellite signals, as long as it did not involve "force" or permanent damage to the spacecraft. Again, the United States provided no specific response to the Russian proposal and continued its programs for research on all types of space defensive systems.

Meanwhile, however, some Pentagon perspectives had begun to become more nuanced since the early part of the Bush administration when nuclear

---

[65] For the full text of Amb. Hu's remarks on June 9, 2005, see the "Current and Future Space Security" Web site of the Center for Nonproliferation Studies at the Monterey Institute of International Studies at <http://cns.miis.edu/research/space/pdf/cdpv988.pdf> (accessed August 11, 2006).

[66] Ibid.

[67] For the text of Amb. Skotnikov's speech on June 9, 2005, to the Conference on Disarmament, see the "Current and Future Space Security" Web site of the Center for Nonproliferation Studies at the Monterey Institute of International Studies at <http://www.cns.miis.edu/research/space/pdf/cdpv984.pdf> (accessed August 11, 2006).

[68] Ibid.

weapons and kinetic-kill concepts had dominated discussions. Instead, this more evolved stance now emphasized interruption rather than destruction. But the strategy's adoption had less to do with efforts to assuage international critics than it did with growing U.S. military concerns with orbital debris. As Vice Commander of Air Force Space Command Lieutenant General Daniel P. Leaf explained in an interview, "Our priority is on temporary and reversible means, not destruction. But we also know that there could be the potential for such a significant threat that destroying it [a satellite] might merit the resultant debris."[69] Thus, while the air force did not rule out kinetic-kill strikes, it had begun to move them to the back burner in terms of its priorities.

Bush administration political appointees in the Pentagon retained a more skeptical view of space restraint. Indeed, their frustration with the ongoing efforts within the United Nations to pass a PAROS resolution[70] finally came to a head in the fall of 2005, when senior civilian officials in the Pentagon forced a change on the traditional U.S. policy (since the Clinton years) of abstaining from the yearly PAROS resolution.[71] Instead, the U.S. delegation for the first time voted against this otherwise unanimous U.N. resolution that called on states to refrain from using space for weapons purposes and urged the CD to take up the issue of a new treaty at its earliest convenience.[72] In doing so, the Bush administration set down an undeniable marker of strong U.S. opposition to further space arms control efforts.

Most European allies of the United States—with the exception of the United Kingdom—opposed the harder and more nationalistic U.S. line on space security. As Washington moved increasingly to reject negotiated forms of security, the states of the European Union (EU) had begun to shift markedly in the opposite direction, instituting a nearly universal common currency, forming new defense relationships, and seeing an increased number of mergers among leading aerospace companies, thus blurring their national identities in space.

[69] Interview with Lt. Gen. (USAF) Daniel P. Leaf, on the Web site of *Foreign Policy* magazine, August 2005, available at <http://www.foreignpolicy.com/story/cms.php?story_id=3141>; excerpted on the "Current and Future Space Security" Web site of the Center for Nonproliferation Studies at the Monterey Institute of International Studies at <http://www.cns.miis.edu/research/space/us/arms.htm> (accessed August 11, 2006).

[70] Among its various clauses, the PAROS resolution "emphasizes the necessity of further measures . . . to prevent an arms race in outer space" and "calls upon all States . . . to contribute actively to the objective of the peaceful use of outer space . . . and to refrain from actions contrary to that objective."

[71] Author's interview with U.S. Department of State official (name withheld upon request), Washington, D.C., January 20, 2006.

[72] Only Israel voted with the United States in rejecting the measure.

Although some EU defense analysts spoke of possible future conventional arms applications based on the Galileo positioning system, no EU country had a space weapons program and few even had military satellites. Instead, EU governments expressed a primary interest in using space to promote their commercial interests and in cooperating to pool resources for more effective space surveillance capabilities, including for collision avoidance, navigation, and debris management. According to French analyst Xavier Pasco, the whole mindset in Europe differed from that in the United States regarding space. For Europe, he noted, the "way forward will be to consider security as a more general concept that will be much more broadly construed than having only a military dimension."[73]

But there was one area where the United States continued to promote international cooperation and led the process of improving collective space security: orbital space debris. Here, in contrast to almost all other areas, Washington emerged as the leader in pushing for strict civilian limits on the release of harmful debris during the course of space operations. Domestically, the United States had already passed commercial rules under the Clinton administration governing the behavior of any operator releasing a satellite that incorporated U.S. technology, thus providing a considerable head start to international efforts. The Bush administration continued these efforts both internationally and at home. In October 2005, the U.S. Federal Communications Commission (FCC) enacted even stricter new guidelines to require any space provider under its jurisdiction to submit a debris mitigation plan for approval by the FCC. Within the U.N. Committee on the Peaceful Uses of Outer Space (COPUOS), the United States pressed for the development of similar guidelines during the 2005–2007 period,[74] calling for such measures as boosting satellites past their service lives into higher parking orbits or de-orbiting them to burn up in the atmosphere. This effort now aimed at convincing all users of space to adopt best practices and relevant technologies and procedures across the lifetime of their spacecraft. However, new space-faring countries like China and India initially opposed these regulations, describing them as "cultural imperialism," given the large amounts of existing orbital debris generated by the U.S. and Soviet programs. Moreover, each had experienced incidents in the past few years

[73] Xavier Pasco, "Enhancing Space Security in the Post Cold War Era: What Contribution from Europe," in John M. Logsdon and Audrey M. Schaffer, eds., *Perspectives on Space Security* (Washington, D.C.: Space Policy Institute, George Washington University, December 2005), p. 57.

[74] Author's interview with Nicholas Johnson, NASA's chief scientist for debris migration and the U.S. team leader in these meetings, Johnson Space Center, Houston, Texas, October 3, 2007 (conducted via e-mail).

where their spacecraft had broken up in orbit, creating hundreds of orbiting fragments because of their failure to adopt debris mitigation techniques.[75] Russia also initially opposed passage of the guidelines, fearing that they would impose new costs on its already cash-strapped space operations and because they would halt its past practice of blowing up sensitive Russian military satellites before they de-orbited.[76] To win support from these space "polluters," the United States eventually accepted the notion that the guidelines would be non-binding. However, some circles in the Bush administration pushed for looser restrictions also because they feared that binding regulations on debris might restrict U.S. efforts to deploy space weapons.

Human Spaceflight and the Beginning of a Paradigm Shift

NASA finally resumed manned spaceflights with the successful launch of the shuttle *Discovery* in the summer of 2005. Its completion of a fourteen-day mission in July marked an important step in the U.S. recovery from the *Columbia* accident,[77] but critics continued to recommend the shutdown of the costly shuttle program. Still, NASA's management remained firmly behind the shuttle and expressed its commitment to fulfill its obligations to the *ISS*—albeit at a bare minimum—before moving on to a next-generation spacecraft for returning to the Moon. By this time, public attention to human spaceflight already had begun to turn in another direction: private space ventures.

Before 2004, no entity besides state-run national programs had ever placed human beings into space. But a small group of entrepreneurs and flight enthusiasts had coalesced in the mid-1990s to sponsor a space version of the prizes that had stimulated and supported Charles Lindbergh's flight in the *Spirit of St. Louis* in 1927 across the Atlantic Ocean. In 1995, they formed the X Prize Foundation in Rockville, Maryland, with the intention of raising support for the first private venture to launch a person into space (which they defined as 100 km in altitude).[78] With growing support, they soon moved the organization

[75] On these incidents, see Orbital Debris Program Office (NASA), *History of On-Orbit Satellite Fragmentations* (Houston: Johnson Space Center, NASA, May 2004), 13th edition, available online at <http://www.orbitaldebris.jsc.nasa.gov/library/SatelliteFragHistory/13thEditionofBreakupBook. pdf> (accessed August 16, 2006).

[76] Ibid.

[77] For details on the *Discovery* mission, see the NASA Web site at <http://www.nasa.gov/mission_pages/shuttle/shuttlemissions/archives/sts-114.html> (accessed August 4, 2006).

[78] For more details on this history, see the X Prize Web site at <http://www.xprizefoundation.com/about_us/history.asp> (accessed August 11, 2006).

to St. Louis to benefit from its obvious symbolic value. The X Prize Foundation announced its "challenge" in 1996 to offer a $10 million prize for the first team to place a person into space twice with the same spacecraft within a two-week period (to discourage simple, nonreusable spacecraft and to encourage the crew's return). Eventually, thanks to major corporate and individual donations (particularly from the Ansari family), the X Prize Foundation raised the necessary funds well before any group could complete the challenge and claim the prize. Finally, in late September 2004, a team headed by aircraft designer Burt Rutan successfully launched the so-called *SpaceShipOne* out of the atmosphere from a specially made aircraft, safely returning its crew member, and repeating the flight on October 4, 2004, to win the competition.[79] Suddenly, the world of space exploration looked very different. Although *SpaceShipOne* and its launch system represented a far cry from the space shuttle or the massive Saturn V rockets that had boosted the Apollo capsules to the Moon, the fact that it had been built at a tiny fraction of the cost (and with private funding) over a period of only eight years opened up a completely new road to space. Thanks to tremendous media attention and coverage on the Internet, the flight also opened a wholly new perspective on space for the public, especially younger people who might eventually be able to visit it. Most important, from now on, states would not control access to space, thus opening this new environment to the general public for the first time. In July 2005, Virgin Atlantic airlines founder Sir Richard Branson and Rutan announced plans to form the jointly held Spaceship Company to build a follow-on vehicle to *SpaceShipOne* and offer the first commercial flights into space through a service called Virgin Galactic.[80] Now, the price of admission to space had dropped from Russia's $20 million to a "mere" $200,000 for the first Virgin Galactic tickets, although Branson promised future economies of scale to bring the cost down considerably. Other entrepreneurial companies—such as Bigelow Aerospace—were already well on their way toward the private construction of prototype orbital space stations.[81]

Meanwhile, in the launch sector, producers of space hardware and related services continued to look for cheaper access to space. A private venture founded by a young Internet multimillionaire, Elon Musk, aimed at delivering payloads of up to 1,000 kilograms into space within twenty-four hours of an order

[79] See the Web site of Burt Rutan's company, Scaled Composites, at <http://www.scaled.com/projects/tierone/041004_spaceshipone_x-prize_flight_2.html> (accessed August 11, 2006).

[80] See the Virgin Galactic Web site at <http://www.virgingalactic.com/en/news.asp> (accessed August 11, 2006).

[81] Leonard David, "Bigelow Aerospace Sets a Business Trajectory," *Space News*, March 26, 2007, p. 24.

at a cost expected to be in the low 10s of millions of dollars, compared to a go-ing rate of between $50–$110 million per launch. The maiden flight of Musk's Falcon booster in early 2006 exploded shortly into its flight. But Falcon's sec-ond flight achieved stage separation and accomplished a half-orbit in space before losing telemetry and subsequently failing. Still, most experts saw these growing pains as an anticipated part of developing a new rocket.

To deal with the expected demand from tourists and from commercial hard-ware providers, a variety of state and local organizations began to vie for com-ing contracts as "spaceports"—that is, ready-to-use launch sites for small com-mercial providers. Groups in New Mexico, California, and Oklahoma began to invest significant sums in setting aside large areas where activities like rocket launches could occur without endangering major population centers.[82] Clearly, the organizers believed that space prosperity was just around the corner.

But the dominant space money-maker continued to be international com-munications. Stagnant demand for space-based services, stemming in part from the cost of building and launching satellites, remained the most significant im-pediment to further development of this sector. Other impediments included frustration over delays in the allocation and actual use of geostationary orbital slots by the International Telecommunications Union, problems in space traf-fic control, and restrictions on exports, particularly in the United States. All of these issues required increasing international attention and negotiation. Still, space communications now led an international commercial space sector that could boast a healthy $115 billion in yearly revenues.[83]

Among American satellite producers, a major complaint continued to be the U.S. International Traffic in Arms Regulations (ITAR), which governed ex-port commerce in dual-use items. Although many satellite technologies were by now available on the world market, U.S. producers remained hamstrung by regulations aimed at preventing China and other targeted states from acquiring military technology. Despite the call by many leading industry officials for new guidelines that would recognize changes that had taken place in the market-place over the past decade, the administration and Congress remained reluc-tant to loosen these restrictions. The net effect served to strengthen relations between other satellite producers (such as Russia and the United Kingdom) and a growing list of clients in East Asia, South Asia, and the Middle East.

[82] See, for example, the Web site of the Mojave Airport and Spaceport at <http://www.mojaveairport.com/> (accessed October 21, 2006).

[83] Spacesecurity.org, *Space Security: 2006* (Waterloo, Canada: University of Waterloo, 2006), p. 18.

China, Russia, and the United States: Toward
Space Conflict or Cooperation?

In its first term, the Bush administration had followed a policy of keeping China at arm's length in space. China had been an early bête noire for the administration, and relations had gone poorly until 9/11, when Beijing began to identify common ground with Washington in combating Islamic extremists in its western territories. Moreover, the passage of time since the EP-3 incident and the increasing U.S. desire for Chinese cooperation against North Korea's nuclear and missile programs had caused senior U.S. officials to reconsider its standoffish policy. In 2004, representatives from NASA and China's National Space Administration held their first official meetings, although the two sides failed to announce any specific progress.

In April 2005, new Chief Administrator Michael Griffin took the helm at NASA. Compared to Sean O'Keefe, Dr. Griffin—a Ph.D. in aerospace engineering with prior NASA service and work in the Strategic Defense Initiative Organization—indicated a new emphasis on scientific qualifications.[84] Although Griffin proved unable to stem the pressure on the NASA budget within the administration, given competing national security demands, his appointment was met with guarded praise among space enthusiasts after years of having NASA run by an "outsider." Griffin's conservative credentials as a former weaponeer, however, also carried over to skepticism that he would break new ground in U.S. space relations with China.[85]

Outside the government experts on space policy continued to emphasize the value of cooperation with China, if only to learn more about its ultimate intentions in space. As Joan Johnson-Freese of the Naval War College had written in mid-2004, "While competition has primarily characterized U.S.-China space relations to date, it is time to consider cooperation as well."[86]

On the eve of China's second manned mission in October 2005, the United States offered an olive branch of sorts to the Chinese government by providing its space agency with tracking data for space debris to assist its technicians in planning a safe orbital trajectory for *Shenzhou VI*.[87] After the successful launch

---

[84] See his biography on the NASA Web site at <http://www.nasa.gov/about/highlights/griffin_bio.html> (accessed July 31, 2006).

[85] Author's interviews with various Washington-based space analysts and unnamed U.S. government officials, spring 2006.

[86] Joan Johnson-Freese, "Scorpions in a Bottle: China and the U.S. in Space," *Nonproliferation Review*, Vol. 11, No. 2 (Summer 2004), p. 168.

[87] "Chinese Experts Welcome US Offer of Warning Datum for Spacecraft Launch" (FBIS document number CPP20051015052021), Xinhua (Beijing), October 15, 2005.

and return of the two Chinese taikonauts, these talks continued, largely at the behest of high-level Bush administration officials in the National Security Council.[88]

In April 2006, Vice Administrator Luo Ge of the Chinese space agency met with NASA chief Griffin during a visit to the United States. Although the two sides announced no immediate progress, NASA soon revealed that Dr. Griffin would travel to China in the fall to meet with his Chinese counterpart, setting the stage for a possible breakthrough in previously tense relations between the two possible twenty-first-century space rivals. However, the meetings went poorly due to Chinese military sensitivities about sharing space technology and about making certain sites available to the NASA chief. Griffin later showed his apparent irritation at this less-than-open-arms reception by the Chinese space administration, dampening hopes for future cooperation in manned spaceflight. Nevertheless, the NASA administrator indicated that he expected further talks and the establishment of an ongoing relationship.

Although largely unsuccessful, the trip came off despite opposition from congressional conservatives, who believed that any space cooperation with China could lead to dangerous technology transfer useful for military purposes.[89] In this regard, a 2006 defense department report to Congress on China's evolving military capabilities stated that "Beijing continues to pursue an offensive anti-satellite system."[90] To support this finding, National Reconnaissance Office Director Donald Kerr revealed in the fall recent Chinese use of a ground-based laser to illuminate a U.S. satellite.[91] While the laser did not damage the spacecraft, for many critics this exposed Chinese duplicity in its arms control stance regarding space. However, a nongovernmental report observed that China was hardly alone, since "as many as 30 states may already have the capability to use low-power lasers to degrade unhardened satellite sensors," with the United States leading in anti-satellite capabilities of various types.[92] One former U.S.

[88] Author's interviews with various Washington-based space analysts and unnamed U.S. government officials, spring 2006.

[89] Comments by Senator Wayne Allard, dinner address to the conference "2006 Forum on Space and Defense," Center for Space and Defense Studies, U.S. Air Force Academy, Colorado Springs, Colorado, January 14, 2006. Senator Allard also noted his opposition to space cooperation with Russia.

[90] Department of Defense, *Military Power of the People's Republic of China, 2006*, Annual Report to Congress, available online at <http://www.defenselink.mil/pubs/pdfs/China%20Report%202006.pdf> (accessed August 13, 2006).

[91] On this event, for example, Vago Muradian, "China Tried to Blind U.S. Sats with Laser," *Defense News*, September 25, 2006, p. 1.

[92] Spacesecurity.org, *Space Security: 2006*, p. 23.

government analyst described China's overall military space capabilities as "sporadic," noting particularly gaps in sensor technology.[93] He concluded that Beijing "does not have a coherent military space architecture."[94] Thus, the nature of the Chinese space threat depended on where one sat, how one evaluated its emerging capabilities, and how one judged its ultimate intentions.

As experts took stock of China's accomplishments in space to this time, its military capabilities still seemed to rank comparatively low. Arguably, more important had been its creation of a reliable space-tracking network, a prerequisite to effective manned, commercial, and future military programs. Previously, Beijing had been limited to a fleet of ships for such tracking (much as the Soviet Union had). But it had used skillful diplomacy to establish a remote station in Namibia and agreements with Sweden and Italy to assist in tracking from other locations. These arrangements supplemented forty other cooperative space agreements with a range of countries, including work with Brazil on high-resolution photo-reconnaissance satellites. By 2006, China's official civilian space budget stood at about $500 million per year,[95] although this figure clearly understated a number of costs. Figures for the military budget were still unknown but likely totaled somewhat less, given the low number of military launches. The working estimate among foreign experts for actual, overall Chinese space spending was approximately $2.2 billion, with civilian expenditures likely still leading those in the military sector.[96] Of course, given the military's role in space launches and its provision of infrastructure support, these figures (as in the early U.S. and Soviet programs) were difficult to separate and involved many dual-use capabilities.

Technology represented only part of U.S. concerns. Indeed, according to some authors, the biggest challenge posed to the United States by China's civilian space program by the end of the first fifty years of spaceflight was not its military assets but its *demographics*.[97] China had the youngest cadre of space scientists and engineers of any major space power, with the average age about two decades younger than their counterparts in the United States and Russia. The

[93] James A. Lewis, "China as a Military Space Competitor," in John M. Logsdon and Audrey M. Schaffer, eds., *Perspectives on Space Security* (Washington, D.C.: Space Policy Institute, George Washington University, December 2005), p. 107.

[94] Ibid.

[95] Budget data provided by Chinese space agency vice administrator Luo Ge in Frank Davies, "China Plans Unmanned Moon Landing by 2012, Official Says," *Orlando Sentinel*, April 4, 2006, p. 1.

[96] Spacesecurity.org, *Space Security: 2006*, p. 110.

[97] See Brian Harvey, *China's Space Program: From Conception to Manned Spaceflight* (Chichester, England: Praxis, 2004), p. 237.

real threat to other space powers was the potential that this dynamic younger generation of Chinese specialists might in the future become oriented not in a civilian direction—as currently—but instead into military space efforts. In this context, political relations among various space powers had emerged at a critical tipping point, something that leaders in Washington might have considered in their decision to at least begin discussions on civilian cooperation through NASA.

In its military program, China's reconnaissance and remote-sensing satellites still lacked the ability to transmit digital images (relying instead on electro-optical transmissions) and suffered from less-than-optimal resolution. Therefore, as it likely worked to correct these deficits, it supplemented its own data with information provided to Chinese ground stations by French, European Union, U.S. (Landsat), and other foreign satellites. China also possessed a small fleet of navigational satellites and planned to orbit a complete constellation to facilitate development of its own locational systems, which could also be employed for precision-guided munitions. A major concern in Western military writings was China's experience—often with foreign experts, such as British specialists in Surrey—in building and launching microsatellites and nanosatellites. While the military capabilities of such spacecraft remained unclear, their dual-use potential might lead to low-cost inspection missions and anti-satellite weapons. However, these programs seemed to be proceeding in a gradual manner at this point because of China's limited resources and preference for indigenous technology (although China continued to purchase and learn from foreign technological experience in a number of fields). As of 2006, gaps in China's military capabilities also included an absence of dedicated electronic intelligence satellites.

Compared to early American reticence to engage China in space cooperation, the EU had readily embraced Beijing. China had pledged some $236 million to participate in the European global positioning system (Galileo), which might allow Beijing to influence how the system would eventually be operated.[98] Such cooperation remained an issue of concern to the U.S. military, as it could potentially provide Beijing with an "unjammable" signal in wartime.[99] But questions remained in the fall of 2007 about Galileo's funding base and about technology transfer and management issues with other partners, including China.

---

[98] Lt. Col. (USAF) Scott W. Beidleman, "GPS vs. Galileo: Balancing for Position in Space," *Astropolitics*, Vol. 3, No. 2 (Summer 2005), p. 140.
[99] Ibid.

Looking forward, China had announced plans for a steady, step-by-step development of its space capabilities. The launch of a Moon orbiter in 2007 would be followed by a lunar rover in 2012 and other unmanned vehicles aimed at returning samples to Earth by 2015 or 2017.[100] Statements by Chinese officials still mentioned no manned Moon exploration and emphasized Beijing's desire for restrictive space arms control measures. Many U.S. defense officials remained skeptical.

As for Russia, by 2005, the RSA had finally begun to regain its footing. The boom in world oil prices had filled Moscow's coffers with hard currency and boosted its already significant trade surplus. Although some of these funds were used to make early payments on Russia's state debts from the 1990s, their presence also provided the prospect of extended funding for space and military modernization for the first time since 1991. By 2007, financing for the RSA totaled over $2 billion, with $1.4 billion coming from the state budget and the rest from foreign contracts.[101] As RSA Director Anatoly Perminov explained, "It's obvious that the scale of international cooperation in space will only continue to grow."[102]

The RSA's plan for the next decade included deployment of a six-person space ferry (the *Kliper*, a winged replacement for the Soyuz spacecraft originally planned for joint development with European partners), two new launch vehicles (the midsize Angara and the heavy-launch Soyuz-2), and a deep-space mission to the Mars moon Phobos.[103] Nevertheless, RSA representatives did not expect a return to the salad days of the Cold War. As Perminov noted, the *Kliper*'s development would depend on its commercial viability and would only be built if the project succeeded in attracting financial support from foreign partners.[104] In the end, the European Space Agency's 2006 decision not to proceed with the project caused its cancellation, although the Energiya enterprise planned to redesign the program and resubmit it in the future. More

---

[100] On China's plans, see Craig Covault, *Aviation Week & Space Technology*, May 23, 2005, p. 37; "PRC: Article Examines Preparations for Shenzhou VI, Plans for Moon Flight," *Xinwen Chenbao* (Shanghai), September 28, 2005 (FBIS Document #CPP20050929510002); and Davies, "China Plans Unmanned Moon Landing by 2012, Official Says."

[101] Konstantin Lantratov, "Affordable Space Projects from Russia," *Kommersant* (Daily), August 21, 2007, online at <http://www.kommersant.com/t795701/r_3/n_25/Space_Rocket_Launch/> (accessed October 8, 2007).

[102] Roskosmos Director Perminov, as quoted in Aleksey Shcheglov, "Ambitsii Roskosmosa rastut" [Roskomos's aims are growing], Strana.ru Web site, posted November 11, 2005.

[103] "Russia Approves 10-Year Space Plan," Spacetoday.net, posted October 26, 2005.

[104] Perminov, as quoted in Shcheglov, "Ambitsii Roskosmosa rastut."

importantly, Russian oil revenues buoyed the overall RSA budget and provided significant recent increases. In 2007, funding to support improvement of the long-ailing GLONASS network doubled, and total civilian space expenditures were expected to jump by 20 percent for 2008.[105]

The Russian military also began to focus again on upgrading its space component by announcing plans for an Aerospace Defense Command by 2008. To support this effort, the Russian government began to move forward with a major upgrade of the military's once-secret northern launch facility at Plesetsk. The plan called for expanding its capabilities and increasing the yearly number of launches by using the new Angara booster. Military officials revealed that Russia eventually planned to conduct all military launches from this facility, thus finally erasing vulnerabilities stemming from its reliance on Kazakhstan to conduct certain types of launches. Although currency issues and hidden aspects of the military budget made exact figures difficult to calculate, the Russian military space budget likely now amounted to slightly less than $1 billion per year, compared to a U.S. space military budget of over $12 billion.[106] Russia retained the technological foundation to surge in this direction if political changes required as much, though not likely to the level of U.S. forces considering the major contractions in the industry during the 1990s and the aging of Russian space scientists and engineers. As President Putin noted in a January 2007 address, the "most important priority" for Russia's space program would be attracting a new cadre of young specialists.[107]

New Critics of the U.S. Military Space Policy

One of the more remarkable developments of the second Bush term was the increase of criticism of its space policies, not from traditional opponents of space weapons but instead from hard-line supporters. These advocates claimed that the administration had lost its momentum and now faced the risk of leaving office without having provided even a thin line of space-based defenses.

[105] Lantratrov, "Affordable Space Projects from Russia."

[106] On the U.S. military space budget, see Jeffrey Lewis, "Selected Space Programs in the 2005 Appropriations Process," Center for Defense Information, Washington, D.C., August 2004, available on the CDI Web site at <http://www.cdi.org/PDFs/FY05Appropriations.pdf> (accessed October 21, 2006).

[107] "Vystupleniye Prezidenta Rossii V.V. Putina na torzhestvennom vechere pocvyashchennom 100-letiyu co dnya pozhdeniya S.P. Koroleva" [Speech by President V. V. Putin at the ceremony to commemorate the 100th anniversary of the birth of S. P. Korolev], available on the Web site of the Russian Federal Space Agency at <http://www.roscosmos.ru/NewsDoSele.asp?NEWSID=2050> (accessed October 8, 2007).

To this point, the Pentagon had conducted a number of experiments, ranging from laser tests on a specially equipped 747 aircraft (precursor work for the Space-Based Laser), microsatellite close fly-bys (to examine the possibilities for anti-satellite use), and various tracking and interception tests in space associated with the ground-based interceptor, THAAD, and the Aegis missile defense systems. The record had proved mixed, at best, in contrast to the confident claims of supporters at the early stage of the administration. As during the SDI period, the problem proved to be making this complex technology work. The Airborne Laser had continued to experience problems in defining its beam and providing adequate power and range to destroy actual missiles. Congressional critics had finally canceled funding for its space-based version until the difficulties could be worked out first with the airborne variant. The April 2005 test of the Demonstration of Autonomous Rendezvous Technology (DART), a potential dual-use satellite inspection system, had ended with its range-finding system malfunctioning, causing an embarrassing collision with the target satellite.[108] However, the Experimental Spacecraft System 11 (XSS-11) had engaged in a series of successful maneuvers in space during 2005 and continued to follow a schedule for close visits of satellites over a two-year mission plan. In addition, Congress had approved research aimed at deployment of a "test-bed" of up to six space-based interceptors by 2011–12, although at low levels of funding.[109] Meanwhile, the planned Near-Field Infrared Experiment had been the object of considerable political jockeying. This project—a tracking satellite potentially armed with a "kill vehicle" to hit ballistic missiles or satellites—had seen the kill vehicle removed by Congress in fiscal years 2004 to 2006, although MDA Director Lieutenant General Trey Obering mentioned that it might return in possible future tests after 2007.[110] Finally, the air force's Common Aero Vehicle (CAV) had run into similar congressional limits. Out of concern that this Mach 20 system might be used to carry nuclear weapons into space, the program remained tethered to the research phase with a strict congressional prohibition against developing any weapons, at least until the CAV "bus" itself had been perfected.

[108] See Dawn Stover, "Battlefield Space Military Hardware Has Orbited Earth for Decades, but No Actual Weapons Have Ever Been Deployed in Space. That May Change Soon—And It May Launch a Major Arms Race," *Popular Science*, Vol. 267, No. 5 (November 2005).

[109] Spacesecurity.org, *Space Security: 2006*, p. 24.

[110] See Center for Defense Information, "NFIRE's Kill Vehicle Is Gone Again, Unless, of Course, It's Back Again," *CDI Space Security Update*, March 10, 2006, online at <http://www.cdi.org/program/document.cfm?documentid=3350&programID=68&from_page=../friendlyversion/printversion.cfm> (accessed August 14, 2006).

In the face of these delays, advocates of U.S. "space power" theories criticized what they perceived as continuing inattention to China and other possible threats. As an article in the influential U.S. Naval Institute magazine *Proceedings* stated, "We are at risk of relinquishing our military space dominance to competitors."[111] The proper course, according to hard-core advocates of space-based defenses, aimed at fulfillment of SDI-era plans for deployment of a large-scale, layered, space defense framework to deal with both missiles and hostile satellites.

In early 2006, a group of influential missile defense supporters calling themselves the Independent Working Group (many of them former Reagan administration officials and SDI laboratory personnel) issued a voluminous report emphasizing the urgent need for space-based defenses.[112] Specifically, they described a threat environment characterized by rising risks to the United States from ballistic missiles and WMD—an environment that "no longer allows the luxury of long lead times for the development and deployment of defenses."[113] Notably, the report sharply criticized the Bush administration's missile and space defense programs for not doing enough, almost as harshly as critics from the left had criticized it for having gone too far. It argued that those defenses that had been deployed by the Bush administration offered only "extremely limited coverage"[114] of the United States. The authors called for a return to SDI-era systems, including deployment of one thousand Brilliant Pebble interceptors by 2010, which the report stated could be accomplished for $16.4 billion (drawing on estimates from the 1980s). Finally, the authors argued for the creation of a special congressional caucus "to build support for U.S. space primacy, space control, and assured access as well as missile defense in general and space-based anti-missile systems in particular."[115]

Still other analysts continued to support more aggressive, offensive-oriented systems, like the so-called Rods from God. These plans, linked to the Pentagon's evolving concept of Global Precision Strike, aimed at deploying perhaps dozens of orbiting battle stations armed with quivers of 20-foot-long tungsten rods that could be called down on deeply buried targets anywhere on the Earth within a few moments' notice. Supporters believed that "a few well-placed tungsten

---

[111] Maj. (USMC, ret.) Franz J. Gayl, "Time for a Military Space Service," *Proceedings* (July 2004), p. 44.

[112] See the text of the Independent Working Group's "Missile Defense, the Space Relationship, & the Twenty-First Century: 2007 Report," posted on the IWG's Web site at <http://ifpa.org/pdf/IWGreport.pdf> (accessed July 31, 2006).

[113] Ibid., p. viii.

[114] Ibid.

[115] Ibid., p. xi.

rods . . . would guarantee the destruction of the targeted facilities."[116] However, as even some supporters admitted, the actual physics of the technology's use required slowing the rods down so that they didn't disintegrate on contact with the ground.[117]

However, congressional support for SDI-era programs, offensive space weapons, and grand visions for large-scale space defenses had waned since 2001, if it had really existed then at all. The reasons had to do with growing concerns about costs, political implications, and effectiveness, problems that continued to plague such umbrella-type defenses. The Independent Working Group's report garnered little steam. Instead, mainstream U.S. political thinking on near-term deployment of space defenses seemed to have moved surprisingly close to that of longtime critics by 2006. Support for space defenses within the administration also weakened following President Bush's abrupt release of Secretary of Defense Donald Rumsfeld in November. The secretary's dismissal came the day after the midterm elections, which had yielded a Democratic victory in retaking both houses of Congress, largely resulting from public discontent over the progress of the war in Iraq. Increasingly, the main emphasis within the U.S. military began to shift toward acquiring the *selective* ability to take out individual satellites and stray missiles.

Part of the reason for this evolving shift in many parts of the Pentagon was a growing awareness among air force officials that testing, deploying, and using space weapons would generate harmful space debris, which might be worse than trying to address space threats by other means. Recognizing the U.S. investment in safe access to Earth orbit, Air Force Undersecretary for Space Programs Gary Payton concluded in an interview regarding kinetic-kill weapons, "It would be hugely disadvantageous for the U.S. to get into that game."[118] This perspective was not universally appreciated by political supporters and military backers of more aggressive forms of space power. In contrast, the intelligence community viewed interference with its space sanctuary as a dangerous trend, which threatened to erase U.S. advantages in collection, verification, electronic information, tracking, weather forecasting, and communications.

Among critics of the Bush administration's military space policy, fears con-

---

[116] Michael Goldfard, "Are Kinetic-Energy Weapons the Future of Space Warfare?" June 8, 2006, available on the *Weekly Standard* Web site at <http://www.weeklystandard.com/Content/Public/Articles/000/000/005/700oklkt.asp> (accessed August 13, 2006).

[117] Re-entry into the atmosphere would superheat the rods, making them structurally malleable in collisions with solid objects.

[118] Quoted in Jeremy Singer, "USAF Interest in Lasers Triggers Concerns About Anti-Satellite Weapons," *Space News*, May 1, 2006, p. A4.

tinued that test programs planned for the future would create "facts in orbit."[119] Once deployed, they argued, these weapons would break past taboos against space-based weapons, cause officials to treat space as already weaponized, and change the norms for international space security relations. Some experts believed that missile defense projects, although intended for defensive purposes, could become a backdoor approach to offensive capabilities. As former defense department procurement chief Coyle stated in 2006, "Once you've got space-based interceptors up there, they can just as well be used for offense as defense."[120] In such an environment, trends toward further weaponization would, according to this perspective, be difficult to stop.

With as little fanfare as possible, the Bush administration quietly released its long-awaited "U.S. National Space Policy" in October 2006. Although much of the language remained unchanged from the 1996 policy, the document's tone and points of emphasis revealed a shift in its underlying philosophy.[121] The intelligence community had succeeded in getting continued support for the right of all space systems to enjoy "passage through and operations in space without interference" and for the rejection of "any limitations on the fundamental right of the United States to . . . acquire data from space."[122] But administration neoconservatives had removed prior treaty-based restrictions on U.S. rights to "deny such freedom of action to adversaries" and to exercise "space control." The new policy also added wording on U.S. *opposition* to any new legal restrictions on space activity, with the exception of debris mitigation. In the commercial space arena, the new policy voiced stronger support for U.S. entrepreneurship in space. Overall, the document had walked to the threshold of weaponization but had failed to cross it overtly, thus suggesting not only the strengthened hand of supporters of military space forces since the last Clinton policy in 1996 but also the continued power of opponents of weaponization, who feared it would endanger the existing benefit of safe access to space.

In January 2007, word began to leak out to the news media that space experts had noticed the peculiar absence of China's *Feng Yun 1C* weather satellite from

[119] Theresa Hitchens, Michael Katz-Hyman, and Victoria Samson, "Space Weapons Spending in the FY 2007 Defense Budget," report on the Center for Defense Information Web site at <http://www.stimson.org/space/pdf/FY07SpaceWeapons.pdf> (accessed August 14, 2006).

[120] Philip Coyle quoted in Bryan Bender, "Pentagon Eyeing Weapons in Space: Budget Seeks Millions to Test New Technologies," *Boston Globe*, March 14, 2006, available online at <http://www.boston.com/news/nation/articles/2006/03/14/pentagon_eyeing_weapons_in_space/> (accessed August 14, 2006).

[121] See the full text of the "U.S. National Space Policy," as posted on the Web site of the White House Office of Science and Technology Policy at <http://ostp.gov/html/US%20National%20Space%20Policy.pdf> (accessed October 21, 2006).

[122] Ibid.

space tracking data, suggesting that China had successfully tested a ground-launched, kinetic-energy ASAT. As more information began to come out from governmental sources, it became apparent that the test had taken place on January 11—following two prior proximity tests in 2005 and 2006—and involved the launch of a Dong-Feng 21 missile, which had destroyed the Chinese target satellite at an altitude of 538 miles, generating hundreds of pieces of trackable space debris and creating a major international incident. Arms control supporters felt betrayed by China's apparent disavowal of its past opposition to space weapons. Opponents of new treaties found justification in their past beliefs by stating that China's test "proved" that arms control would not work (although neglecting the fact that such talks had not been conducted for the past decade). Senator Jon Kyl (Rep., Arizona) gave a speech in late January 2007 arguing that China's ASAT capability now required U.S. testing and deployment of advanced space-based defenses.[123] At the international level, U.S. CD Ambassador Christina Rocca indicated no change in U.S. space security policy since early in the Bush administration: "Despite the ASAT test, we continue to believe that there is no arms race in space, and therefore no problem for arms control to solve."[124]

The reaction of the commercial space community proved to be more revealing of a change in thinking and marked the engagement of a major new player in the space security debate. For the first time, commercial space providers broke ranks with the U.S. administration by calling for serious attention to the threat posed by debris-generating weapons activities in space. A number of opinion pieces in various trade publications by industry leaders argued that weapons tests should be halted and debris management moved much further up on the list of urgent space priorities.[125] An editorial in the normally neutral *Space News*, for example, described Bush administration statements rejecting treaty approaches as "premature" and argued instead, "It only makes sense to ban an activity that creates debris that threatens the satellites of many countries."[126] A much-quoted industry impact assessment issued in response to the Chinese test by the Teal Group warned ominously, "An ASAT weapons race

[123] Dave Ahearn, "Senator Urges Funding Space-Based Satellite Defense," *Defense Daily*, January 31, 2007, p. 1.

[124] "Statement to the Conference on Disarmament by Ambassador Christina Rocca, U.S. Permanent Representative," February 13, 2007, on the Web site of the U.S. Mission to the United Nations in Geneva at <http://www.usmission.ch/Press2007/0208CDstatement.htm> (accessed February 28, 2007).

[125] See, for example, opinion piece by Intelsat Chief Executive Officer David McGlade, "Preserving the Orbital Environment," *Space News*, February 19, 2007, p. 27.

[126] "China's Anti-Satellite Test" (Editorial), *Space News*, January 22, 2007, p. 14.

will have the effect of increasing the financial risk of any satellite program, and this will undoubtedly be felt most within the commercial market through decreased investor confidence and(or) high insurance rates."[127] After years on the sidelines, the commercial industry had suddenly engaged itself in the space security debate.

The Chinese government only belatedly issued a statement on January 23 affirming its conduct of the ASAT test. China's Foreign Ministry spokesman tried to tone down the international furor by stating, "This experiment is not targeted at any country, nor will it pose a threat to any country."[128] Not surprisingly, the Chinese explanation proved highly unsatisfactory to foreign observers. Japanese Prime Minister Shinzo Abe took the unusual step of directly condemning China's test and arguing that it constituted a violation of the principle of noncontamination of space embodied in Article IX of the Outer Space Treaty.[129] For others, the notion that the thousands of pieces of trackable and untrackable debris did not constitute a "threat" to others in space seemed at best disingenuous and at worst a complete compromise of China's earlier, principled position. Overall, China had clearly miscalculated the international reaction, perhaps thinking that it had done nothing more than test a technology developed decades before by the Soviet Union and the United States. But the number of spacecraft in low–Earth orbit and the number of countries (and companies) affected had increased by an order of magnitude, making the debris problem caused by this test a much more serious problem. Moreover, international access to space tracking data on the Internet meant that information about the test became highly publicized around the world, instead of remaining hidden in the intelligence files of a few key states, as it might have during the Cold War.

The Chinese ASAT test mobilized and (to some degree) unified the space community in a way it had arguably never been before. The question of the sustainability of this movement—linking commercial, civilian, scientific, and some military space practitioners—remained uncertain. Nevertheless, concern about the space environment had now—after fifty years—jumped from obscure scientific journals and intergovernmental reports to the front pages of major world newspapers and thousands of Internet sites. In this respect, China's ASAT test and the international reaction represented perhaps a fitting

[127] Teal Group Press Release, "Teal Group Assesses Satellite Market Impact of China ASAT Test," Fairfax, Virginia, January 22, 2007.

[128] "Foreign Ministry Spokesperson Liu Jianchao's Regular Press Conference on January 23, 2007," Web site of the Chinese Foreign Ministry at <http://www.fmprc.gov.cn/eng/xwfw/s2510/2511/t291388.htm> (accessed February 28, 2007).

[129] "Japan's Abe Charges China's Satellite Test Illegal," *Agence France Presse*, January 31, 2007.

bookend to close the 1957–2007 period. Clearly, a new discussion about space security—involving new actors—was about to begin.

In June, the countries in the U.N. COPUOS approved the long-awaited Space Debris Mitigation Guidelines, with support from all of the major space-faring countries. While still a voluntary set of measures, the approval of this document and its forwarding to the United Nations for official vote in the fall represented the shifting of space debris to the front burner of international space concerns. No longer would national governments be able to claim ignorance about the effects of debris, and the international community of scientists and policy experts stood ready to condemn any state that violated these new norms. While the effort fell far short of a formal ban on weapons testing, the United Nations' December approval of the debris mitigation guidelines marked significant progress toward the future limitation of acceptable military activities in space. In this sense, space security had reached a new, higher plateau.

As the fiftieth anniversary of *Sputnik I*'s launch neared, Russia reemerged as an active player in the space security debate. At the United Nations, it renewed its sponsorship for a draft resolution on "Transparency and confidence-building measures in outer space activities" and pledged its continued support for a treaty banning weapons in space. At the same time, Colonel General Vladimir Popovkin, commander of Russia's Space Forces, announced that Russia was ready to respond in kind if another country weaponized space first. As Popovkin stated, "We do not want to fight in space, and we do not want to call the shots there either, but we will not permit any other country to do so."[130] In practical terms, Russia was repeating China's statement with its ASAT test in January.

On the U.S. side, the rhetoric and thinking at DOD seemed to have evolved too—in a more positive direction. In sharp contrast to arguments used by senior state department officials as late as May 2007 about the unacceptability of space security discussions,[131] Air Force General Kevin P. Chilton stated in a document released on the eve of congressional hearings on his confirmation as the new commander of U.S. Strategic Command (STRATCOM) that "we should examine the potential utility of a code of conduct or 'rules of the road' for the space domain, thus providing a common understanding of ac-

---

[130] "Russia Promises Retaliation If Weapons Deployed in Space," *RIA Novosti*, September 27, 2007.

[131] See Testimony by Amb. Don Mahley, U.S. State Department, prepared for the House Committee on Oversight and Government Reform's Subcommittee on National Security and Foreign Affairs, hearing on "Weaponizing Space: Is Current U.S. Policy Protecting Our National Security," May 23, 2007, online at <http://nationalsecurity.oversight.house.gov/documents/20070523162935.pdf> (accessed September 20, 2007).

ceptable or unacceptable behavior within a medium shared by all nations."[132] What would become of General Chilton's new policy remained to be seen, but it signaled a new readiness on the part of the United States to consider coopera-tive approaches to space security.

Conclusion

Despite the ambitious plans of the Bush administration on entering office, political and budget factors combined to shift its attention away from the dra-matic change in U.S. space security policy predicted for it. Nevertheless, U.S. policy moved initially toward a strongly nationalistic, military-based approach to space security with the U.S. withdrawal from the ABM Treaty in 2002 and with the surge in funding provided for a range of defensive and offensive space systems after 9/11. But within the U.S. military, the process of studying and as-sessing future scenarios led to growing (although not universal) resistance to kinetic-kill weapons among the panoply of space weapons under consideration, thus providing some grounds for an expectation of continued restraint. The U.S. failure to push forward space arms control of any kind during this criti-cal period of overwhelming U.S. space superiority—besides voluntary civilian debris restrictions—continued to leave the field open to future deployments of space defenses by the U.S. and other militaries as well. Perhaps believing that the arms control route to security had been effectively closed by the long U.S. refusal to talk, China eventually decided to step through this open window.

By the fall of 2007, it seemed that space relations might be on the verge of a step up in the level of consensual knowledge about space reminiscent of 1962 or 1985. While two of the leading space powers had moved to the edge of space's weaponization, threatening to throw international space security back into a nationalist, military-led direction, both had signed onto new debris-mitigation measures and had put active testing of space weapons (with the exception of U.S. ground- and sea-based missile defense interceptors) on hold. The Penta-gon had quietly dropped provocative rhetoric about "space dominance," and China had restated its rejection of any intentions to engage in a space arms race, pledging instead its desire to move ahead to new forms of space arms control. While Russia had warned of its readiness to confront any U.S. moves toward space weapons, it had offered again to lead efforts at the United Nations

[132] "Advance Questions for General Kevin P. Chilton, USAF Nominee for Commander, United States Strategic Command," Daily Report, *Air Force Magazine*, September 27, 2007, available on the Air Force Association Web site at <http://dailyreport.afa.org/NR/rdonlyres/2AECDCE4-DBB0-4908-9E18-7E85A7A95E56/0/092707Chilton.pdf> (accessed October 8, 2007).

to prevent such a direction in space. In Washington, a newly Democratic-led Congress raised doubts about at least near-term support for space weapons. Perhaps most importantly, the emergence of the commercial space industry as a player in the space security debate seemed to indicate a possible power shift on these issues in the future, particularly if any country's military experiments began to threaten the industry's $120 billion of assets in low–Earth orbit.[133]

[133] Figure provided in Teal Group, "Teal Group Assesses Satellite Market Impact of China ASAT Test."

CHAPTER EIGHT

Alternative Futures for Space Security

The task of looking ahead to predict the future in space is one that can only be taken up with some trepidation. The lessons of history are subtle and contradictory, offering no easy explanations or determinate outcomes. In considering the evidence presented by the first fifty years of human space activity, there are instances of close cooperation and cutthroat competition, although fortunately with no direct conflict. The good news is that any country could have defected from the norm of space restraint and yet none has strayed very far.

As we have seen, the restraint exercised by the superpowers in dealing with space security during the Cold War resulted in large part from the impact of technological and environmental factors. Bilateral negotiations moderated this learning under the influence of change-inducing trigger events, which shifted actors' original intentions. Still, despite this difficult process and the reluctance of many actors to break out of past patterns, military restraint took place and endured over time. This "learning against one's will" highlights the fragility of space restraint and its dependence on political relationships as well as a consensus on the adequacy of verification. Unfortunately, the evidence of weapons restraint during the first fifty years of space history—and particularly the years since 2001—shows no decisive or final value change by actors in regard to collective forms of space security, despite conditions of functional (and therefore strategic) interdependence. The changing perspective of each incoming national leadership remains a factor that can challenge and even destroy previously established space norms, as seen in 1981 and 2001. China's 2007 ASAT test is another example, where a national leadership that had not been part of existing space understandings violated prior norms against high-altitude testing, where debris takes decades to de-orbit.

Looking ahead, what matters is whether core environmental understandings among leading space-faring countries can be expanded and maintained—thanks to the consensual knowledge stored in their executive branches, legislatures, militaries, and treaties—or instead whether these norms are superseded by new, short-sighted, and aggressive approaches. As we have seen, while the physical environment of space is vulnerable to offensive and debris-generating activities, cooperation to restrain such programs still depends on decisions by political and military leaderships and is influenced by factors outside the space relationship.

The Cold War superpowers cooperated to preserve a valuable collective good—safe access to space. Such policies served them well, even (and perhaps especially) during periods of great political hostility. A key reason for this cooperative restraint was the benefit the powers derived from peaceful uses of space, including military support and, recently, precision force application.

Taking Stock of Recent Trends

In terms of the substance of the agreements formed, some aspects of military space restraint to date may appear to be an indicator of complex learning, as there is an implication of the abandonment of past goals of competition in some areas. However, we must also take into consideration the fact that certain cooperative bans involved *prospective* limitations, such as in regard to the Moon. For this reason, the restraint that emerged during the first fifty years of space security should instead be seen as indicating cognitive changes that lie midway between simple and complex learning, dependent on time. In other words, *Cold War efforts to limit space competition can best be understood as an "option to learn" at some point in the future.*[1] However, since learning is dependent on cognitive factors and ongoing consensus between the relevant actors, these changes could also lead to the unraveling of space restraint and cause states to "unlearn" the environmental lessons of the Cold War. A related problem is the failure of new space actors, to date, to adopt these lessons. China's 2007 ASAT test and India's hostile response suggest that "copycat" testing of such debris-causing weapons—by states seeking to demonstrate their space military capabilities and resolve—might emerge. The February 2008 U.S. shootdown of a hydrazine-laden reconnaissance satellite by a sea-based Standard Missile 3, albeit at very low altitude and without any long-lasting debris consequences, could make binding norms more difficult to achieve. Such efforts will likely

---

[1] This notion, therefore, accepts the traditional two-tiered learning framework but adds a variation. Space restraint during the Cold War offered a form of cooperation that fulfills the definition of simple learning, but which could potentially lead to "complex" learning at a future time if technologies and policies evolve sufficiently to enact its more restrictive tenets.

require close international consultations to specify "acceptable" and "unacceptable" criteria for future use of kinetic weapons. Otherwise, future space actors may be tempted to bend debris-mitigation rules to accommodate their own specific military needs. Unfortunately, given the context of ongoing competition among states in space, restraint and consensual knowledge must constantly be reaffirmed. Improved communication among states, widespread opposition to weapons, and growing integration of national space programs and industries are powerful indicators that this consensual knowledge may expand rather than contract in the twenty-first century. However, that outcome is far from inevitable.

Yet environmentally directed definitions of space security could increase in acceptability as states and other actors realize that (1) unilateral policies create problems that restrict safe access to space for all players; and (2) the costs of expanded weapons competition—in both financial and strategic terms—are likely to be unusually high in space compared to other environments. It is precisely this possibility of future learning among states that makes it plausible that structural obstacles might continue to be overcome, despite the increasing number of space actors.[2]

On the negative side, the international consensus for collective forms of space security has been weakened considerably in recent years by mistrust and new assumptions about technological trends, including doubts about the human ability to stop certain threats by nonmilitary means. Yet differences of opinion on these points exist internationally. Despite and indeed because of recent ASAT activities, most states still come down firmly in support of efforts to manage space collectively and without resort to destabilizing weapons. In the U.S. Congress and among other seats of U.S. power and influence, the weight of opinion is still not clear. Divided government in Washington from 2006 to the present may be bringing a new perspective to these problems—as Congress has cut back funds for possible orbital weapons systems. But the outcome of this and other shifting national kaleidoscopes of security calculations and political power is what will determine the future course of space security relations—not some predetermined fate.

[2] Game theory suggests that restraint is less likely in n-person games than in contests involving only two players. As Cornes and Sandler conclude, "noncooperative strategy becomes more prevalent when group size increases." (See Richard Cornes and Todd Sandler, *The Theory of Externalities, Public Goods, and Club Goods* [New York: Cambridge University Press, 1986], p. 13.) However, as political scientist John G. Ruggie has found from his examination of a number of historical cases of interstate cooperation, cooperative agreements are influenced not only by the number of actors but also by the establishment of a certain "*social* milieu" among the participants. Such conditions can sometimes overcome structural constraints. (See John Gerard Ruggie, "International Responses to Technology: Concepts and Trends," *International Organization*, Vol. 29, No. 3 [Summer 1975], pp. 559–60.)

The rest of this chapter attempts to take stock of the lessons of the first fifty years of space security and to look ahead to the challenges facing all users over the next several decades in space. It starts by highlighting the *reasons* for key accomplishments and changes during the five periods of space history covered in this book. Next, it discusses the new factors that may recast the space arena in the twenty-first century, creating what may be a paradigm shift in the nature of space activities from the early years, which were dominated by the two Cold War competitors. Finally, it surveys the prescriptions of the four schools of space thinking regarding the future (from Chapter One), taking into account their varying reliance on national military strength, military-technical dynamics, commercial activity, and international cooperation. The conclusion offered here is less a single prediction than an analysis of trade-offs and trends, including a careful weighing of the strengths and weaknesses of the different approaches. As mentioned earlier, what is likely to differ—compared to the first fifty years of space security—is the mix among the actors in space and their objectives. While national militaries and national space agencies will continue to play major roles, they will have to deal with a rapidly increasing number of commercial space actors and other nongovernmental participants, who may change the interests and priorities of the collective space community.

Historical Trends and "Lessons"

As we have seen, the United States and the Soviet Union made no progress toward space restraint or cooperation until mid-1962. This was not for lack of effort, especially on the part of the United States. In the area of space science, Soviet mistrust and security concerns stalled initial negotiations, as the Khrushchev leadership sought to isolate the Soviet program from foreign exposure of its weaknesses, while providing no unnecessary "leg up" to the American program. In the security realm, technical issues and mistrust also separated the two sides. For the United States, problems of verification prevented arms control progress; for the Soviets, there was an unwillingness to sacrifice technologies that might provide new means of overcoming crucial U.S. strategic advantages in other military areas. In the course of the first few years of space activity, however, both sides acquired considerable new information about the specific characteristics of the space environment and the problems of conducting traditional military activities there. The effects of nuclear tests in space, in particular, proved to be severely damaging to both satellite activities and human spaceflight. This shared knowledge about the unique hazards of space within the two national leaderships preceded the ability of their political systems to accept cooperation and overcome the inertia of Cold War hostility.

Only at the end of this period, following the devastating effects on satellites of nuclear testing in space and its threats to the respective manned programs did the two governments recognize the need to push past domestic obstacles to meaningful space cooperation. This shift required a broader definition of security than had previously been entertained by the U.S. and Soviet defense establishments, and it required treating this new environment with greater restraint. In the short period before the death of President Kennedy, the two leaders achieved important progress in solidifying space's first cooperative norms and in creating its first legal rules. What had been a completely unregulated realm in 1962 evolved into a region of significant mutual restrictions by the end of 1963. While the competitive goals of both superpowers remained intact, their policies changed to prevent threats to the space environment from damaging higher-value activities than weapons, namely, high-prestige civilian programs and mostly secret military reconnaissance missions.

With the establishment of a precedent, the next step forward in space restraint came not with the help of exogenous political pressure but from factors created by repeated interactions between the major space powers. Talks over the course of 1965–66 between the U.S. and Soviet governments, as well as within the U.N. Committee on the Peaceful Uses of Outer Space, culminated in the 1967 Outer Space Treaty, which was based on the existing 1963 U.N. Legal Principles Agreement. The possibility that one side or the other might try to assert unilateral, territorial claims on the Moon helped stimulate this progress. Interestingly, in a step *backward* from the largely competitive aims that continued to characterize their programs, the two sides showed that stability and a desire to prevent losses in the passive military sector overrode military aims of keeping future weapons options open.

With the Moon race won and Vietnam War costs increasing, the United States abandoned its "racing" mentality in space. The development of space-based means for verification facilitated the emergence of U.S.-Soviet détente and played a critical role in opening the door to new forms of arms control. Civilian and space science cooperation benefited from the accompanying political rapprochement. Trends toward space cooperation now seemed inevitable, and theorists of superpower integration heralded a new future of U.S.-Soviet "condominium" in dealing with the world's security problems. However, space politics proved more volatile than the rules of orbital physics.

With the decline of U.S.-Soviet détente, the fragility of *civilian* space cooperation was exposed. Space science projects, particularly, became a victim of the deteriorating course of superpower relations. But this political pressure did not affect all areas of activity equally, owing to differences in interdependence across the various fields of space activity. Specifically, existing agreements to protect space access through military restraint proved relatively immune to the sharp deterioration in U.S.-Soviet political relations. Although some military research and test programs took place during this period—including Soviet and U.S. ASAT tests and the U.S. Strategic Defense Initiative—they did so largely within the limits set by the new treaties and by treaty supporters in both governments, who helped preserve the consensual understandings regarding space security. In the end, thanks to U.S. Secretary of State George Shultz and Soviet General Secretary Mikhail Gorbachev (among others), policies of unilateral advantage in space eventually lost out to those of cooperation. By the late 1980s, the drive toward military-led space security fizzled with the change of political direction in Moscow and the rising costs and technical problems of U.S. space defenses. A new appreciation for the environmental threat posed by space debris—thanks to data from the U.S. F-15 ASAT test in 1985—also emerged from this period and formed the basis for domestic policy changes in the United States. Later, these understandings would stimulate collective action within the U.N. Committee on the Peaceful Uses of Outer Space.

The Cold War's end in 1991 brought with it the political prerequisites for a variety of new cooperative U.S.-Russian projects in space science and manned spaceflight. In the military field, the warming of military-to-military relations led to a dramatic scaling back of prior weapons programs and official acceptance of the desirability of bilateral consultations before deployment of any future space defenses by either side. Commercial cooperation flourished, and the two sides essentially merged their capabilities in the space launch field, while signing up for long-term integration of their manned programs in the *International Space Station* (*ISS*). But the United States also moved far ahead of other actors in the development of space-enabled warfare[3] capabilities to increase the precision and corresponding effectiveness of its weapons, while reducing casualties and collateral damage. This trend challenged other countries and also heightened the priority to protect space assets. The development of such weapons capabilities by other states will promote either stability or instability in space, depending on how they are handled by governments.

After 2000, the new U.S. administration moved space security relations away from negotiated mechanisms and toward renewed space nationalism. While international commercial and space science cooperation continued after 2001, the official U.S. view of collective security approaches in space turned negative because of increasing sensitivity to possible space threats. The George W. Bush administration's withdrawal from the Anti-Ballistic Missile (ABM) Treaty and

---

[3] I thank Pete Hays for suggesting this terminology.

enunciation of an assertive strategy of "space control" put all other space powers on notice that the United States no longer intended to play by the old rules of military space restraint. Instead, it looked for a technical fix to a potential technical vulnerability, largely ignoring the *political* roots of the problem. Russia initially indicated that an arms race in space might follow the U.S. ABM Treaty decision. It eventually simply kept its multiple-warhead missiles (rather than destroying them under the now-bypassed START II agreement) and began to develop selective new retaliatory capabilities.

Near the end of President Bush's second term, however, it had become clear that there would be an imperfect match between the administration's early space defense rhetoric and its eventual accomplishments. Initial confusion about program priorities and the long lead times for the development of effective technologies meant that little would be accomplished besides research and limited deployment of ground-based interceptors in Alaska and California. Thus, the Bush administration would not remain in power long enough to see its vision for reshaping space security implemented, although its use of the Aegis missile defense system for ASAT purposes laid down a marker of its intention to make certain limited uses of space weapons acceptable. The responsibility for fulfilling this vision will soon pass to another administration, one not necessarily supportive of such an aggressive approach. Funding for expanded ASAT and space-based missile defense programs will have to come from a new Congress, which is unlikely to return to Republican control until at least 2010.

Outside the purview of the U.S. military, China's ASAT test and its effect in energizing the commercial community against the threat of military-created space debris has introduced new dynamics into the space security debate. By the early twenty-first century, the impact of globalization on space industry integration has blurred the boundaries separating national programs and technologies. Unlike in other environmental fields, these trends may actually improve chances for successful management given the interests of commercial entities in space being free of debris. The emergence of small commercial providers—ranging from Surrey Satellite Technology, Ltd., to SpaceX, to Virgin Galactic—means that a wave of new entrepreneurs will have to be taken into account in any attempt to "dominate" space with a military-led strategy, raising questions about implications and how space commerce (including domestic) might be affected. Such new realities put potential space "defenders"—including China—on notice that their behavior may be judged by new criteria.

As the world marks the end of the first fifty years of space history, the future course of space security remains nearly as murky as when that history began. Critical lessons have been learned about space and the challenges of operating in this simultaneously hostile and fragile environment. Moreover, while states and their leaders are free to accept or reject existing agreements, what differs about the present period is that a history of restraint has been established and, mostly, maintained. The fact that the space commons is becoming an increasingly heavily trafficked region of human activity also means that new forms of coordination will be required to allow space development to continue to move forward.

Emerging Issues in International Space Management

In order to highlight emerging trends, it is worthwhile discussing some of the nonmilitary management problems related to the space environment that are likely to affect space security. While certain environmental factors (including particularly man-made radiation and orbital space debris) have been discussed in various places in this book, an emerging set of environment-related problems require further attention. Many of these constraints did not exist during the Cold War, when fewer actors occupied critical orbits and regions of space. But they are becoming serious problems today because of an increase of space actors and the expansion of space activities, especially in the area of communications. The success or failure of international efforts to prevent conflicts over these issues will have a significant impact on the ability of states to manage international space security.

Space "Traffic Control"

A key challenge of the increase in international space activity and the concomitant growth of space debris is the need for space "traffic control" and collision avoidance. The problem stems not only from the high speed and complex orbital characteristics of spacecraft and debris but also from the fact that states and other space operators do not provide full data on their activities. Most critical in efforts at collision avoidance is data on maneuvers by individual spacecraft that change their orbital tracks. A spacecraft that maintains an established track is relatively easy to follow, but movements out of this orbit (using on-board thrusters and stored fuel supplies) to achieve a new mission, improve existing functions, or de-orbit can create hazards for other operators expecting the spacecraft to remain in its prior orbit. A collision involving active spacecraft is one of the worst nightmares of space operators, since the damage to the spacecraft themselves might be only the start of a cascade of ensuing problems. As described by a team of U.S. Air Force Academy space faculty, "A collision between two-medium sized spacecraft would result in an enormous amount of high velocity debris. The resulting cloud would expand as it orbited and greatly increase the likelihood of impacting another spacecraft. The domino effect could ruin a band of space for decades."[4] One of the more active bands of space activity is within the increasingly crowded geostationary orbital (GEO) realm at 22,300 miles in altitude, an area of space populated heavily by commercial and military operators alike. Space collision expert William Ailor puts the probability of a collision involving GEO satellites over the next decade as 1 in 10.[5] Avoiding such problems will require enhanced communication and cooperation among a variety of players in space to improve transparency and the accuracy of time-sensitive tracking data. Conditions of international hostility and high mistrust that would likely characterize an active weapons competition in space could jeopardize such cooperation. Yet, as Ailor concludes, "Collision avoidance and services to lower collision and other hazards to operating spacecraft . . . as space traffic increases . . . will require international cooperation to make it truly effective in reducing risks."[6] The current blanket exclusion of military activities from multilateral efforts through the U.N. Committee on the Peaceful Uses of Outer Space to control space debris at best only postpones a difficult negotiation and, at worst, creates a gaping loophole that could make future space traffic and debris management impossible.

Another emerging traffic issue is the possible future use of nuclear reactors in space, particularly—and unlike in the Soviet RORSAT program—for propulsion. Old concepts like NERVA are beginning to be discussed in the context of deep-space exploration plans. Increased transparency and rule making will be necessary to address legitimate concerns (and protect against possible accidents) should states decide to move forward with this technology. Alternatively, an international consensus may emerge on the need to develop safer, state-of-the-art power sources for such missions using both conventional or nuclear-electric (nonreactor) technologies.[7]

Space Surveillance

Related also to the international management of the space environment is the challenge of developing space surveillance systems capable of collating and distributing information on orbital spacecraft and debris. Currently, only

---

[4] Jerry Jon Sellers (with contributions by William J. Astore, Robert B. Giffen, and Wiley J. Larson), *Understanding Space: An Introduction to Astronautics* (New York: McGraw-Hill, 2004), p. 85.

[5] William Ailor, "Collision Avoidance and Improving Space Surveillance," *Astropolitics*, Vol. 2, No. 2 (Summer 2004), p. 110.

[6] Ibid., p. 199.

[7] On these issues, see Simon P. Worden and Jess Sponable, "Access to Space: A Strategy for the Twenty-First Century," *Astropolitics*, Vol. 4, No. 1 (Spring 2006), pp. 78–80.

the United States has an effective system in place for this purpose, and largely for military reasons. However, Washington shares this information with other states and operators to promote safe practices and reduce chances of catastrophic accidents. Before China's launch of its *Shenzhou VI* manned spacecraft in October 2005, for example, the United States provided China with updated data on existing objects in orbit to help prevent threats to this spacecraft. But new restrictions in the 2004 U.S. Defense Authorization Act require the North American Aerospace Defense Command to refrain from freely distributing information on sensitive U.S. military objects without specific permission from the Pentagon.[8] While the concerns of Pentagon planners are understandable, the criteria for restricting this data could be politicized in the future, making collision avoidance more difficult and raising international tensions. For these reasons, other states and *groups* of countries working together are trying to reduce reliance on the U.S. military for space surveillance. Russia maintains a Space Surveillance System that is nominally effective (with some apparent gaps), while the European Union, China, and Japan all have gradually expanding systems for debris and spacecraft monitoring.[9] What is unclear is how well these systems will be able to incorporate larger amounts of debris data and the degree to which these systems could meet other emerging demands, such as monitoring arms control treaties or weapons testing. To the extent that capable new systems can be developed to increase transparency, such efforts could improve chances both for collision avoidance and for catching states that violate existing norms against debris generation, including weapons tests that create new orbital hazards.

Radio Frequency and GEO Slot Allocations

Another set of emerging problems created by technological and environmental limitations of space is related to the decreasing availability of useful radio frequencies for space communications and the correspondingly limited availability of slots in GEO for communications satellites. Currently, the International Telecommunications Union (ITU)—a body of the United Nations—is responsible for preventing such conflicts through an allocation system involving a complex set of technical and political factors. In practice, the system has not worked perfectly, because states not capable of using various slots or frequencies have sometimes "squatted" in allocations granted for geographical or political reasons rather than capability for actual use. Nevertheless, the pressure

---

[8] Simon Collard-Wexler, Jessy Cowan-Sharp, Sarah Estabrooks, Thomas Graham Jr., Robert Lawson, and William Marshall, *Space Security 2004* (Toronto: Northview Press, 2005), p. 18.

[9] Ibid., p. 17.

for change has been managed by the ability of states to make deals within the system and the fact that demand has not yet overwhelmed supply. As these conditions change, however, there is likely to be increasing pressure for new forms of management. The demand for specific frequencies is influenced by their capacity to carry information, with certain high-frequency bands (Ultra-High Frequency, X-band, and K-band) currently being claimed exclusively by U.S. military users.[10] Growing numbers of GEO satellites are putting pressure on the availability of usable frequencies, with the added complication that the U.S. military (the largest single user of bandwidth) turns frequently to commercial satellites for surge capacity during high-use periods, such as during the invasion of Iraq in the spring of 2003. Technological improvements have assisted in the management of conflicts up to the present by giving receivers a greater ability to parse bandwidths without interference. The U.S. development of laser communications may be another way of overcoming emerging spectrum crowding through the pulse delivery of large amounts of information in specific frequencies,[11] but technical gaps could keep some users from being able to exploit such new capabilities. Moreover, transparency remains a major problem due to the sensitivity of military (and some commercial) communications. Military pressures could increase in space if particular states believed that their national security interests "require" the opening of certain frequencies for their use, including perhaps a revised allocation of satellite slots in GEO. It is not clear how these conflicts would be resolved.

Alternative Scenarios for Future
Space Security Relations

In examining the impact of these changes on future space security, any serious analysis needs to take into account an array of factors: military developments, threat perceptions, commercial pressures, environmental issues, globalization trends, and national and international political pressures. Given the complexity of these factors, it would be absurd to argue that one can offer a deterministic prediction or that any single set of options is comprehensive. Still, it is important to try to bring order to this apparent chaos.

To consider the dominant trends in contemporary thinking about future space security, we should return to the four schools of space analysis and re-examine their predictions about the future: (1) the deployment of space defenses and efforts by individual countries to assert "space control" (space na-

[10] Ibid., p. 9.
[11] Ibid., p. 10.

tionalism); (2) the slower emergence of space weapons and possible presence of multiple actors with limited weapons in orbit (technological determinism); (3) the loose coupling of national and international goals for safe access to space, with new commercial and other nonstate actors gradually becoming major actors and joining international efforts to make and establish space "rules" (social interactionism); and (4) a new form of consensus about space security and the empowerment of international organizations and new treaties to govern space (global institutionalism). Each of these options is an ideal type, meaning that it is not likely to be seen in its purest form. Yet these four basic directions express the main alternatives extant in the current international space debate and offer *different* perspectives on the tools for possible future space management, the role of cooperation, and the prevalence of different types of environmental learning.

Space Nationalism

Focusing on the notion of future "space control," recent space nationalists have predicted that the natural consequence of state competition will be the dominance of one state over others.[12] Its more extreme advocates see a kind of global showdown taking place in space between the forces of evil and good. For this reason, Steven Lambakis argues that current U.S. policies of space restraint need to be reversed, before it is too late: "History supports the belief that hostile foreign governments and nongovernmental entities will endeavor to impair America's space capabilities or use satellites to their own advantage."[13] Everett Dolman goes further, lamenting the lack of the necessary competitive spirit among American decision makers, saying, "That the space race is over and the Space Age is in decay seems dismally obvious."[14] Nevertheless, this school's argument makes the assumption that competition will dominate the future in space rather than trends toward the diminution of conflict, expanded cooperation, and eventual global integration.

Dolman and other space nationalists expect that other countries will challenge the United States in seeking to win this struggle. Notably, they do not see a U.S. victory in this conflict as predetermined, thus leading Dolman to argue for an aggressive U.S. strategy to "control" space access militarily and, in the commercial sector, for freeing up U.S. businesses to claim and take over the

[12] See, for example, Everett C. Dolman, *Astropolitik: Classical Geopolitics in the Space Age* (London: Frank Cass, 2002).

[13] Steven Lambakis, *On the Edge of the Earth: The Future of American Space Power* (Lexington: University Press of Kentucky, 2001), p. 1.

[14] Dolman, *Astropolitik*, p. 182.

Moon by withdrawing from the Outer Space Treaty.[15] His aim is to let loose all forces—military, economic, and political—that might facilitate U.S. success in this final, fatal struggle for global dominance. As Dolman predicts, "Who controls low-Earth orbit controls near-Earth space. Who controls near-Earth space dominates Terra [Earth and its atmosphere]. Who dominates Terra determines the destiny of humankind."[16] But, for Dolman, war itself is not necessarily inevitable, as he believes that weaker states may be cowed into submission. For others in the space nationalist school, actual conflict is seen as more likely. As Lambakis says, "We can state with certainty that we will be challenged in space ... simply because it makes military sense to do so."[17] Dean Cheng, a more moderate analyst within this school, suggests that China's development of space weapons will occur "because it is consistent with what the Chinese leadership perceives their national interests to require."[18] But what about Indian weapons, or new Russian systems? The level and diffusion of these technologies, as well as a decision to switch them to hair-trigger alert for maximum effectiveness, could be highly risky for all.

As a prescription, most space nationalists support a significant deployment of space defenses and offensive systems. There are presumably advocates for similar stances in China, India, Russia, and perhaps within other space powers. The Independent Working Group's solution—discussed in the previous chapter—is the U.S. deployment of one thousand Brilliant Pebbles interceptors, which Dolman describes as "the optimum solution for military planners."[19] Indeed, such an action by the United States could result in short-term U.S. space hegemony.

In analyzing this school's predictions, we can observe that history is not kind to military hegemons. Such policies lead to loss of reputation and increased efforts by enemies to subvert such strategies, particularly in ways that are hard to control. As John Lewis Gaddis argues, the notion of military hegemony is "a nineteenth-century vision that plays badly at the beginning of the twenty-first."[20] In space, more specifically, the environment is vulnerable to ruination (such as from EMP or debris) by the application of asymmetric power more than in any other past frontier. Thus, since a cycle of action-reaction arming is the virtually unavoidable conclusion once space is weaponized by one state,

---

[15] Ibid., p. 165.

[16] Ibid., p. 8.

[17] Lambakis, *On the Edge of the Earth*, p. 137.

[18] Dean Cheng, "China's A-Sat Test: Of Interceptors and Inkblots," *Space News*, February 12, 2007, p. 19.

[19] Everett Carl Dolman, "China's 'Shot Across the Bow'," *Space News*, February 12, 2007, p. 18.

[20] John Lewis Gaddis, *Surprise, Security, and the American Experience* (Cambridge, Mass.: Harvard University Press, 2004), p. 67.

there are bound to be serious problems created for safe access. As we look at historical cases, we can also observe that attempted discriminatory security regimes regarding weapons have the disadvantage of being hard to maintain. Evidence from the Treaty on the Nonproliferation of Nuclear Weapons shows that when a few states are allowed to have weapons and refuse to give them up, others will eventually seek to emulate them. This means that the logical conclusion of such a strategy is a U.S. commitment to undertake a ceaseless preventive war against possible military space rivals.

Can such a struggle be supported? In contrast to the predictions of this school, it is difficult to see how the visceral, ideologically driven competition of the Cold War will be repeated in space, at least among major countries active today. Even a rising China, given its scarcely communist system, close integration with the U.S. capitalist economy, and lack of global territorial ambitions, seems to offer a pale ideological substitute for the Soviet Union—not that China could not stage a significant space race. Iran's recent test of a space launch vehicle could presage the emergence of a more potent ideological foe, but Tehran's space capabilities are likely to remain far behind those of the United States for decades to come, suggesting low motivation for any kind of sustained U.S. space domination strategy.

Similarly, the assumption that past weapons restraint "hurt" U.S. space policy seems to belie the tremendous advantages the United States enjoys today in space, albeit coupled with the vulnerabilities that all users in space face with conditions of environmental interdependence. Could these conditions have been avoided by military means? It seems doubtful. Moreover, as the counterfactual space history offered in Chapter Two argues, the failure of Washington and Moscow to constrain space weapons during the Cold War could have led to far *worse* results in terms of U.S. space security. Just as other states would undoubtedly have objected to the cluttering up of near-Earth space by the proliferation of U.S. military platforms and debris from weapons test programs, for the United States to fail to try diplomatic means for halting deployments by others would at best be irresponsible and at worst very dangerous.

The space nationalist school might counter such an argument by saying that a future "Pearl Harbor" in space would quickly change the perception of restraint-based models. Fortunately, the U.S. military is already moving to harden satellites and reduce dependency on individual spacecraft—the kind of threat that China's ASAT now represents. It is also gradually moving in the direction of reducing its reliance on single, very expensive, multiuse satellites and diversifying responsibilities across a larger number of cheaper, smaller, and therefore less-vulnerable spacecraft. Other ideas for so-called nonoffensive defenses are taking root as well, thanks to their likely greater reliability and lower cost than weapons. These preventive measures make the concept of an across-the-board "Pearl Harbor" in space hard to visualize short of a massive, multiple-missile debris or electromagnetic pulse (EMP) attack on low–Earth orbit, which could happen only at great cost to China or another aggressor and would be highly transparent in its preparation and launch, taking days to put into place and putting launch facilities at great risk of destruction once the first shot was fired. (Of course, such an attack could have happened at any time during the Cold War as well, but it did not.) Moreover, such an attack would also affect all actors as much as the United States and would likely reap scorn for its perpetrator, as happened with China after its ASAT test. As military strategist Barry Posen observes regarding the next logical step, namely, stationing weapons in space, "Ultimately the United States has more to lose than to gain from such a competition."[21] Such restraint worked to both sides' advantage during the Cold War and after 1991.

The prior existence of core U.S. defensive systems on land, at sea, and in the air and the multiplicity of delivery systems that use space, including the advanced precision-guided conventional weapons that have revolutionized the way the United States conducts wars, make space critical in a support and force enhancement role. Weapons testing and deployment in space in a multilateral context may jeopardize these advantages. It may be far better to redress space vulnerabilities through non-space means. As military expert Norman Friedman observes, "All of this suggests that the only really effective way to conduct warfare against space assets would be to attack their ground-based control and downlink facilities."[22]

Finally, the question of who will constitute the future actors in space matters. Although the space nationalist school accepts and even endorses international cooperation, it rejects the notion of possible changes in space security decision making stemming from the influence of new nonstate actors, including those in the commercial sector. This prediction may be shortsighted.

Technological Determinism

The predictions of the second school of space policy analysis are similar to the space nationalists in suggesting that weapons are inevitable, but the parameters they portray are considerably more measured and their implications far less apocalyptic because they do not assume virulent political hostility. As

[21] Barry R. Posen, "Command of the Commons: The Military Foundation of U.S. Hegemony," *International Security*, Vol. 28, No. 1 (Summer 2003), p. 20.

[22] Norman Friedman, *Seapower and Space: From the Dawn of the Missile Age to Net-Centric Warfare* (Annapolis, Md.: Naval Institute Press, 2000), p. 312.

Michael O'Hanlon writes, "Even if weaponization is indeed inevitable . . . when and how it happens may matter a great deal."[23] Harkening back to the Cold War experience, O'Hanlon suggests that a variety of systems still make no sense from a military-technical point of view, particularly space-to-Earth weapons: "In addition to being politically very provocative, they offer few benefits to a global military power already capable of rapid intercontinental strike. The technologies within reach, such as tungsten rods or a common aero vehicle . . . do not warrant advanced development and deployment."[24] Within the U.S. military some space experts have noticed how power is moving away from states and instead to nonstate actors. As Marine Lieutenant Colonel Maurice Perdomo points out, "Washington wants to control access to space, but it is getting harder to do so when all the nations of the world are clamoring to use commercial space services that are available from multinational corporations and consortiums."[25] These potential problems lead members of this school to advocate a "go slow" approach regarding space defenses. Using an analogy drawn from the study of emerging markets, Air Force Lieutenant Colonel Wesley Halman, for example, argues that the United States would be smarter to employ a "fast-following" strategy regarding space weapons, somewhat akin to the Soviet approach in copying U.S. nuclear weapons in the late 1940s and 1950s.[26] That is, he recommends letting other states develop the initial technologies first and—if they do—learning from their mistakes in testing, research, and development and then outdoing them with second-generation weapons.[27]

While technological determinists argue that weapons are unavoidable, the logic of their position also contends that countervailing technologies are possible too. As Peter Hays points out, "Hardening of LEO satellites to withstand HAND [High-Altitude Nuclear Detonations]-induced radiation stands out as one of the least costly and potentially effective means of addressing this threat."[28] Such defensive and nonaggressive measures may also strengthen deterrence

[23] Michael E. O'Hanlon, *Neither Star Wars nor Sanctuary: Constraining the Military Uses of Space* (Washington, D.C.: Brookings Institution, 2004), p. 19.

[24] Ibid., p. 28.

[25] See Lt. Col. (USMC) Maurice Perdomo, "United States National Space Security Policy and the Strategic Issues for DOD Space Control," Master's thesis, U.S. Army War College, March 2005, available at <http://www.strategicstudiesinstitute.army.mil/pdffiles/ksil8.pdf > (accessed August 25, 2006).

[26] On the Soviet effort, see Richard Rhodes, *Dark Sun: The Making of the Hydrogen Bomb* (New York: Touchstone, 1995).

[27] Lt. Col. (USAF) Wesley Hallman, "A Fast-Following Space Control Strategy," *Astropolitics*, Vol. 3, No. 1 (Spring 2005).

[28] Lt. Col. (USAF) Peter L. Hays, *United States Military Space: Into the Twenty-First Century*, Institute for National Security Studies, Occasional Paper No. 42 (Colorado Springs: Institute for National Security Studies, U.S. Air Force Academy, September 2002), p. 124.

in space. Similarly, efforts to develop other types of nonoffensive defensives (decoys, shielding, maneuver) could also be described as inevitable once states move down this road.[29] Such measures could mitigate chances of war and make the work of aggressors much more difficult. Critical to this process, however, will be the nature and extent of offensive deployments, norms against possible use, and the effectiveness of simultaneously advancing defensive or avoidance technologies.

Another analyst who sees space weapons as inevitable, particularly after the Chinese test, suggests that they be focused in areas where they will do the least damage to other uses of space. The Hoover Institution's Bruce Berkowitz argues that the United States should create guidelines for space similar to the Incidents at Sea Agreement signed between Washington and Moscow in 1972.[30] In his view, building space weapons is not especially difficult given available technology, but that does not excuse a policy of sitting on one's hands and not trying to prevent debris problems and the risks they pose. For this reason, he suggests a similar technological analogy for space: "Minimizing space debris is just like protecting intellectual property."[31] Berkowitz concludes, "We'll never demilitarize space. But we can make it a safer place to work."[32] These ideas and the prospect of preventing space weapons altogether through rule-based mechanisms are developed even further by the next school.

Social Interactionism

The third main school of thought on space's future focuses on the social framework created by the interdependence among multiple actors in the space environment. In contrast to the previous two schools, however, it views the deployment of weapons in space and even the future testing of Earth-to-space weapons as preventable through "rules of the road" and other arrangements. One relevant model is a multiple-play and communicative collective goods environment where actors have an overriding self-interest in protecting safe access to space for activities of higher value than the deployment of weapons. For that reason, its proponents argue, leading space actors should be capable

---

[29] For a detailed technical study on the feasibility of these systems, see Phillip J. Baines, "Prospects for 'Non-Offensive' Defenses in Space," in James Clay Moltz, ed., *New Challenges in Missile Proliferation, Missile Defense, and Space Security*, CNS Occasional Paper No. 12 (Monterey: Calif.: Center for Nonproliferation Studies, Monterey Institute of International Studies, July 2003), online at <http://www.cns.miis.edu/pubs/opapers/op12/index.htm> (accessed October 11, 2007).

[30] Bruce Berkowitz, "Debris in Space," *Wall Street Journal*, January 25, 2007, p. A18.

[31] Ibid.

[32] Ibid.

of forming agreements to engage in similar mutual restraint relationships as existed during the Cold War. In this context, China's ASAT test might even have had a salutary role. As Michael Krepon and Michael Katz-Hyman argue, "Now all stakeholders in space are keenly aware that debris constitutes an indiscriminate, lethal hazard,"[33] thus promoting learning.

In current conditions involving increasing numbers of commercial actors, this school argues, the behavior of governments will be influenced and constrained by other national interests, such as commercial aims, consumer safety, and attention to market stability. For these reasons, the social interactionists argue that an arms race in space is unlikely and that states will instead become more conservative about testing, deploying, and using space weapons in circumstances where they must consider the impact of these activities on commercial interests, rather than only because they might destabilize arms control. This kind of paradigm shift toward "soft" power definitions of security and commercial leadership of space activity may already be happening and can be expected to shape new directions in space security.[34] As Krepon and Katz-Hyman note, "China's space program is also intimately connected to its economic goals and status consciousness. Beijing's status has been damaged by creating an enduring hazard to space operations in low Earth orbit."[35] From a broader perspective, Intelsat's Chief Executive Officer David McGlade argues, "The future preservation of the space environment will rely on every nation's appreciation that its own self-interest lies in preserving this precious common good."[36] In this context, simple cost-benefit analysis—even absent complex learning—should steer states away from such harmful outcomes as testing and deploying weapons. As strategic analyst John Pike prescribes colorfully on this issue, "People who live in glass houses should not organize rock-throwing contests."[37]

The social element of space relations is also relevant. For these reasons, Krepon's prescription for space is "assurance" rather than dominance.[38] His analy-

---

[33] Michael Krepon and Michael Katz-Hyman, "An Arms Race in Space Isn't the Problem," *Space News*, February 12, 2007, p. 18.

[34] On this concept, see Joseph S. Nye, *Soft Power: The Means to Success in World Politics* (New York: PublicAffairs, 2004). On its application to space, see Joan Johnson-Freese, *Space as a Strategic Asset* (New York: Columbia University Press, 2007), p. 24.

[35] Krepon and Katz-Hyman, "An Arms Race in Space."

[36] David McGlade, "Preserving the Orbital Environment," *Space News*, February 19, 2007, p. 27.

[37] Pike quote in Dawn Stover, "Battlefield Space Military Hardware Has Orbited Earth for Decades, but No Actual Weapons Have Ever Been Deployed in Space. That May Change Soon—And It May Launch a Major Arms Race," *Popular Science*, Vol. 267, No. 5 (November 2005), p. 50.

[38] Michael Krepon (with Christopher Clary), *Space Assurance or Space Dominance? The Case Against Weaponizing Space* (Washington, D.C.: Henry L. Stimson Center, 2003), p. 88.

sis of what is of value in space amplifies themes from an increasingly globalized world economy, making space a "transmission belt for global commerce and a realm of exploration for the benefit of all of humanity."[39] From this perspective, it is also increasingly clear that new commercial entities will resist harmful interference from all countries' space weapons, particularly if their testing puts at risk next-generation satellites in low–Earth orbit, raises launch costs, and restricts access by tourists to favorable locations from which to view the Earth.

The logic of this school's argument suggests that commercial actors will eventually need to be brought in as players in space security decision making. The question these analysts raise is whether governments will embrace this process or seek to deny it. If the latter approach is taken, the results could be dire. French space analyst Alain Dupas notes that "bold moves in the purely military space sector—such as the development of debris-creating anti-satellite weapons or other space-based offensive systems—could compromise the stability of the still fragile commercial space sector."[40] For this reason, Dupas calls instead for "public-private partnerships" in space development as a desirable model for the United States and Europe, with an emphasis on limiting harmful military activities.[41]

U.S. analyst Charles Peña makes a similar point, stating that "one thing is abundantly clear: the U.S. military greatly benefits from using commercial space systems."[42] Therefore, he argues that Washington should be very careful about harming commercial vitality and entrepreneurship. Peña concludes that "commercial space . . . should be the driving force of space and shape space policy," not military interests.[43] Yet, unlike global institutionalists, his focus is on creating a laissez-faire model for international space development rather than an international organization led by governments.

One example of an increasingly market-oriented approach even by the U.S. government is NASA's recent decision to turn to small, private launch-services technology.[44] Such a private-sector emphasis is now percolating through NASA's regional centers, which are reaching out to industry for innovative

---

[39] Ibid., p. 30.

[40] Alain Dupas, "Commercial-Led Options," in James Clay Moltz, ed., *Future Security in Space: Commercial, Military and Arms Control Trade-Offs*, Occasional Paper No. 10 (Monterey, Calif.: Center for Nonproliferation Studies, Monterey Institute of International Studies, July 2002), p. 60, available online at <http://www.cns.miis.edu/pubs/opapers/op10/op10.pdf> (accessed August 25, 2006).

[41] Ibid., pp. 59–60.

[42] Charles V. Peña, "U.S. Commercial Space Programs: Future Priorities and Implications for National Security," in Moltz, *Future Security in Space*, p. 8.

[43] Ibid., p. 10.

[44] On some of these projects, see Worden and Sponable, "Access to Space."

approaches to stimulate "bottom-up" pressures to expand the U.S. presence and public participation in space.[45] Such trends may gather steam once these launchers, new spaceports, and more venture-capital funding are available. Alternatively, if private demand for space services, new forms of energy, tourism, and construction contracts becomes self-sustaining, government support may be superfluous, except perhaps for deep-space exploratory missions (where the financial payoffs are less clear). Even in this area, however, the possible prestige value of deep-space exploration cannot be completely discounted from a commercial point of view.[46]

A possible drawback to this informal, more market-led space security approach is the risk of inadequate attention to potential problems: traffic/debris control, frequency allocation, and possible Moon resource conflicts. While commercial actors have an interest in these areas, governments may be required to force companies to pay for management costs and to abide by restrictions. Similarly, this approach leaves open the question of who will play the role of "enforcer" if agreements are ignored, in the absence of a space hegemon or an empowered international organization. The challenge seems to be one of providing at least minimum costs for enforcement mechanisms while still allowing commercial actors the freedom to develop space industries and services to their full potential.

For these reasons, the concept of "developing" space (rather than defending or securing it) provides a powerful theme for those who see space primarily from the social context of repeated international interactions to protect and improve a kind of economic conveyor belt. Progress in space toward the solution of problems on Earth by entrepreneurs, public-private coalitions, or transnational groups could stimulate creative approaches to space security. The transformative nature of these contacts—with or without formal organizations—could have a powerful impact on reducing incentives for military space conflict.

Global Institutionalism

The fourth school argues that the fundamentally transborder nature of space activity requires *international* responses and the formation of more powerful institutions and treaties. By and large, states continue to be the main actors,

---

[45] Author's interviews with staff members at NASA-Ames, August 10, 2006.

[46] The tremendous popularity of NASA's Mars Rover missions, for example, could have easily been tapped for commercial value, even if only via advertising. While this notion might seem like shameless exploitation of space (making Mars spacecraft a new form of billboard), it is likely to be an unavoidable—and not especially harmful—part of the next era in space exploration.

although the emergence of important commercial players, intergovernmental consortia, and other nongovernmental entities (such as universities) plays a larger role here than in the space nationalism school. Indeed, this neoliberal institutionalist model does not reject the notion that current American global influence could not be used to create a new legal and moral order in space, rather than a military-dominated realm.

One of the more influential advocates of this perspective has been Bruce DeBlois. In examining the history of military space activity from his perspective as a military intelligence officer in the late 1990s, then-Air Force Lieutenant Colonel DeBlois spelled out an argument for continuing positive U.S. leadership in space, saying, "Forty years of cold war history show a successful pattern of US policy aimed at supporting space as a sanctuary. The reason is that we have more to lose if space is weaponized."[47] The vision that DeBlois outlined in his study portrays space more as a resource to be preserved for use by all rather than as a territory to be seized and protected.

Not surprisingly, this approach rejects the notion of unilateralism in space and the concepts of space dominance or control as even potentially successful strategies. The University of Maryland's Nancy Gallagher notes the need to maintain the Outer Space Treaty and reach "one or more supplemental accords."[48] She suggests starting with bans on destruction of peaceful space assets or interference with those conducting "legitimate" activities. She would follow with an agreement prohibiting space-based anti-ballistic missile defenses because of their possible utility for threatening satellites in geostationary orbits.[49] Theresa Hitchens calls for a prohibition on space testing of debris-causing weapons to stem the harmful effects on other space activity and their tendency to lead to an arms race.[50] Critics would argue that such agreements could unduly restrict U.S. freedom of action. Gallagher counters, "Space policy is but one of many security problems that illustrate the fallacies of assuming that the ascendance of the United States as the sole information-age superpower offers perpetual military dominance . . . regardless of other countries' interests or concerns."[51]

[47] Lt. Col. (USAF) Bruce DeBlois, "Space Sanctuary: A Viable National Strategy," *Aerospace Power Journal*, Vol. 12, No. 4 (Winter 1998), p. 44, online at <http://www.airpower.maxwell.af.mil/airchronicles/apj/apj98/win98/deblois.pdf> (accessed August 24, 2006).

[48] Nancy Gallagher, "Towards a Reconsideration of the Rules for Space Security," in John M. Logsdon and Audrey M. Schaffer, eds., *Perspectives on Space Security* (Washington, D.C.: Space Policy Institute, George Washington University, December 2005).

[49] Ibid., p. 36.

[50] Theresa Hitchens, "Safeguarding Space: Building Cooperative Norms to Dampen Negative Trends," *Disarmament Diplomacy*, No. 81 (Winter 2005), on the Acronym Institute Web site at <http://www.acronym.org.uk/dd/dd81/81th.htm> (accessed August 25, 2006).

[51] Ibid., p. 34.

Most European space security analysts share this perspective.[52] No European Union country today is actively considering space weapons, and statements by European government leaders routinely emphasize the importance of international cooperation. NATO countries in Europe have been unanimous in rejecting U.S. opposition to the yearly U.N. resolutions on the Prevention of an Arms Race in Outer Space. As former French and U.N. space official Gerard Brachet observes, "We need to collectively reflect on possible measures to improve the safety of space operations, measures that we can all agree to and would not adversely affect the economics of space operations."[53] These are the priorities that Europe hopes to bring into play in space negotiations and international space management.

A key difference between the space nationalist and global institutionalist space policy schools can be seen in most vivid relief in the distinction between Dolman's notion of U.S. moral *superiority* and DeBlois's of American moral *leadership* in space. In the case of the global institutionalists, the United States is best served by using the attractiveness of its ideas to build *voluntary* support for its positions and associated consensual institutions. As DeBlois argues, "The idea of putting weapons in space to dominate the globe is simply not compatible with who we are and what we represent as Americans."[54]

Two possible disadvantages of the global institutionalist school are associated with enforcement costs and the risks of free riders. That is, while states often find it easy to put up funds for national defense or domestic public works programs, they are often less likely to fund international organizations, such as those that may be required for the management of space. European countries, for example, will not be able to "free ride" for their space security as in the past. As the European Commission's Eneko Landaburu explains, "The EU's citizens should be aware that they will never get the ability to shape world events that most of them say they want unless they are prepared to pay the extra cost, either in financial terms, or in terms of institutional and political reforms that will give them the kind of hard power enabling the EU to act entirely independent of the US security umbrella."[55] This means developing either EU-wide or other international capabilities to monitor space traffic and debris, verify arms

---

[52] See, for example, Gerard Brachet, "Collective Security in Space: A Key Factor for Sustainable Long-Term Use of Space," and Detlev Wolter, "Common Security in Outer Space and International Law: An International Perspective," both in John M. Logsdon, James Clay Moltz, and Emma S. Hinds, eds., *Collective Security in Space: European Perspectives* (Washington, D.C.: Space Policy Institute, George Washington University, January 2007).

[53] Brachet, "Collective Security in Space," p. 8.

[54] DeBlois, "Space Sanctuary," p. 46.

[55] Eneko Landaburu, "Hard Facts About Europe's Soft Power," *Europe's World*, Vol. 1, No. 3 (Summer 2006), p. 33.

control agreements, and operate human spaceflight activities. None of these capabilities will come cheaply.

A third possible disadvantage of the global institutionalist approach is its currently limited political support in the largest and most active space-faring country: the United States. Indeed, even before the George W. Bush administration, President Bill Clinton mostly followed a state-centric policy of opposing limits on U.S. military freedom in space through new treaties. Although China's ASAT test and the new Democratic leadership in Congress may lead to some new U.S. thinking, skepticism about formal international controls remains significant in the United States. Enhanced transparency by foreign space powers and expanded U.S. cooperation with potential space rivals could mitigate this opposition over time, although again, such efforts with critical countries like China are still in their infancy. The U.S. commercial space industry would be the one actor capable of carrying out such a transformation in the minds of members of Congress, particularly if it could exhibit both the economic benefits and the limited security risk of expanding commercial ties with China in space, at least in certain areas.

Finally, in light of the asymmetries involved in commercial space activities to date, we might ask whether overly assertive global institutionalist approaches might *harm* international relations in space. For example, the attempts by non-space powers to regulate potential great-power mining on the Moon through the 1979 United Nations Moon Treaty have threatened to force the will of a technologically underdeveloped majority on a technologically advanced minority.[56] Such strategies are unlikely to succeed in the future. But the concern of these disadvantaged states may simply suggest the need for improved international consultations and perhaps the development of a compromise system based on limited user fees within the context of the Outer Space Treaty. On balance, it is unlikely to prevent the Moon's development or lead inevitably to conflict.

Current Trends and Possible Signposts Ahead

At the end of the first fifty years of space history, many questions related to future space security remain undecided. After the first experiments of the two superpowers with military means of achieving space security, the historical evidence from 1962 to the present shows surprising restraint compared to

---

[56] By attempting to claim that these resources should benefit all states, the signatory countries sought to stake a claim to the potentially vast mineral resources in space that are as yet untapped. This management system, however, has not come to fruition, because the major space powers have refused to sign the treaty. For a recent analysis of these issues, see Andrew Brearley, "Mining the Moon: Owning the Night Sky?" *Astropolitics*, Vol. 4, No. 1 (Spring 2006).

the record of nuclear and conventional arms deployments and actual interstate conflict in other environments. This did not mean that the states involved in space have altered their views of one another or become fundamentally cooperative. Instead, it signified a realization that no better means to achieve security in space had been devised, and that protecting this fragile environment for civilian, commercial, and passive military uses served their interests better. But this required reining in a series of planned military technologies, which proved neither easy nor inevitable.

Despite the 2002 U.S. withdrawal from the ABM Treaty and the 2006 U.S. National Space Policy, no irreversible decisions have been made regarding the deployment of space defenses. Thus, both directions for space security—unilateral and collective—remain very much in play. Returning to the introduction of this book, the question is where the line on military restraint will be drawn. As noted above, this struggle is becoming less of a dialogue and more of a global conversation, with emerging international, commercial, and nongovernmental actors influencing a realm traditionally dominated by only two states.

Herbert York's concept of military-led "technological exuberance," raised in Chapter One and seen throughout the history of space activity, remains relevant to any discussion of space security. But there are also purely civilian-oriented space enthusiasts struggling to stimulate and "capture" the space industrial base. Similarly, members of Congress are more interested in protecting high-wage jobs than in whether funds flow only into military technologies. The bulk of new spending on space may instead begin to come from small and moderate-sized companies and private entrepreneurs, transforming the old "iron triangle" (Congress, the Pentagon, and large aerospace firms) of space finance. In this sense, today's public and private investments in space infrastructure and civilian technologies—spaceports, launchers, manned spacecraft, satellites, and tracking networks—are likely to become the most reliable measure of tomorrow's space power, more than which countries have developed extensive space defenses. The latter technologies are most likely to become expensive sunk costs, while the former investments could stimulate more usable forms of power through innovative technological capabilities and possible new industries.

As to what makes states "secure" in space, those militaries that are able to add capacity across the spectrum of passive, military-support technologies while diversifying their assets across low-cost and replaceable platforms are likely to be in the best shape to prevent future threats to their spacecraft. The surrounding context will matter. This means that the strength or weakness of international mechanisms for space management and the level of transparency and monitoring will make a difference in the *sustainability* of nonweapons approaches.

By restraining their competitive impulses during the Cold War, the United States and the Soviet Union raised the chances that future actors—with fewer competitive geopolitical ambitions—might be able to move beyond traditional military-dominated scenarios in space, even in the face of *technological* possibilities for new weapons. If the normative goals of states eventually become rooted in the desirability of space's "development," we might look back on the mutual restrictions placed on space weapons by the United States and the Soviet Union in the first three decades of space activity and say that they served a useful purpose in helping to safeguard this new frontier for future generations.

Given the historical record recounted here, any prediction of future collective learning by states regarding space must be viewed as both tentative and contingent. Just as certain trigger events—like the 1962 Starfish Prime nuclear test, 1967 U.S.-Soviet space accidents, and the 1985 *Solwind* test—helped motivate the two countries into cooperating in space, trends could move in the other direction. A hostile attack on the ISS or its successor, a state-sponsored attempt to shoot down a *foreign* spacecraft in orbit, or repeated instances of action-reaction ASAT tests by various countries generating large amounts of debris might move the international community in one direction or the other. The key factors in determining how states will react will be both national and international perspectives on how best to move forward with securing the safe use of space. History shows that lessons learned by officials in one period can be unlearned by their successors. Still, the lowered degree of political hostility among most main space-faring entities since the end of the Cold War has already created new forms of space cooperation in both the military and nonmilitary areas. It has also promoted growing global recognition of shared environmental problems in space, such as orbital debris. The possibility, then, is that the potentially harmful effects of an expanded number of actors in space may be managed successfully, particularly if the political and technological changes accompanying globalization continue to integrate the space activities of various states, causing them to alter or limit their past competitive goals.

Conclusion

Despite the predictions of a variety of analysts, the future course of international relations in space remains unwritten. For this reason, it is important to consider carefully the four alternative futures outlined here, which are based on very different organizing principles: international hostility (space nationalism), selective military restraint (technological determinism), commercial coopera-tion (social interactionism), and international management (global institu-tionalism).

In regard to the issue of space weapons (or defenses), it is easy for analysts using purely deductive analysis to develop threatening scenarios about space warfare. But such analysis cannot categorically state either where such threats will arise or if they will arise. Similarly, analysts focusing on the preventability of military space threats and the promise of international management cannot prove that threats will not emerge or that cooperative international approaches to stop them will be effective. The main difference, however, is that moving forward with a weapons-based approach toward security is more difficult to reverse than an approach based on international cooperation. The latter is also less costly if states can successfully reduce their vulnerability to possible attacks and increase the sanctions on those who might be tempted to violate noninter-ference and antidebris norms.

Effective U.S. decision making will require that the Congress and the presi-dent exercise their constitutional roles in bringing to the table the widest range of possible political, economic, security, and even moral problems the country is likely to face in space and try to balance the military risks against the pos-sibility of damaging space, or the U.S. reputation. Such balance and restraint have not always been possible in space decision making. Both the actors and the international setting will affect these decisions, as will the inclination of key players to invest either in collective security or, alternatively, national defenses for space. Since the end of the U.S.-Soviet Cold War, the multipolar dynam-ics of military competition—if unleashed—could be more difficult to manage. Contemporary international relations in the context of globalization suggest, however, that collaborative outcomes (in whatever form) will be hard to avoid. As environmental analyst Ronnie Lipschutz observes, "Rules, laws, norms, practices, and customs are the articulation of this scheme of [international] self-management from which there appears to be no escape."[57]

Given the ability of multiple actors to damage and even render unusable critical regions of the space environment such as low–Earth orbit, this book has highlighted the importance of various forms of military restraint in space. In this sense, it has focused on such unique space-related problems as man-made EMP radiation and orbital debris, issues over which the United States has reached out to the international community for cooperation at critical junc-tures over the past fifty years. To date, the United States has left out efforts to control military-generated debris, preferring to maintain its independence. But

---

[57] Ronnie D. Lipschutz, *Global Environmental Politics: Power, Perspectives, and Practice* (Wash-ington, D.C.: Congressional Quarterly Press, 2004), p. 237.

China's recent behavior shows that it too is capable of unilateralism in space. This poses a problem the United States cannot ignore.

Historian John Lewis Gaddis comments, "The history of American grand strategy during the Cold War is remarkable for the *infrequency* with which the United States acted unilaterally."[58] In regard to space, whose environmental conditions affect all equally, these words seem especially relevant. By pursuing military-led security in space, the United States, China, and other potential unilateralists risk weakening past norms of space restraint and providing incentives for others to do the same, including perhaps by damaging the orbital environment with pellets, sand, or other asymmetric "collective bads." Such outcomes will be more costly and self-defeating for all countries over the long run than any gains to be achieved by short-sighted attempts at space hegemony. Future power in space, if trends outlined in the previous section are true, is likely to be judged by the weight of transnational partnerships, not by individual countries trying to operate alone in disregard of the security interests of others.

With the constantly expanding knowledge base on space, its unique environmental characteristics, and ongoing processes of space industry integration, current factors seem to favor continued national restraint over unlimited military competition. As noted above, one reason is that individual entrepreneurs, transnational companies, and international space consortia are likely to play more significant roles than individual states in what William Burrows has called the "second" space age.[59] These trends could steer outcomes away from the predictions of nationalist and technologically determinist arguments about space's weaponization. If these new actors do become more influential in space, the "demand" for space weapons may be reduced while the perceived need for stronger restrictions against such systems, which might interfere with new forms of space commerce, tourism, and exploration, is likely to rise.

But this bright future is still contingent on human initiative and the domestic and international willingness to support multilateral cooperation, which cannot be taken for granted in either the United States or other countries, and will require investments of funds, technology, and political capital. Mistrust, hypernationalism, exclusionist religious beliefs, arms races (which may or may not involve the United States), or simply unexpected global crises could derail progress in this integrative direction and result in hostile space dynamics. Trends toward knee-jerk weapons-based responses to security threats are pow-

---

[58] Gaddis, *Surprise, Security, and the American Experience*, p. 58.

[59] William E. Burrows, *This New Ocean: The Story of the First Space Age* (New York: Random House, 1998).

erfully ingrained in the strategic cultures of many societies. As a former defense department scientist involved in U.S.-Soviet arms control verification noted near the end of the Cold War:

> When faced with being accountable through the ages to come for the safety of their respective countries during their few years of tenureship, should a President or Premier take the route of the untried, or should he follow the oft-tried and oft-failed, but always defensible and always strongly felt, route of the mailed fist? He is certain that no one can accuse him of weakness if he opts for guns and bombs. Maybe stupidity, but not weakness. That such conduct actually is a sign of much more profound weakness is a point too subtle for his anxious mind to comprehend.[60]

In light of the U.S. and Chinese decisions to explore at least limited space weapons programs, these words underlie the need for a renewed and expanded international dialogue about space security. The goals of such efforts should be to reduce misperceptions and improve the chances of identifying mutually beneficial outcomes for space, and then to develop realistic pathways to ensure that these routes are actually taken. The main lesson we can draw from the past is that space will continue to be a highly interactive environment. Great care will be required to manage this important experiment in environmental security, technological development, and human conflict prevention.

[60] Jack F. Evernden, quoted in William E. Burrows, *Deep Black: Space Espionage and National Security* (New York: Random House, 1986), p. 345.

REFERENCE MATTER

Selected Bibliography

Books and Major Studies

Allison, Graham. *Essence of Decision*. New York: Little, Brown, 1971.

Amann, Ronald, and Julian Cooper, eds. *Industrial Innovation in the Soviet Union*. New Haven, Conn.: Yale University Press, 1982.

Argyris, Chris, and Donald A. Schon. *Organizational Learning: A Theory of Action Perspective*. Reading, Mass.: Addison-Wesley, 1978.

Arquilla, John. *The Reagan Imprint: Ideas in American Foreign Policy from the Collapse of Communism to the War on Terror*. Chicago: Ivan R. Dee, 2006.

Atkinson, William C. *A History of Spain and Portugal*. Baltimore: Penguin, 1960.

Bailes, Kendall. *Technology and Society Under Lenin and Stalin: Origins of the Soviet Technical Intelligentsia, 1917–1941*. Princeton, N.J.: Princeton University Press, 1978.

Barrett, Scott. *Environment and Statecraft: The Strategy of Environmental Treaty-Making*. New York: Oxford University Press, 2003.

Basiuk, Victor. *Technology, World Politics, and American Policy*. New York: Columbia University Press, 1977.

Baucom, Donald R. *The Origins of SDI: 1944–1983*. Lawrence: University of Kansas Press, 1992.

Beier, J. Marshall, and Steven Mataija, eds. *Arms Control and the Rule of Law: A Framework for Peace and Security in Outer Space*. Toronto: Center for International and Security Studies, York University, 1998.

Bergaust, Erik. *The Next Fifty Years in Space*. New York: MacMillan, 1964.

Boorstin, Daniel J. *The Discoverers*. New York: Vintage (Random House), 1983.

Breslauer, George W., and Philip E. Tetlock, eds. *Learning in U.S. and Soviet Foreign Policy*. Boulder, Colo.: Westview Press, 1991.

Broad, William J. *Star Warriors: A Penetrating Look into the Lives of the Young Scientists Behind Our Space Weaponry*. New York: Simon & Schuster, 1985.

Bulkeley, Rip. *The Sputniks Crisis and Early United States Space Policy: A Critique of the Historiography of Space*. Bloomington: Indiana University Press, 1991.

Bunn, George. *Arms Control by Committee: Managing Negotiations with the Russians*. Stanford, Calif.: Stanford University Press, 1992.

Burrows, William E. *Deep Black: Space Espionage and National Security*. New York: Random House, 1986.

Burrows, William E. *This New Ocean: The Story of the First Space Age*. New York: Random House, 1998.

Caidin, Martin. *Red Star in Space*. New York: Crowell-Collier Press, 1963.

Chasek, Pamela S. *Earth Negotiations: Analyzing Thirty Years of Environmental Diplomacy*. New York: United Nations University Press, 2001.

Chertok, B. Ye. *Rakety i Lyudi* [Rockets and people]. Moscow: Mashinostroenie, 1999.

Child, Jack. *Antarctica and South American Geopolitics: Frozen Lebensraum*. New York: Praeger, 1988.

Christman, Albert B. *Sailors, Scientists, and Rockets: Origins of the Navy Rocket Program and of the Naval Ordnance Test Station, Inyokern* (*History of the Naval Weapons Center, China Lake, California*, Volume I). Washington, D.C.: U.S. Government Printing Office, 1971.

Chun, Lt. Col. (USAF) Clayton K. S. *Shooting Down a "Star": Program 437, the US Nuclear ASAT System and Present-Day Copycat Killers*. Cadre Paper No. 6. Maxwell Air Force Base, Ala.: Air University Press, April 2000.

Clarke, Arthur C. *The Exploration of Space*. New York: Harper, 1959.

Collard-Wexler, Simon, Jessy Cowan-Sharp, Sarah Estabrooks, Thomas Graham Jr., Robert Lawson, and William Marshall. *Space Security 2004*. Toronto: Northview Press, 2005.

Congressional Research Service. *Soviet Space Programs: 1976–80*, Part I. Report prepared for the Senate Commerce Committee. Washington, D.C.: U.S. Government Printing Office, 1982.

Cornes, Richard, and Todd Sandler. *The Theory of Externalities, Public Goods, and Club Goods*. New York: Cambridge University Press, 1986.

Day, Dwayne A., John M. Logsdon, and Brian Latell, eds. *Eye in the Sky: The Story of the Corona Spy Satellites*. Washington, D.C.: Smithsonian Institution, 1998.

Deutsch, Karl M. *The Nerves of Government: Models of Political Communication and Control*. New York: Free Press, 1963.

Dokos, Thanos P. *Negotiations for a CTBT 1958–1994: Analysis and Evaluation of American Policy*. Lanham, Md.: University Press of America, 1995.

Dolman, Everett C. *Astropolitik: Classical Geopolitics in the Space Age*. London: Frank Cass, 2002.

Ducrocq, Albert. *The Conquest of Space: Moon Probes and Artificial Satellites: Their Impact on Human Destiny*. London: Putnam, 1961.

Eisenhower, Dwight D. *The White House Years: Waging Peace (1956–61)*. New York: Doubleday, 1965.

Eisenhower, Susan, ed. *Partners in Space: US-Russian Cooperation After the Cold War*. Washington, D.C.: Eisenhower Institute, 2004.

Emme, Eugene M., ed. *The History of Rocket Technology: Essays on Research Development and Utility*. Detroit: Wayne State University Press, 1964.

Etheredge, Lloyd S. *Can Governments Learn? American Foreign Policy and Central American Revolutions*. New York: Pergamon Press, 1985.

Evangelista, Matthew. *Innovation and the Arms Race*. Ithaca, N.Y.: Cornell University Press, 1988.

Fawcett, J. E. S. *Outer Space: New Challenges to Law and Policy*. Oxford, England: Clarendon Press, 1984.

Feiwel, G. F., ed. *Game-Theoretic Models of Bargaining*. New York: Cambridge University Press, 1985.

Fensch, Thomas, ed. *The Kennedy-Khrushchev Letters: Top Secret*. The Woodlands, Tex.: New Century, 2001.

Feoktistov, Konstantin, and I. Bubnov. *O Kosmoletakh* [About spaceflight]. Moscow: Molodaya Gvardiya, 1982.

Fitzgerald, Frances. *Way Out There in the Blue: Reagan, Star Wars and the End of the Cold War*. New York: Simon & Schuster, 2000.

Freedman, Lawrence. *The Evolution of Nuclear Strategy*. London: Palgrave, 2003.

Friedman, George, and Meredith Friedman. *The Future of War: Power, Technology and American World Dominance in the Twenty-First Century*. New York: St. Martin's Griffin, 1996.

Friedman, Norman. *Seapower and Space: From the Dawn of the Missile Age to Net-Centric Warfare*. Annapolis, Md.: Naval Institute Press, 2000.

Friedman, Thomas L. *The Lexus and the Olive Tree: Understanding Globalization*. New York: Anchor, 2000.

Frutkin, Arnold W. *International Cooperation in Space*. Englewood Cliffs, N.J.: Prentice-Hall, 1965.

Gaddis, John Lewis. *The Long Peace: Inquiries into the History of the Cold War*. New York: Oxford University Press, 1987.

Gaddis, John Lewis. *Surprise, Security, and the American Experience*. Cambridge, Mass.: Harvard University Press, 2004.

Gallagher, Nancy W. *The Politics of Verification*. Baltimore: Johns Hopkins University Press, 1999.

George, Alexander L., Philip J. Farley, and Alexander Dallin, eds. *U.S.-Soviet Security Cooperation: Achievements, Failures, Lessons*. New York: Oxford University Press, 1988.

Gibney, Frank, and George J. Feldman. *The Reluctant Space-Farers: A Study in the Politics of Discovery*. New York: New American Library, 1965.

Gilpin, Robert. *American Scientists and Nuclear Weapons Policy*. Princeton, N.J.: Princeton University Press, 1962.

Goldstein, Judith, and Robert O. Keohane, eds. *Ideas and Foreign Policy: Beliefs, Institutions, and Political Change*. Ithaca, N.Y.: Cornell University Press, 1993.

Goodchild, Peter. *The Real Dr. Strangelove: Edward Teller*. Cambridge, Mass.: Harvard University Press, 2004.

Gorbachev, Mikhail. *Memoirs*. New York: Doubleday, 1995.

Green, Constance McLaughlin, and Milton Lomask. *Vanguard: A History*. Washington, D.C.: Smithsonian Institution Press, 1971.

Haas, Ernst B. *When Knowledge Is Power: Three Models of Change in International Organizations*. Berkeley: University of California Press, 1990.

Hardin, Russell. *Collective Action*. Baltimore: Johns Hopkins University Press, 1982.

Harford, James. *Korolev: How One Man Masterminded the Soviet Drive to Beat America to the Moon*. New York: Wiley, 1997.

Harvey, Brian. *Russia in Space: The Failed Frontier?* Chichester, England: Praxis, 2001.

Harvey, Brian. *China's Space Program: From Conception to Manned Spaceflight*. Chichester, England: Praxis, 2004.

Harvey, Dodd L., and Linda C. Ciccoritti. *U.S.-Soviet Cooperation in Space*. Miami, Fla.: University of Miami Press, 1974.

Hays, Lt. Col. (USAF) Peter L. *United States Military Space: Into the Twenty-First Century*. Occasional Paper No. 42. Colorado Springs: U.S. Air Force Academy, Institute for National Security Studies, September 2002.

Hirsh, Lester M., ed. *Man and Space: A Controlled Research Reader*. New York: Pitman, 1966.

Hitchens, Theresa. *Future Security in Space: Charting a Cooperative Course*. Washington, D.C.: Center for Defense Information, September 2004.

Hobbes, Thomas. *Leviathan, Parts One and Two*. Indianapolis, Ind.: Liberal Arts Press, 1977 (original 1651).

Hoffmann, Erik P., and Robbin F. Laird. *The Scientific-Technological Revolution and Soviet Foreign Policy*. New York: Pergamon Press, 1982.

Hoffmann, Erik P., and Robbin F. Laird. *Technocratic Socialism: The Soviet Union in the Advanced Industrial Era*. Durham, N.C.: Duke University Press, 1985.

Holloway, David. *Stalin and the Bomb: The Soviet Union and Atomic Energy, 1939–1956*. New Haven, Conn.: Yale University Press, 1994.

Independent Working Group. *Missile Defense, the Space Relationship, and the Twenty-First Century: 2007 Report*. Washington, D.C.: Institute for Foreign Policy Analysis, 2006.

Jankowitsch, Peter. "International Cooperation in Outer Space." Occasional Paper No. 11. Muscatine, Iowa: Stanley Foundation, 1976.

Jervis, Robert. *Perception and Misperception in International Politics*. Princeton, N.J.: Princeton University Press, 1976.

Jessup, Phillip, and Howard J. Taubenfeld. *Controls for Outer Space and the Antarctic Analogy*. New York: Columbia University Press, 1959.

Johnson, Nicholas. *Soviet Military Strategy in Space*. London: Jane's, 1987.

Johnson, Nicholas L., and Darren S. McKnight. *Artificial Space Debris*. Malabar, Fla.: Orbit, 1987.

Johnson-Freese, Joan. *Space as a Strategic Asset*. New York: Columbia University Press, 2007.

Joyner, Christopher C. *Governing the Frozen Commons: The Antarctic Regime and Environmental Protection*. Columbia: University of South Carolina Press, 1998.

Kahn, Herman. *On Thermonuclear War*. Princeton, N.J.: Princeton University Press, 1960.

Kash, Don. *Cooperation in Space*. Lafayette, Ind.: Purdue University Press, 1967.

Kennedy, Paul. *The Rise and Fall of the Great Powers*. New York: Vintage, 1989.

Keohane, Robert O. *After Hegemony: Cooperation and Discord in the World Political Economy*. Princeton, N.J.: Princeton University Press, 1984.

Khorev, A., et al. *Reabilitirovan Posmertno* [Rehabilitated after death]. Moscow: Uridicheskaya Literatura, 1989.

Killian, James R., Jr. *Sputnik, Scientists, and Eisenhower: A Memoir of the First Special Assistant to the President for Science and Technology*. Cambridge, Mass.: MIT Press, 1977.

Kissinger, Henry. *White House Years*. New York: Little, Brown, 1979.

Kistiakowsky, George B. *A Scientist at the White House: The Private Diary of President Eisenhower's Special Assistant for Science and Technology*. Cambridge, Mass.: Harvard University Press, 1976.

Klein, Cmdr. (USN) John J. *Space Warfare: Strategy, Principles and Policy*. London: Routledge, 2006.

Knox, MacGregor, and Williamson Murray, eds. *The Dynamics of Military Revolution, 1300–2050*. New York: Cambridge University Press, 2001.

Krepon, Michael, with Christopher Clary. *Space Assurance or Space Dominance? The Case Against Weaponizing Space*. Washington, D.C.: Henry L. Stimson Center, 2003.

Kull, Steven. *Minds at War: Nuclear Reality and the Inner Conflicts of Defense Policymakers*. New York: Basic, 1988.

Lakoff, Sanford, and Herbert F. York. *A Shield in Space? Technology, Politics, and the Strategic Defense Initiative*. Berkeley: University of California Press, 1989.

Lambakis, Steven. *On the Edge of Earth: The Future of American Space Power*. Lexington: University Press of Kentucky, 2001.

Langton, N. H., ed. *The Space Environment*. New York: American Elsevier, 1969.

Larson, Deborah Welch. *Origins of Containment*. Princeton, N.J.: Princeton University Press, 1985.

Lettow, Paul. *Ronald Reagan and His Quest to Abolish Nuclear Weapons*. New York: Random House, 2005.

Ley, Willey. *Harnessing Space*. New York: Macmillan, 1963.

Lipschutz, Ronnie D. *Global Environmental Politics: Power, Perspectives, and Practice*. Washington, D.C.: Congressional Quarterly Press, 2004.

Livermore, H. V. *Portugal: A Short History*. Edinburgh, Scotland: Edinburgh University Press, 1973.

Logsdon, John M. *The Decision to Go to the Moon: Project Apollo and the National Interest*. Cambridge, Mass.: MIT Press, 1970.

Logsdon, John M., ed. *Exploring the Unknown: Selected Documents in the History of the U.S. Civil Space Program*, Volume 1: *Organizing for Exploration*. Washington, D.C.: NASA, 1995.

Logsdon, John M., and Gordon Adams, eds. *Space Weapons: Are They Needed?* Washington, D.C.: Space Policy Institute, George Washington University, October 2003.

Logsdon, John M., James Clay Moltz, and Emma S. Hinds. *Collective Security in Space: European Perspectives*. Washington, D.C.: Space Policy Institute, George Washington University, January 2007.

Logsdon, John M., and Audrey M. Schaffer, eds. *Perspectives on Space Security*. Washington, D.C.: Space Policy Institute, George Washington University, December 2005.

Lukin, P. I., et al. *Kosmos i pravo* [Space and law]. Moscow: Institute of State and Law, 1980.

Lupton, Lt. Col. (USAF, ret.) David E. *On Space Warfare: A Space Power Doctrine*. Maxwell Air Force Base, Ala.: Air University Press, June 1998.

Mahan, Capt. (USN) A. T. *The Interest of America in Sea Power: Present and Future*. Port Washington, N.Y.: Kennikat Press, 1897.

Manno, Jack. *Arming the Heavens: The Hidden Military Agenda for Space*. New York: Dodd, Mead, 1984.

March, James G., and Johan P. Olsen, eds. *Ambiguity and Choice in Organizations*. Oslo, Norway: Universitetsforlaget, 1976.

McAleer, Neil. *The Omni Space Almanac*. New York: World Almanac, 1987.

McDougall, Walter A. . . . *the Heavens and the Earth: A Political History of the Space Age*. New York: Basic, 1985.

McIntyre, John R., ed. *International Space Policy: Legal, Economic, and Strategic Options for the Twentieth Century and Beyond*. New York: Quorum, 1987.

Merriman, Roger Bigelow. *The Rise of the Spanish Empire in the Old World and in the New*. New York: Macmillan, 1918.

Moltz, James Clay. "Managing International Rivalry on High Technology Frontiers: U.S.-Soviet Competition and Cooperation in Space." Ph.D. dissertation, University of California at Berkeley, 1989.

Moltz, James Clay, ed. *Future Security in Space: Commercial, Military, and Arms Control Trade-Offs*. Occasional Paper No. 10. Monterey, Calif.: Center for Nonproliferation Studies, Monterey Institute of International Studies, July 2002.

Moltz, James Clay, ed. *New Challenges in Missile Proliferation, Missile Defense, and Space Security*. Occasional Paper No. 12. Monterey, Calif.: Center for Nonproliferation Studies, Monterey Institute of International Studies, July 2003.

NASA. *Orbital Debris*. Proceedings of a workshop held at NASA Lyndon B. Johnson Space Center, Houston, Texas, July 27–29, 1982 (compiled by Donald J. Kessler and Shin-Yi Su). Washington, D.C.: NASA, Scientific and Technical Information Branch, 1985.

National Academy of Sciences. *Antarctic Treaty System: An Assessment, National Research Council*. Washington, D.C.: National Academy Press, 1986.

Neufeld, Michael J. *The Rocket and Reich: Peenemunde and the Coming of the Ballistic Missile Era*. Cambridge, Mass.: Harvard University Press, 1995.

Neustadt, Richard E., and Ernest R. May. *Thinking in Time*. New York: Free Press, 1986.

Nisbet, Robert, ed. *Social Change*. Oxford, England: Blackwell, 1972.

Nisbet, Robert. *Social Change and History: Aspects of the Western Theory of Development*. New York: Oxford University Press, 1969.

Nowell, Charles E. *A History of Portugal*. Princeton, N.J.: Van Nostrand, 1952.

Nye, Joseph S. *Soft Power: The Means to Success in World Politics*. New York: PublicAffairs, 2004.

Oberg, James. *Red Star in Orbit*. New York: Random House, 1981.

Office of Technology Assessment. *U.S.-Soviet Cooperation in Space*. Washington, D.C.: U.S. Congress, Office of Technology Assessment, OTA-TM-STI–27, July 1985.

O'Hanlon, Michael E. *Neither Star Wars nor Sanctuary: Constraining the Military Uses of Space*. Washington, D.C.: Brookings Institution, 2004.

Papp, Daniel S., and John R. McIntyre, eds. *International Space Policy*. New York: Quorum, 1987.

Pardoe, Geoffrey K. C. *The Future for Space Technology*. London: Frances Pinter, 1984.

Parry, Albert. *Russia's Rockets and Missiles*. Garden City, N.Y.: Doubleday, 1960.

Paterson, Matthew. *Understanding Global Environmental Politics: Domination, Accumulation, Resistance*. London: Macmillan Press, 2000.

Peebles, Curtis. *Battle for Space*. New York: Beaufort, 1983.

Perdomo, Lt. Col. (USMC) Maurice. "United States National Space Security Policy and the Strategic Issues for DOD Space Control." Master's thesis, U.S. Army War College, Carlisle, Penn., March 2005.

Peterson, M. J. *Managing the Frozen South: The Creation and Evolution of the Antarctic Treaty System*. Berkeley: University of California Press, 1988.

Podvig, Pavel, ed. *Russian Strategic Nuclear Forces*. Cambridge, Mass.: MIT Press, 2001.

Raushenbakh, V., ed. *Iz istorii sovetskoy kosmonavtiki (sbornik pamyati akademika S. P. Koroleva)* [From the history of Soviet cosmonautics (memories of Academician S. P. Korolev)]. Moscow: Nauka, 1983.

Reiss, Edward. *The Strategic Defense Initiative.* New York: Cambridge University Press, 1992.

Rhea, John, ed. *Roads to Space: An Oral History of the Soviet Space Program.* New York: Aviation Week Group (McGraw-Hill), 1995.

Rhodes, Richard. *Dark Sun: The Making of the Hydrogen Bomb.* New York: Touchstone, 1996.

Riabchikov, Evgeny. *Russians in Space.* Moscow: Novosti Press, 1971.

Richelson, Jeffrey T. *America's Space Sentinels: DSP Satellites and National Security.* Lawrence: University of Kansas Press, 1999.

Richelson, Jeffrey T. *Spying on the Bomb: American Nuclear Intelligence from Nazi Germany to Iran and North Korea.* New York: Norton, 2006.

Roberts, J. M. *The Pelican History of the World.* New York: Penguin, 1980.

Rosecrance, Richard. *The Rise of the Trading State: Commerce and Conquest in the Modern Age.* New York: Basic, 1986.

Ruzic, Neil P. *Where the Winds Sleep: Man's Future on the Moon, A Projected History.* Garden City, N.Y.: Doubleday, 1970.

Sagan, Scott D. *The Limits of Safety: Organizations, Accidents, and Nuclear Weapons.* Princeton, N.J.: Princeton University Press, 1993.

Schauer, William H. *The Politics of Space: A Comparison of the Soviet and American Programs.* New York: Holmes and Meier, 1976.

Schelling, Thomas C. *The Strategy of Conflict.* Cambridge, Mass.: Harvard University Press, 1960.

Schirra, Wally, with Richard N. Billings. *Schirra's Space.* Annapolis, Md.: Naval Institute Press, 1988.

Schlesinger, Arthur M., Jr. *A Thousand Days: John F. Kennedy in the White House.* New York: Houghton Mifflin, 1965.

Schwartz, Stephen, ed. *Atomic Audit: The Costs and Consequences of U.S. Nuclear Weapons Since 1940.* Washington, D.C.: Brookings Institution Press, 1998.

Seaborg, Glenn T. *Kennedy, Khrushchev, and the Test Ban.* Berkeley: University of California Press, 1981.

Seller, Jerry Jon, with contributions by William J. Astore, Robert B. Giffen, and Wiley J. Larson. *Understanding Space: An Introduction to Astronautics.* New York: McGraw-Hill, 2004.

Siddiqi, Asif A. *Sputnik and the Soviet Space Challenge.* Gainesville: University of Florida Press, 2003.

Siddiqi, Asif A. *The Soviet Race with Apollo.* Gainesville: University of Florida Press, 2003.

Simpson, John W. *Nuclear Power from Underseas to Outer Space.* La Grange Park, Ill.: American Nuclear Society, 1995.

Smolders, Peter L. *Soviets in Space: The Story of Salyut and the Soviet Approach to Present and Future Space Travel.* London: Lutterworth Press, 1973.

Snyder, Jack. *Myths of Empire: Domestic Politics and International Ambition.* Ithaca, N.Y.: Cornell University Press, 1991.

Sorensen, Theodore C. *Kennedy.* New York: Harper & Row, 1965.

Spacesecurity.org. *Space Security 2006*. (Waterloo, Canada: University of Waterloo, 2006).

Stares, Paul. *The Militarization of Space: U.S. Policy 1945–1984*. Ithaca, N.Y.: Cornell University Press, 1985.

Stares, Paul B. *Space and National Security*. Washington, D.C.: Brookings Institution, 1987.

Steinbruner, John. *The Cybernetic Theory of Decision*. Princeton, N.J.: Princeton University Press, 1974.

Stern, Paul C., Robert Axelrod, Robert Jervis, and Roy Radner, eds. *Perspectives on Deterrence*. New York: Oxford University Press, 1989.

Stoiko, Michael. *Soviet Rocketry: Past, Present, and Future*. New York: Holt, Rinehart and Winston, 1970.

Suter, K. D. *World Law and the Last Wilderness*. Sydney: Friends of the Earth, 1980.

Swenson, Loyd S., Jr., James M. Grimwood, and Charles C. Alexander. *This New Ocean: A History of Project Mercury*. Washington, D.C.: NASA, 1966.

Talbott, Strobe, trans. and ed. *Khrushchev Remembers: The Last Testament*. New York: Little, Brown, 1974.

Taubman, Philip. *Secret Empire: Eisenhower, the CIA, and the Hidden Story of America's Space Espionage*. New York: Simon & Schuster, 2003.

Teller, Edward. *Memoirs: A Twentieth-Century Journey in Science and Politics*. Cambridge, Mass.: Perseus, 2001.

Tolba, Mostafa K., with Iwona Rummel-Bulska. *Global Environmental Diplomacy: Negotiating Environmental Agreements for the World, 1973–1992*. Cambridge, Mass.: MIT Press, 1998.

Trento, Joseph J. *Prescription for Disaster: From the Glory of Apollo to the Betrayal of the Shuttle*. New York: Crown, 1987.

Triggs, Gillian D. *The Antarctic Treaty Regime: Law, Environment and Resources*. New York: Cambridge University Press, 1987.

Turner, Frederick Jackson. *The Frontier in American History*. New York: Henry Holt, 1921.

U.S. Congress. *Final Report of the Select Committee on U.S. National Security and Military/Commercial Concerns with the People's Republic of China*. May 25, 1999 (unclassified version).

U.S. Congress. *Report of the Commission to Assess United States National Security Space Management and Organization*. Pursuant to Public Law 108–65, January 11, 2001.

U.S. Department of Defense, *Military Power of the People's Republic of China, 2006*. Annual Report to Congress, 2006.

Van Creveld, Martin. *Technology and War: From 2000 B.C. to the Present*. New York: Free Press, 1991.

Vereshchetin, V. S. *Mezdunarodone sotrudnichestvo v kosmose* [International cooperation in space]. Moscow: Nauka, 1977.

Vladimirov, Leonid. *Sovetskiy kosmicheskiy blef* [The Soviet space bluff]. Frankfurt, Germany: Possev-Verlag, 1973.

Von Bencke, Matthew J. *The Politics of Space: A History of U.S.-Soviet/Russian Competition and Cooperation*. Boulder, Colo.: Westview Press, 1997.

Waltz, Kenneth N. *Theory of International Politics*. Reading, Mass.: Addison-Wesley, 1979.

Weber, Steve. *Cooperation and Discord in U.S.-Soviet Arms Control.* Princeton, N.J.: Princeton University Press, 1991.

Westermeyer, William E. *The Politics of Mineral Resource Development in Antarctica: Alternative Regimes for the Future.* Boulder, Colo.: Westview Press, 1984.

Wilensky, Harold L. *Rich Democracies: Political Economy, Public Policy, and Performance.* Berkeley: University of California Press, 2002.

Wirtz, James J. *The Tet Offensive: Intelligence Failure in War.* Ithaca, N.Y.: Cornell University Press, 1991.

Wirtz, James J., and Jeffrey A. Larsen, eds. *Rockets' Red Glare: Missile Defenses and the Future of World Politics.* Boulder, Colo.: Westview Press, 2001.

Wolter, Detlev. *Common Security in Outer Space and International Law.* Geneva: U.N. Institute for Disarmament Research, 2006.

Wright, David, Laura Grego, and Lisbeth Gronlund. *The Physics of Space Security: A Reference Manual.* Cambridge, Mass.: American Academy of Arts and Sciences, 2005.

Yevsikov, Victor. *Re-Entry Technology and the Soviet Space Program (Some Personal Observations).* Falls Church, Va.: Monograph Series on the Soviet Union, Delphic Associates, December 1982.

York, Herbert F. *Race to Oblivion.* New York: Simon & Schuster, 1970.

York, Herbert F. *The Advisors: Oppenheimer, Teller, and the Superbomb.* Stanford, Calif.: Stanford University Press, 1976.

York, Herbert F. *Making Weapons, Talking Peace: A Physicist's Odyssey from Hiroshima to Geneva.* New York: Basic, 1987.

Young, Hugo, Bryan Silcock, and Peter Dunn. *Journey to Tranquility: The History of Man's Assault on the Moon.* London: Jonathan Cape, 1969.

Zhukov, G. P. *Kosmos i mir* [Space and peace]. Moscow: Nauka, 1985.

Zhukov, Gennady, and Yuri Kolosov. *International Space Law.* New York: Praeger, in cooperation with Novosti Press, 1984.

Archival Sources

National Security Archives, George Washington University, Washington, D.C.

Ronald Reagan Presidential Library and Archives, Simi Valley, Calif.

Russian Foreign Ministry, Moscow

Journals and Trade Publications

Aerospace Daily

Aerospace Power Journal

Arms Control Today

Astropolitics

Aviation Week and Space Technology

Bulletin of the Atomic Scientists

Defense Daily

Defense News

Disarmament Diplomacy

Foreign Affairs

Foreign Policy

International Security

The Nonproliferation Review

Orbital Debris Quarterly

Popular Science

Space News

Space Policy

Wired

Index

spaceflight; Scientific cooperation;
Space science
Clarke, Arthur C., 27
Clinton administration: China policies,
228; commercial space policy, 237;
commitment to ABM Treaty, 255; co-
operation with Russians, 234; missile
defense program, 229, 235, 239, 247,
248–50, 253–56; orbital debris problem,
246; relations with Russia, 239, 242,
252–53, 254–55; space cooperation with
Russia, 242; space security policy, 229,
234, 235–40, 245–47, 252–53, 324
Cochran, Thad, 249
CoCom, see Coordinating Committee for
Multilateral Export Controls
Cognitive breaks, 63
Cognitive change, see Learning
Cohen, William, 248, 255
Cold War: Berlin issues, 98, 109–10, 121;
counterfactual history of space security, 48–50; end of, 223, 225, 226–27, 307;
environmental factors in space policy,
46–47, 55–64; expectations of nuclear
war, 110, 120; "missile gap," 106; multi-
lateralism, 328; space cooperation, 29,
115–16, 126, 182–83, 305–7; space security issues, 11–12, 64–65; U-2 spy planes,
101, 118. See also Nuclear arms race;
Post-Cold War period; United States-
Soviet relations
Collective bads, 58
Collective goods, 34–35, 61
Commerce department, U.S., 189, 219, 237
Commercial space services: Bush administration policies, 296; as constraint
on military use of space, 319, 320;
counterfactual history, 49; European,
244, 261, 283; expansion, 177–78; export controls, 286; future of, 6–7, 317,
320–21; globalization, 308, 317; human
spaceflight, 284–85, 308; orbital debris
threat and, 283, 297–99, 308; Reagan
policies, 188–89, 198–99; relations
with governments, 321; relations with
military, 320; remote sensing, 237; rev-

enues, 286; rocket builders, 243; Russian enterprises, 250–51; Russian-U.S.
partnerships, 233, 243–45, 250, 260–61,
307; scientific experiments on satellites,
204; Soviet organization, 205, 210–11;
support of international agreements,
324; U.S. policies, 114–15, 211, 237.
See also Communications satellites;
Launch services; Navigation systems
Commission to Assess the Ballistic Mis-
sile Threat to the United States, 229,
248
Commission to Assess United States
National Security Space Management
and Organization, 256, 261–62
Committee on Space Research, 148
Committee on the Peaceful Uses of
Outer Space, see United Nations Com-
mittee on the Peaceful Uses of Outer
Space
Common Aero Vehicle (CAV), 262, 293
Commons, tragedy of, 34–35
Communications Satellite Act, 114
Communications satellites: demand for
services, 286; effects of electromagnetic
radiation, 51, 119, 123; international
cooperation, 57–58, 102, 126, 145, 166,
311–12; military, 119, 162, 170–71, 184;
radio frequency allocations, 311–12;
Soviet, 170–71; U.S., 104, 114–15, 119,
143–44, 162, 184. See also Satellites
Comprehensive Test Ban Treaty, 252
COMSAT, 114, 145
Conference on Disarmament (CD):
member countries, 128; negotiations,
246; Russian policies, 245; space arms
control discussions, 273–74, 279, 280,
281, 282; support of nuclear test ban,
128; U.S. policies, 265, 297
Congress, U.S.: ABM deployment opponents, 165; ASAT test ban, 202; debates
on Chinese space competition, 275–77;
Democratic control, 200–201, 295;
interest in space program, 95, 215, 236;
military space program supporters,
26; missile defense opponents, 65, 293;